WHY ZEBRAS DON'T GET ULCERS

WHY ZEBRAS DON'T GET ULCERS

An Updated Guide to Stress, Stress-Related Diseases, and Coping

ROBERT M. SAPOLSKY

W. H. Freeman and Company
New York

Cover and Text Design: Victoria Tomaselli

Library of Congress Cataloging-in-Publication Data

Sapolsky, Robert M.
Why zebras don't get ulcers : an updated guide to stress, stress-related diseases, and coping / Robert M. Sapolsky.
p. cm.
Includes bibliographical references and index.
ISBN 0-7167-3210-6 (alk. paper)
1. Stress (Physiology) 2. Stress (Psychology) 3. Stress management I. Title.
[DNLM: 1. Stress—physiopathology. 2. Stress, Psychological—prevention & control. 3. Adaptation, Psychological. QT 162.S8 S241w 1998]
QP82.2.S8S266 1998
616′.001′9—dc21
DNLM/DLC 98–5578
CIP

Printed in the United States of America
Fourth printing, 1999

CONTENTS

PREFACE

Perhaps you're reading this while browsing in a bookstore. If so, glance over at the guy down the aisle when he's not looking, the one pretending to be engrossed in the Stephen Hawking book. Take a good look at him. He's probably not missing fingers from leprosy, or covered with smallpox scars, or shivering with malaria. Instead, he probably appears perfectly healthy, which is to say he has the same diseases that most of us have—cholesterol levels that are high for an ape, hearing that has become far less acute than in a hunter-gatherer of his age, a tendency to dampen his tension with Valium. We in our westernized society now tend to get different diseases than we used to. But what's more important, we tend to get different *kinds* of diseases now, with very different causes and consequences. A millennium ago, a young hunter-gatherer inadvertently eats a reedbuck riddled with anthrax and the consequences are clear—he's dead a few days later. Now, a young lawyer unthinkingly decides that red meat, fried foods, and a couple of beers per dinner constitute a desirable diet, and the consequences are anything but clear—a half-century later, maybe he's crippled with cardiovascular disease, or maybe he's taking bike trips with his grandkids. Which outcome occurs depends on some obvious factors, like what his liver does with cholesterol, what levels of certain enzymes are in his fat cells, whether he has any congenital weaknesses in the walls of his blood vessels. But the outcome will also depend heavily on such surprising factors as his personality, the amount of emotional stress he experiences over the years, whether he has someone's shoulder to cry on when those stressors occur.

There has been a revolution in medicine concerning how we think about the diseases that now afflict us. It involves recognizing the interactions between the body and the mind, the ways in which emotions and personality can have a tremendous impact on the functioning and health of virtually every cell in the body. It is about the role of stress in making some of us more vulnerable to disease, the ways in which some of us cope with stressors, and the critical notion that you cannot really understand a disease *in vacuo*, but rather only in the context of the person suffering from that disease.

This is the subject of my book. I begin by trying to clarify the meaning of the nebulous concept stress and to teach, with a minimum of pain, how various hormones and parts of the brain are mobilized in response to stress. I then focus on the links between stress and increased risk for certain types of disease, going chapter by chapter through the effects of stress on the circulatory system, on energy storage, on growth, reproduction, the immune system, and so on. Next I describe how the aging process may be influenced by the amount of stress experienced over the lifetime. I then examine the link between stress and the most common and arguably most crippling of psychiatric disorders, major depression. As part of updating the book for this revised edition, I have added three new chapters: one on the effects of stress on memory, one a tour of certain personality types associated with maladaptive stress-responses, and one on what the stressor of being poor has to do with health.

Some of the news in this book is grim—sustained or repeated stress can disrupt our bodies in seemingly endless ways. Yet most of us are not incapacitated by stress-related disease. Instead, we cope, both physiologically and psychologically, and some of us are spectacularly successful at it. For the reader who has held on until the end, the final chapter reviews what is known about stress management and how some of its principles can be applied to our everyday lives. There is much to be optimistic about.

I believe that everyone can benefit from some of these ideas and can be excited by the science on which they are based. Science provides us with some of the most elegant, stimulating puzzles that life has to offer. It throws some of the most provocative ideas into our arenas of moral debate. And occasionally, it improves our lives. I love science, and it pains me to think that so many are terrified of the subject or feel that choosing science means that you cannot also choose compassion, or the arts, or be awed by nature. Science is not meant to cure us of mystery, but to reinvent and reinvigorate it.

Thus I think that any science book for nonscientists should attempt to convey that excitement, to make the subject interesting and accessible even to those who would normally not be caught dead near the subject. That has been a particular goal of mine in this book. Often, that has meant simplifying complex ideas, and as a counterbalance to this, I include copious references at the end of the book, often with annotations concerning controversies and subtleties about material presented in the main text. These references are an excellent entrée for those readers who want something more detailed on the subject.

Many sections of this book contain material about which I am far from expert, and over the course of the writing, a large number of savants have been called for advice, clarification, and verification of facts. I thank them all for their generosity with their time and expertise: Nancy Adler, Robert Axelrod, Alan Baldrich, Marcia Barinaga, Alan Basbaum, Justo Bautisto, Tom Belva, Anat Biegon, Vic Boff (whose brand of vitamins graces the cupboards of my parents' home), Carlos Camargo, Matt Cartmill, M. Linette Casey, Richard Chapman, Cynthia Clinkingbeard, Felix Conte, George Daniels, Regio DeSilva, Irven DeVore, James Doherty, John Dolph, Leroi DuBeck, Richard Estes, Michael Fanselow, David Feldman, Caleb Tuck Finch, Paul Fitzgerald, Gerry Friedland, Meyer Friedman, Rose Frisch, Roger Gosden, Ray Hintz, Allan Hobson, Robert Kessler, Bruce Knauft, Mary Jeanne Kreek, Stephen Laberge, Emmit Lam, Jim Latcher, Richard Lazarus, Helen Leroy, Jon Levine, Seymour Levine, John Liebeskind, Ted Macolvena, Jodi Maxmin, Peter Milner, Gary Moberg, Terry Muilenburg, Ronald

Myers, Carol Otis, Daniel Pearl, Ciran Phibbs, Jenny Pierce, Ted Pincus, Virginia Price, Gerald Reaven, Sam Ridgeway, Carolyn Ristau, Paul Rosch, Ron Rosenfeld, Aryeh Routtenberg, Paul Saenger, Saul Schanburg, Kurt Schmidt-Nielson, Carol Shively, J. David Singer, Bart Sparagon, David Speigel, Ed Spielman, Dennis Styne, Steve Suomi, Jerry Tally, Carl Thoresen, Peter Tyak, David Wake, Michelle Warren, Jay Weiss, Owen Wolkowitz, Carol Worthman, and Richard Wurtman.

I am particularly grateful to the handful of people—friends, collaborators, colleagues, and ex-teachers—who took time out of their immensely busy schedules to read chapters. I shudder to think of the errors and distortions that would have remained had they not tactfully told me I didn't know what I was writing about. I thank them all sincerely: Robert Ader of the University of Rochester; Marvin Brown of the University of California, San Diego; Laurence Frank at the University of California, Berkeley; Jay Kaplan of Bowman Gray Medical School; Charles Nemeroff of Emory University; Seymour Reichlin of Tufts/New England Medical Center; Robert Rose of the MacArthur Foundation; Wylie Vale of the Salk Institute; Jay Weiss of Emory University; and Redford Williams of Duke University.

A number of people were instrumental in getting this book off the ground and into its final shape. Much of the material in these pages was developed in continuing medical education lectures I have given for health professionals over the years. These have been presented under the auspices of the Institute for Cortex Research and Development, and its director, Will Gordon, who has given me much freedom and support in exploring this material. Bruce Goldman of the Portable Stanford series first planted the idea for this book in my head, and Kirk Jensen recruited me for W. H. Freeman and Company; both helped in the initial shaping of the book. Finally, my secretary, Patsy Gardner, has been of tremendous help in all the logistical aspects of pulling this book together. I thank you all, and look forward to working with you in the future.

I received tremendous help with organizing and editing the first edition of the book, and for that I thank Audrey Herbst, Tina Hastings, Amy Johnson, and Meredyth Rawlins of Freeman. Liz Meryman, who selects the art for *The Sciences*, helping to merge the two cultures in that beautiful publication, graciously consented to read the manuscript and gave splendid advice on appropriate artwork. In addition, I thank Alice Fernandes-Brown, who was responsible for making my idea for the cover such a pleasing reality. To this list I must add the following people, who worked on this updated edition: John Michel, Amy Trask, Georgia Lee Hadler, Victoria Tomaselli, Bill O'Neal, Kathy Bendo, Paul Rohloff, Jennifer MacMillan, and Sheridan Sellers.

This book has been, for the most part, a pleasure to write and I think it reflects one of the things in my life for which I am most grateful—that I take so much joy in the science that is both my vocation and avocation. I thank the mentors who taught me to do science and, even more so, taught me to enjoy science: the late Howard Klar, Howard Eichenbaum, Mel Konner, Lewis Krey, Bruce McEwen, Paul Plotsky, and Wylie Vale.

A band of research assistants have been indispensable to the writing of this book. Steve Balt, Roger Chan, Mick Markham, Michelle Pearl, Serena Spudich, and Paul Stasi have wandered the basements of archival libraries, called strangers all over the world with questions, distilled arcane articles into coherency. In the line of duty, they have sought out drawings of opera castrati, the daily menu at Japanese-American internment camps, the causes of voodoo death, and the history of firing squads. All of their research was done with spectacular competence, speed, and humor. I am fairly certain this book could not have been completed without their help and am absolutely certain its writing would have been much less enjoyable. And finally, I thank my editor at Freeman, Jonathan Cobb. He has taught me about writing style, reminded me that commas cannot be randomly distributed, and supplied a clear and correct vision of what is needed, whether on the scale of fixing a single awkward phrase or an awkward concept permeating the entire book. And, somewhere amid this process, he has also become a friend.

Parts of the book describe work carried out in my own laboratory, and these studies have been made possible by funding from the National Institutes of Health, the National Institute of Mental Health, the National Science Foundation, the Sloan Foundation, the Klingenstein Fund, the Alzheimer's Association, and the Adler Foundation. The African fieldwork described herein has been made possible by the long-standing generosity of the Harry Frank Guggenheim Foundation. Finally, I heartily thank the MacArthur Foundation for supporting all aspects of my work.

There is a tradition among stress physiologists who dedicate their books to their spouses or significant others. It seems an unwritten rule that you are supposed to incorporate something cutesy about stress in the dedication: To Madge, who attenuates my stressors; for Arturo, the source of my eustress; for my wife who, over the course of the last umpteen years, has put up with my stress-induced hypertension, ulcerative colitis, loss of libido, and displaced aggression. I will forgo that style here in dedicating this book to my wife; I have something simpler to say:

For Lisa, my best friend,
who has made my life complete

WHY ZEBRAS DON'T GET ULCERS

WHY DON'T ZEBRAS GET ULCERS?

It's two o'clock in the morning and you're lying in bed. You have something immensely important and challenging to do that next day—a critical meeting, a presentation, an exam. You have to get a decent night's rest, but you're still wide awake. You try different strategies for relaxing—take deep, slow breaths, try to imagine restful mountain scenery—but instead you keep thinking that unless you fall asleep in the next minute, your career is finished. Thus you lie there, more tense by the second.

If you do this on a regular basis, somewhere around two-thirty, when you're really getting clammy, an entirely new, disruptive chain of thoughts will no doubt intrude. Suddenly, amid all your other worries, you begin to contemplate that nonspecific pain you've been having in your side, that sense of exhaustion lately, that frequent headache. The realization hits you—I'm sick, fatally sick! Oh, why didn't I recognize the symptoms, why did I have to deny it, why didn't I go to the doctor?

When it's two-thirty on those mornings, I always have a brain tumor. They're very useful for that sort of terror, because you can attribute every conceivable nonspecific symptom to a brain tumor and justify your panic. Perhaps you do, too; or maybe you lie there thinking that you have cancer, or an ulcer, or that you've just had a stroke.

Even though I don't know you, I feel confident in predicting that you don't lie there thinking, "I just know it; I have leprosy."

True? You are exceedingly unlikely to obsess about getting a serious case of dysentery if it starts pouring. And few of us lie there feeling convinced that our bodies are teeming with intestinal parasites or liver flukes.

Of course not. Our nights are not filled with worries about smallpox, scarlet fever, malaria, or bubonic plague. Cholera doesn't run rampant through our communities; river blindness, black water fever, and elephantiasis are third-world exotica. Few female readers will die in childbirth, and even fewer of those reading this page are likely to be malnourished.

Thanks to revolutionary advances in medicine and public health, our patterns of disease have changed, and we are no longer kept awake at night worrying about infectious diseases (except, of course, AIDS or tuberculosis) or the diseases of poor nutrition or hygiene. As a measure of this, consider the leading causes of death in the United States in 1900: pneumonia, tuberculosis, and influenza (or, if you were young, female, and inclined toward risk-taking, childbirth). When is the last time you heard of someone under age eighty dying of the flu? Yet in 1918, one of the most barbaric years of World War I, a soldier was far more likely to die of the flu or pneumonia than of battle wounds.

Our current patterns of disease would be unrecognizable to our great-grandparents or, for that matter, to most mammals. Put succinctly, we get different diseases and are likely to die in different ways from most of our ancestors (or from most humans currently living in the less privileged areas of this planet). Our nights are filled with worries about a different class of diseases; we are now living well enough and long enough to slowly fall apart.

The diseases that plague us now are ones of slow accumulation of damage—heart disease, cancer, cerebrovascular disorders. While none of these diseases is particularly pleasant, they certainly mark a big improvement over succumbing at age twenty after a week of sepsis or dengue fever. Along with this relatively recent shift in the patterns of disease have come changes in the way we perceive the disease process. We have come to recognize the vastly complex intertwining of our biology and our emotions, the endless ways in which our personalities, feelings, and thoughts both reflect and influence the events

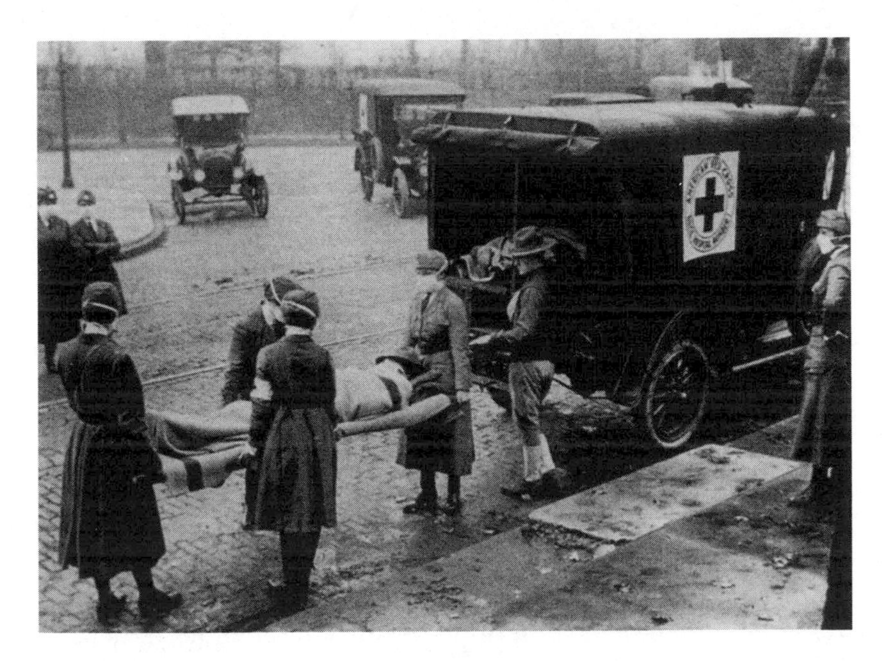

Influenza pandemic, 1918.

in our bodies. One of the most interesting manifestations of this recognition is understanding that extreme emotional disturbances can adversely affect us. Put in the parlance with which we have grown familiar, *stress can make us sick,* and a critical shift in medicine has been the recognition that many of the damaging diseases of slow accumulation can be either caused or made far worse by stress.

In some respects this is nothing new. Centuries ago, sensitive clinicians intuitively recognized the role of individual differences in vulnerability to disease. Two individuals could get the same disease, yet the courses of their illness could be quite different and in vague, subjective ways might reflect the personal characteristics of the individuals. Or a clinician might have sensed that certain types of people were more likely to contract certain types of disease. But since early in this century, the addition of rigorous science to these vague clinical perceptions has made stress physiology—the study of how the body responds to stressful events—a real discipline. As a result, there is today an extraordinary amount of physiological, biochemical, and molecular information available as to how all sorts of intangibles in our lives—emotional turmoil, psychological characteristics, our place in society, and the sort of society in which we live—can affect very real bodily events: whether cholesterol gums up our blood vessels or is safely cleared from the circulation, whether a cell that has become cancerous will be detected in time by the immune system, whether neurons in our brain will survive five minutes without oxygen during a cardiac arrest.

This book is a primer about stress, stress-related disease, and the mechanisms of coping with stress. How is it that our bodies can adapt to some stressful emergencies, while other ones make us sick? Why are some of us especially vulnerable to stress-related diseases, and what does that have to do with our personalities? How can purely psychological stressors make us sick? What might stress have to do with our vulnerability to depression, the speed at which we age, or how well our memories work? What do our patterns of stress-related diseases have to do with where we stand on the rungs of society's ladder? Finally, how can we increase the effectiveness with which we cope with the stressors that surround us?

Perhaps the best way to begin is by making a mental list of the sorts of things we find stressful. No doubt you would immediately come up with some obvious examples—traffic, deadlines, family relationships, money worries. But what if I said, "You're thinking like a speciocentric human. Think like a zebra for a second." Suddenly, new items might appear at the top of your list—serious physical injury, predators, starvation. The need for that prompting illustrates something critical—you and I are more likely to get an ulcer than a zebra is. For animals like zebras, the most upsetting things in life are *acute physical stressors*. You are that zebra, a lion has just leapt out and ripped your stomach open, you've managed to get away, and now you have to spend the next hour evading the lion as it continues to stalk you. Or, perhaps just as stressfully, you are that lion, half-starved, and you had better be able to sprint across the savanna at top speed and grab something to eat or you won't survive. These are extremely stressful events, and they demand imme-

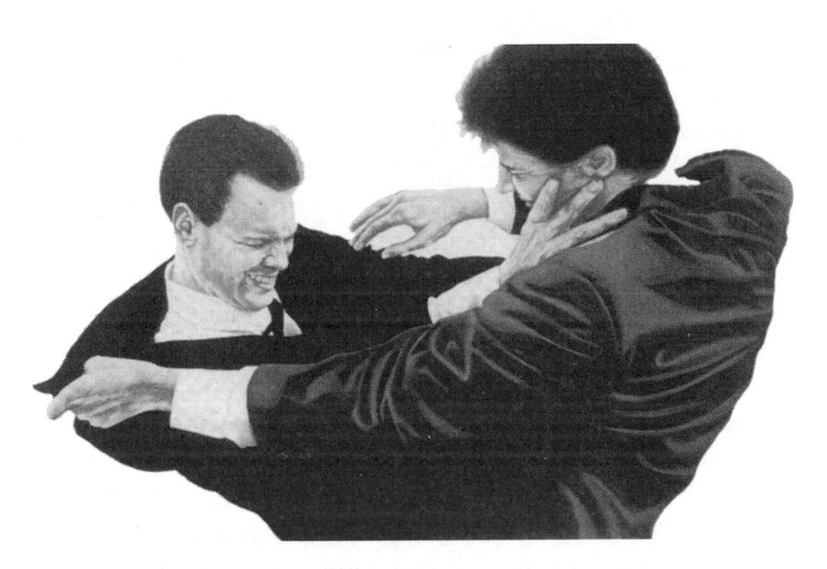

Robert Longo, 1981: untitled work on paper. (Two yuppies contesting the last double latte at a restaurant?)

diate physiological adaptations if you are going to live. Your body's responses are brilliantly adapted for handling this sort of emergency.

An organism can also be plagued by *chronic physical stressors*. The locusts have come and eaten your crops; for the next six months, you have to wander a dozen miles a day to get enough food. Drought, famine, parasites, that sort of unpleasantness—not the sort of experience we have often, but central events in the lives of nonwesternized humans and most other mammals. The body's stress-responses are reasonably good at handling these sustained disasters.

Critical to this book is a third category of ways to get upset—*psychological and social stressors*. Regardless of how poorly we are getting along with a family member or how incensed we are about losing a parking spot, we rarely settle that sort of thing with a fistfight. Likewise, it is a rare event when we have to stalk and personally wrestle down our dinner. Essentially, we humans live well enough and long enough, and are smart enough, to generate all sorts of stressful events purely in our heads. How many hippos worry about whether Social Security is going to last as long as they will, or even what they are going to say on a first date? Viewed from the perspective of the evolution of the animal kingdom, psychological stress is a recent invention. We humans can experience wildly strong emotions (provoking our bodies into an accompanying uproar) linked to mere thoughts. Two people can sit facing each other, doing nothing more physically strenuous than moving little pieces of wood now and then, yet this can be an emotionally taxing event: chess grand masters, during their tournaments, can place metabolic demands on their bodies that begin to approach those of athletes during the peak of a competitive event.* Or a person can do nothing more exciting than sign a piece of paper: if she has just signed the order to fire a hated rival after months of plotting and maneuvering, her physiological responses might be shockingly similar to those of a savanna baboon who has just lunged and slashed the face of a competitor. And if someone spends months on end twisting his innards in anxiety, anger, and tension over some emotional problem, this might very well lead to illness.

This is the critical point of this book: if you are that zebra running for your life, or that lion sprinting for your meal, your body's physiological response mechanisms are superbly adapted for dealing with such short-term physical emergencies. When we sit around and worry about stressful things, we turn on the same physiological responses—but they are potentially a disaster when provoked chronically. A large body of evidence suggests that stress-related disease emerges, predominantly, out of the fact that we so often activate a physiological system that has evolved for responding to acute physical emergencies, but we turn it on for months on end, worrying about mortgages, relationships, and promotions.

This difference between the ways that we get stressed and the ways a zebra does lets us begin to wrestle with some definitions. To start, I must call forth a concept that you were tortured with in ninth-grade biology and probably have not had to think about since—homeostasis. Ah, that dimly remembered concept, the idea that the body has an ideal level of oxygen that it

---

* Perhaps journalists are aware of this fact; consider this description of the Kasparov-Karpov tournament of 1990: "Kasparov kept pressing for a murderous attack. Toward the end, Karpov had to oppose threats of violence with more of the same and the game became a melee."

needs, an ideal degree of acidity, an ideal temperature, and so on. All these different variables are maintained in homeostatic balance, the state in which all sorts of physiological measures are being kept at the optimal level for a particular time of day, season, age of organism, and so on.

The homeostatic concept has been modified in recent years, producing a new term that I steadfastly tried to ignore until recently. The original homeostatic concept was grounded in the idea that there was a single optimum for any measure in your body, and that some local mechanism maintained that optimum. *Allostasis* refers to the notion that different circumstances demand different homeostatic set points (after all, the ideal blood pressure when you are sleeping is likely to be quite different than when you are bungee-jumping), and that maintaining whatever an optimal set point might be typically demands far-flung regulatory changes throughout the body instead of just local adjustments.

Within that framework, a stressor can be defined as anything that throws your body out of allostatic balance—for example, an injury, an illness, subjection to great heat or cold. The stress-response, in turn, is your body's attempt to restore balance. This consists of the secretion of certain hormones, the inhibition of others, the activation of particular parts of the nervous system, and other physiological changes that will be described in subsequent chapters. These are definitions that would suffice for that lion or zebra.

But when we consider ourselves and our human propensity to worry ourselves sick, we have to expand on the notion of stressors merely being things that knock you out of allostatic balance. A stressor can also be the *anticipation* of that happening. Sometimes we are smart enough to see things coming and, based only on anticipation, can turn on as robust a stress-response as if the event had actually occurred. Some aspects of anticipatory stress are not unique to humans—whether you are a human surrounded by a bunch of threatening teenagers in a deserted subway station or a zebra face to face with a lion, your heart is probably racing, even though nothing physically damaging has occurred (yet). But unlike less cognitively sophisticated species, we can turn on the stress-response by thinking about potential stressors that may throw us out of allostatic balance far in the future. For example, think of the African farmer watching a swarm of locusts descend on his crops. There is some food stored away; he is not about to suffer the allostatic imbalance of starving for months, but that man will still be undergoing a stress-response. Zebras and lions may see trouble coming and mobilize a stress-response in anticipation, but they can't get stressed about things so far in advance.

And sometimes we humans can be stressed by things that simply make no sense to zebras or lions. It is not a general mammalian trait to become anxious about mortgages or the Internal Revenue Service, about public speaking or fears of what you will say in a job interview, about the inevitability of death. Our human experience is replete with psychological stressors, a far cry from the physical world of hunger, injury, blood loss, or temperature extremes. When we activate the stress-response out of fear of something that turns out to be real, we congratulate ourselves that this cognitive skill allows us to mobilize our defenses early. And when we get into a physiological uproar for no reason at all, or over something we cannot do anything about, we call it things like "anxiety," "neurosis," "paranoia," or "needless hostility."

Thus, the stress-response can be mobilized not only in response to physical or psychological insults, but also in expectation of them. It is this generality of the stress-response that is the most surprising—a physiological system activated not only by all sorts of physical disasters but also by just thinking about them as well. This generality was first appreciated about sixty years ago by one of the godfathers of stress physiology, Hans Selye. To be only a bit facetious, stress physiology exists as a discipline because this man was both a very insightful scientist and rather inept at handling laboratory rats.

In the 1930s, Selye was just beginning his work in endocrinology, the study of hormonal communication in the body. Naturally, as a young, unheard-of assistant professor, he was fishing around for something with which to start his research career. A biochemist down the hall had just isolated some sort of extract from the ovary, and colleagues were wondering what this ovarian extract did to the body. So Selye obtained some of the stuff from the biochemist and set about studying its effects. He attempted to inject his rats daily, but apparently not with a great display of dexterity. Selye would try to inject the rats, miss them, drop them, spend half the morning chasing the rats around the room or vice versa, flailing with a broom to get them out from behind the sink, and so on. At the end of a number of months of this, Selye examined the rats and discovered something extraordinary: the rats had peptic ulcers, greatly enlarged adrenal glands (the source of two important stress hormones), and shrunken immune tissues. He was delighted; he had discovered the effects of the mysterious ovarian extract.

Being a good scientist, he ran a control group: rats injected daily with saline alone, instead of the ovarian extract. And, thus, every day they too were injected, dropped, chased, and chased back. At the end, lo and behold, the control rats had the same peptic ulcers, enlarged adrenal glands, and atrophy of tissues of the immune system.

Now your average budding scientist, at this point, might throw up his or her hands and furtively apply to business school. But Selye, instead, reasoned through what he had observed. The physiological changes couldn't be due to the ovarian extract after all, since the same changes occurred in both the control and the experimental groups. What did the two groups of rats have in common? Selye reasoned that it was his less-than-trauma-free injections. Perhaps, he thought, these changes in the rats bodies were some sort of nonspecific responses of the body to generic unpleasantness. To test this idea, he put some on the roof of the research building in the winter, others down in the boiler room. Still others were exposed to forced exercise, or to surgical procedures. In all cases, he found increased incidences of peptic ulcers, adrenal enlargement, and atrophy of immune tissues.

We know now exactly what Selye was observing. He had just discovered the tip of the iceberg of stress-related disease. Legend (mostly promulgated by Selye himself) has it that Selye was the person who, searching for a way to describe the nonspecificity of the unpleasantness to which the rats were responding, borrowed a term from engineering and proclaimed that the rats were undergoing "stress." But in fact, in the 1920s the term had already been introduced to medicine in roughly the sense that we understand it today by a physiologist named Walter Cannon. What Selye did was to formalize the concept with two ideas:

- The body has a surprisingly similar set of responses (which he called the *general adaptation syndrome*) to a broad array of stressors.
- Under certain conditions, stressors will make you sick.

It is this generality that is puzzling. If you are trained in physiology, it makes no sense at first glance. In physiology, one is typically taught that *specific* challenges to the body trigger *specific* responses and adaptations. Warming a body causes sweating and dilation of blood vessels in the skin. Chilling a body causes just the opposite—constriction of those vessels and shivering. Being too hot seems to be a very specific and different physiological challenge from being too cold, and it would seem logical that the body's responses to these two very different states should be extremely different. Instead, what kind of crazy bodily system is this that is turned on whether you are too hot or too cold, whether you are the zebra, the lion, or a terrified adolescent going to a high-school dance? Why should your body have such a generalized and stereotypical stress-response, regardless of the predicament you find yourself in?

When you think about it, it makes some physiological sense, given the adaptations brought about by the stress-response. Regardless of whether you are that zebra or that lion, you need energy if you are going to survive a demanding emergency. And you need it immediately, in the most readily utilizable form, rather than stored away somewhere in your fat cells for some building project next spring. One of the hallmarks of the stress-response is the rapid mobilization of energy from storage sites and the inhibition of further storage. Glucose and the simplest forms of proteins and fats come pouring out of your fat cells, liver, and muscles, all to stoke whichever muscles are struggling to save your neck.

If your body has mobilized all that glucose, it needs also to deliver it to the critical muscles as rapidly as possible. Heart rate, blood pressure, and breathing rate increase, all to transport nutrients and oxygen at greater rates.

Equally logical is another feature of the stress-response. During an emergency, it makes sense that your body halts long-term, expensive building projects. If there is a tornado bearing down on the house, this isn't the day to repaint the kitchen.

Hold off until you've weathered the disaster. Thus, during stress, digestion is inhibited—there isn't enough time to derive the energetic benefits of the slow process of digestion, so why waste energy on it? You have better things to do than digest breakfast when you are trying to avoid being someone's lunch. Similarly, growth is inhibited during stress, and the logic is just as clear. You're sprinting for your life: extending your long bones shouldn't be at the top of your list of priorities. Along with that, reproduction, probably the most energy-expensive, optimistic thing that you can do with your body (especially if you are female), is curtailed—keep your energies on the problem at hand, and worry about eggs and sperm some other time. During stress, sexual drive decreases in both sexes; females are less likely to ovulate or to carry pregnancies to term, while males begin to have trouble with erections and secrete less testosterone.

Along with these changes, immunity is also inhibited. The immune system, which defends against infections and illness, is ideal for spotting the tumor cell that will kill you in a year, or making enough antibodies to protect you in a few weeks, but is it really needed this instant? The logic here appears to be the same—look for tumors some other time; expend the energy more wisely now. (As we will see in Chapter 8, there are some major problems with this idea that the immune system is suppressed during stress in order to save energy.)

Another feature of the stress-response becomes apparent during times of extreme physical pain. With sufficiently sustained stress, our perception of pain can become blunted. It's the middle of a battle; soldiers are storming a stronghold with wild abandon. A soldier is shot, grievously injured, and the man doesn't even notice it. He'll see blood on his clothes and worry that one of his buddies near him has been wounded, or he'll wonder why his innards feel numb. As the battle fades, someone will point with amazement at his injury—didn't it hurt like hell? It didn't. Such stress-induced analgesia is highly adaptive and well documented. If you are that zebra and injured, you still have to escape. Now would not be a particularly clever time to go into shock from extreme pain.

Finally, during stress, shifts occur in cognitive and sensory skills. Suddenly certain aspects of memory improve, which is always helpful if you're trying to figure out how to get out of an emergency (has this happened before? is there a good hiding place?). Moreover, your senses become sharper. Think about watching a terrifying movie on television, on the edge of your seat at the tensest part. The slightest noise—a creaking door—and you nearly jump out of your skin. Better memory, sharper detection of sensations—all quite adaptive and helpful.

Collectively, the stress-response is ideally adapted for that zebra or lion. Energy is mobilized and delivered to the tissues that need them; long-term building and repair projects are deferred until the disaster has passed. Pain is blunted, cognition sharpened. Walter Cannon, the physiologist who, at the beginning of the century, paved the way for much of Selye's work and is generally considered the other godfather of the field, concentrated on the adaptive aspect of the stress-response in dealing with emergencies such as these. He formulated the well-known "fight or flight" syndrome to describe the stress-response, and he viewed it in a very positive light. His books, with titles such as *The Wisdom of the Body*, were suffused with a pleasing optimism about the ability of the body to weather all sorts of stressors.

Yet stressful events can sometimes make us sick. Why?

Selye, with his ulcerated rats, wrestled with this puzzle and came up with an answer that was sufficiently wrong that it is generally thought to have cost him a Nobel prize. He developed a three-part view of how the stress-response worked. In the initial (alarm) stage a stressor is noted; metaphorical alarms go off in your head, telling you that you are hemorrhaging, too cold, low on blood sugar, or whatever. The second stage (adaptation, or resistance) comes with the successful mobilization of the stress-response system and the reattainment of allostatic balance.

It is with prolonged stress that one enters the third stage, which Selye termed "exhaustion," where stress-related diseases emerge. Selye believed that one becomes sick at that point because stores of the hormones secreted during the stress-response are depleted. Like an army that runs out of bullets, suddenly we have no defenses left against the threatening stressor.

It is very rare, however, as we will see, that any of the crucial hormones are actually depleted during even the most sustained of stressors. The army does not run out of bullets. Instead, spending so much on bullets causes the rest of the body's economy to collapse. It is not so much that the stress-response runs out; rather, with sufficient activation, *the stress-response itself can become damaging*. This is a critical concept, because it underlies the emergence of much stress-related disease.

That the stress-response itself can become harmful makes a certain sense when you examine the things that occur in reaction to stress. They are generally shortsighted, inefficient, and penny-wise and dollar-foolish, but they are the sorts of costly things your body has to do to respond effectively in an emergency. If you experience every day as an emergency, you will pay the price.

If you constantly mobilize energy at the cost of energy storage, you will never store any surplus energy. You will fatigue more rapidly, and your risk of developing a form of diabetes will even increase. The consequences of chronically overactivating your cardiovascular system are similarly damaging: if your blood pressure rises to 180/120 when you are sprinting away from a lion, you are being adaptive, but if it is 180/120 every time you see the mess in your teenager's bedroom, you could be heading for a cardiovascular disaster. If you constantly turn off long-term building projects, nothing is ever repaired. For paradoxical reasons that will be explained in later chapters, you become more at risk for peptic ulcers. In kids, growth can be inhibited to the point of a rare but recognized pediatric endocrine disorder—stress dwarfism—and in adults, repair and remodeling of bone and other tissues can be disrupted. If you are constantly under stress, a variety of reproductive disorders may ensue. In females, menstrual cycles can become irregular or cease entirely; in males, sperm count and testosterone levels may decline. In both sexes, interest in sexual behavior decreases.

But that is only the start of your problems in response to chronic or repeated stressors. If you suppress immune function too long, trouble will surely follow. The tragedy of those who have AIDS has taught us that we become susceptible to all sorts of horrendous diseases if we are grossly immunodeficient. In a less dramatic way, the immune suppression brought about by chronic stress may exact a price: individuals are less likely to resist any of a variety of diseases.

Finally, the same systems of the brain that function more cleverly during stress can also be damaged by one class of hormones secreted during stress. As will be discussed, this may have something to do with how rapidly our brains lose cells during aging, and how much memory loss occurs with old age.

All of this is pretty grim. In the face of repeated stressors, we may be able to precariously reattain allostasis, but it doesn't come cheap, and the efforts to reestablish that balance will eventually wear us down. Here's a way to think about it: the "two elephants on a seesaw" model of stress-related disease. Put two little kids on a seesaw, and they can pretty readily balance themselves on it. This is allostatic balance when nothing stressful is going on, with the children representing the low levels of the various stress hormones that will be presented in coming chapters. In contrast, the torrents of those same stress hormones released by a stressor can be thought of as two massive elephants on the seesaw. With great effort, they can balance themselves as well. But if you constantly try to balance a seesaw with two elephants instead of two little kids, all sorts of problems will emerge:

- First, the enormous potential energies of the two elephants are consumed balancing the seesaw, instead of being put to some more useful task, such as having the elephants mow the lawn or paint the house. This is equivalent to diverting energy from various long-term building projects in order to solve short-term stressful emergencies.
- By using two elephants to do the job, damage will occur just because of how large, lumbering, and unsubtle elephants are. They squash the flowers in the process of entering the playground, they strew leftovers and garbage all over the place from the snacks they must eat while balancing the seesaw, they wear out the seesaw faster, and so on. This is equivalent to a pattern of stress-related disease that will run through many of the subsequent chapters: it is hard to fix one major problem in the body without knocking something else out of balance (the very essence of allostasis spreading across systems throughout the body). Thus, you may be able to solve one bit of imbalance brought on during stress by using your elephants (your massive levels of various stress hormones), but such great quantities of those hormones can make a mess of something else in the process.
- A final, subtle problem: when two elephants are balanced on a seesaw, it's tough for them to get off. Either one hops off

and the other comes crashing to the ground, or there's the extremely delicate task of coordinating their delicate, lithe leaps at the same time. This is a metaphor for another theme that will run through subsequent chapters—sometimes stress-related disease can arise from turning off the stress-response too slowly, or turning off the different components of the stress-response at different speeds. When the secretion rate of one of the hormones of the stress-response returns to normal yet another of the hormones is still being secreted like mad, that can be the equivalent of one elephant suddenly being left alone on the seesaw, crashing to earth.*

The preceding pages should allow you to begin to appreciate the two punch lines of this book:

- If you are faced with a physical stressor and you cannot appropriately *turn on* the stress-response, you are in big trouble. To see this, all you have to do is examine someone who cannot activate the stress-response. As will be explained in the coming chapters, two critical classes of hormones are secreted during stress. In one disorder, Addison's disease, you are unable to secrete one class of these hormones, called *glucocorticoids*. In another, called *Shy-Drager syndrome*, it is the secretion of the second class, the hormones epinephrine and norepinephrine (also called *adrenaline* and *noradrenaline*), that is impaired. People with Addison's disease or Shy-Drager syndrome are not more at risk for cancer or heart disease or any other such disorders of slow accumulation of damage. However, people with untreated Addison's disease, when faced with a major stressor such as a car accident or an infectious illness, fall into an "Addisonian" crisis—their blood pressure drops, they cannot maintain circulation, they go into shock. In Shy-Drager's syndrome, it is hard enough simply to stand up, let alone go sprinting after a zebra for dinner—mere standing causes a severe drop in blood pressure, involuntary twitching and rippling of muscles, dizziness, all sorts of

---

* If you find this analogy silly, imagine what it is like to have a bunch of scientists locked up together at a stress conference working with it. I was at a meeting where this analogy was first introduced, and in no time there were factions pushing analogies about elephants on pogo sticks, elephants on monkey bars and merry-go-rounds, sumo wrestlers on seesaws, and so on.

unpleasantness. These two diseases teach something important, namely, that you need the stress-response during physical challenges.

- That first punch line is obviously critical, especially for the zebra who occasionally has to run for its life. But the second punch line is far more relevant to us, sitting frustrated in traffic jams, worrying about expenses, mulling over tense interactions with colleagues. If you *repeatedly turn on* the stress-response, or if you cannot appropriately *turn off* the stress-response at the end of a stressful event, the stress-response can eventually become nearly as damaging as some stressors themselves. A large percentage of what we think of when we talk about stress-related diseases are disorders of excessive stress-response.

A few important qualifications are necessary concerning that last statement, which is one of the central ones of this book. On a superficial level, the message it imparts might seem to be that stressors make you sick or, as emphasized in the last few pages, that chronic or repeated stressors make you sick. It is actually more accurate to say that chronic or repeated stressors can *potentially* make you sick or can increase your *risk* of being sick. Stressors, even if massive, repetitive, or chronic in nature, do not automatically lead to illness, and most of us are free of stress-related diseases at any given time. Much of the final chapter will concentrate on this good piece of news.

There is an additional point that should be emphasized. To state that "chronic or repeated stressors can increase your risk of being sick" is actually incorrect, but in a subtle way that will initially seem like semantic nit-picking. It is never really the case that stress makes you sick, or even increases your risk of being sick. Stress increases your risk of getting *diseases* that make you sick, or if you have such a disease, stress increases the risk of your defenses being overwhelmed by the disease. This distinction is important in a few ways. First, by putting more steps between a stressor and getting sick, there are more explanations for individual differences—why only some people wind up actually getting sick. Moreover, by clarifying the progression between stressors and illness, it becomes easier to design ways to intervene in the process. Finally, it begins to explain why the stress concept often seems so suspect or slippery to many practitioners of more mainstream medicine—clinical medicine is traditionally quite good at being able to make statements like "You feel sick because you have disease X," but is traditionally quite bad at being able to explain how you got disease X in the first place. Thus, mainstream medical practitioners often say, in effect, "You feel sick because you have disease X, not because of some nonsense having to do with stress," which ignores the stressors' role in where the disease came from.

With this framework in mind, we can now begin the task of understanding the individual steps in this system. Chapter 2 introduces the hormones and brain systems involved in the stress-response: which ones are activated during stress, which ones are inhibited? This leads the way to Chapters 3 through 10, which examine the individual parts of your body that are affected. How do those hormones enhance cardiovascular tone during stress, and how does chronic stress cause heart disease (Chapter 3)? How do those hormones and neural systems mobilize energy during stress, and how does too much stress cause energetic diseases (Chapter 4)? And so on. Chapter 11 examines the role of stress in the aging process and the disturbing recent findings that sustained exposure to certain of the hormones secreted during stress may actually accelerate the aging of the brain, including our own. As will be seen, these processes are often more complicated and subtle than they may seem from the simple picture presented in this chapter.

Chapter 12 ushers in a topic obviously of central importance to understanding our own propensity toward stress-related disease: why is psychological stress stressful? This serves as a prelude to the remaining chapters. Chapter 13 reviews major depression, a horrible psychiatric malady that afflicts vast numbers of us and that is often closely related to psychological stress. Chapter 14 discusses what personality differences have to do with individual differences in patterns of stress-related disease. This is the world of anxiety disorders and Type A-ness, plus some surprises about unexpected links between personality and the stress-response. Chapter 15 takes a larger view still, looking at what your place in society, and the type of society in which you live, has to do with patterns of stress-related disease. If you plan to go no further, here's one of the punch lines of that chapter—if you want to increase your chances of avoiding stress-related diseases, make sure you don't inadvertently allow yourself to be born poor.

In many ways, the ground to be covered up to Chapter 16 is all bad news, as we are regaled with the evidence about new and unlikely parts of our bodies and minds that are made miserable by stress. The final chapter is meant to give some hope. Given the same external stressors, certain bodies and certain psyches deal with stress better than others. What are those folks doing right, and what can the rest of us learn from them? We'll look at the main principles of stress management, and some surprising and exciting realms in which they have been applied with stunning success. While the intervening chapters document our numerous vulnerabilities to stress-related disease, the final chapter shows that we have an enormous potential to protect ourselves from many of them. Most certainly, all is not lost.

2

GLANDS, GOOSEFLESH, AND HORMONES

In order to begin the process of learning how stress can make us sick, there is something about the workings of the brain that we have to appreciate. It is perhaps best illustrated in the following rather technical paragraph from an early investigator in the field:

> As she melted small and wonderful in his arms, she became infinitely desirable to him, all his blood-vessels seemed to scald with intense yet tender desire, for her, for her softness, for the penetrating beauty of her in his arms, passing into his blood. And softly, with that marvellous swoon-like caress of his hand in pure soft desire, softly he stroked the silky slope of her loins, down, down between her soft, warm buttocks, coming nearer and nearer to the very quick of her. And she felt him like a flame of desire, yet tender, and she felt herself melting in the flame. She let herself go. She felt his penis risen against her with silent amazing force and assertion, and she let herself go to him. She yielded with a quiver that was like death, she went all open to him.

Now think about this. If D. H. Lawrence is to your taste, there may be some interesting changes occurring in your body. You haven't just run up a flight of stairs, but maybe your heart is beating faster. The temperature has not changed in the room, but you may have just activated a sweat gland or two. And even though certain rather sensitive parts of your body are not being overtly stimulated by touch, you are suddenly very aware of them.

You sit in your chair not moving a muscle, and simply think a thought, a thought having to do with your feeling angry or sad or euphoric or lustful, and suddenly your pancreas secretes some hormone. Your *pancreas*? How did you manage to do that with your pancreas? You don't even know where your pancreas is. Your liver is making an enzyme that wasn't there before, your spleen is faxing a message to your thymus gland, blood flow in little capillaries in your ankles has just changed. All from thinking a thought.

We all understand intellectually that the brain can regulate functions throughout the rest of the body, but it is still surprising to be reminded of how far-reaching those effects can be. The purpose of this chapter is to learn a bit about the lines of communication between the brain and elsewhere, in order to see which sites are activated and which quieted when you are sitting in your chair and feeling severely stressed. This is a prerequisite for seeing how the stress-response can save your neck during a sprint across the savanna, but make you sick during months of worry.

STRESS AND THE AUTONOMIC NERVOUS SYSTEM

The principal way in which your brain can tell the rest of the body what to do is to send messages through the nerves that branch from your brain down your spine and out to the periphery of your body. One dimension of this communication system is pretty straightforward and familiar. The voluntary nervous system is a conscious one. You decide to move a muscle and it happens. This part of the nervous system allows you to shake hands or fill out your tax forms or scratch behind your ear or do a polka. It is another branch of the nervous system that projects to organs besides skeletal muscle, and this part controls the other interesting things your body does—blushing, getting gooseflesh, having an orgasm. In general, we have less control over what our brain says to our sweat glands, for example, than

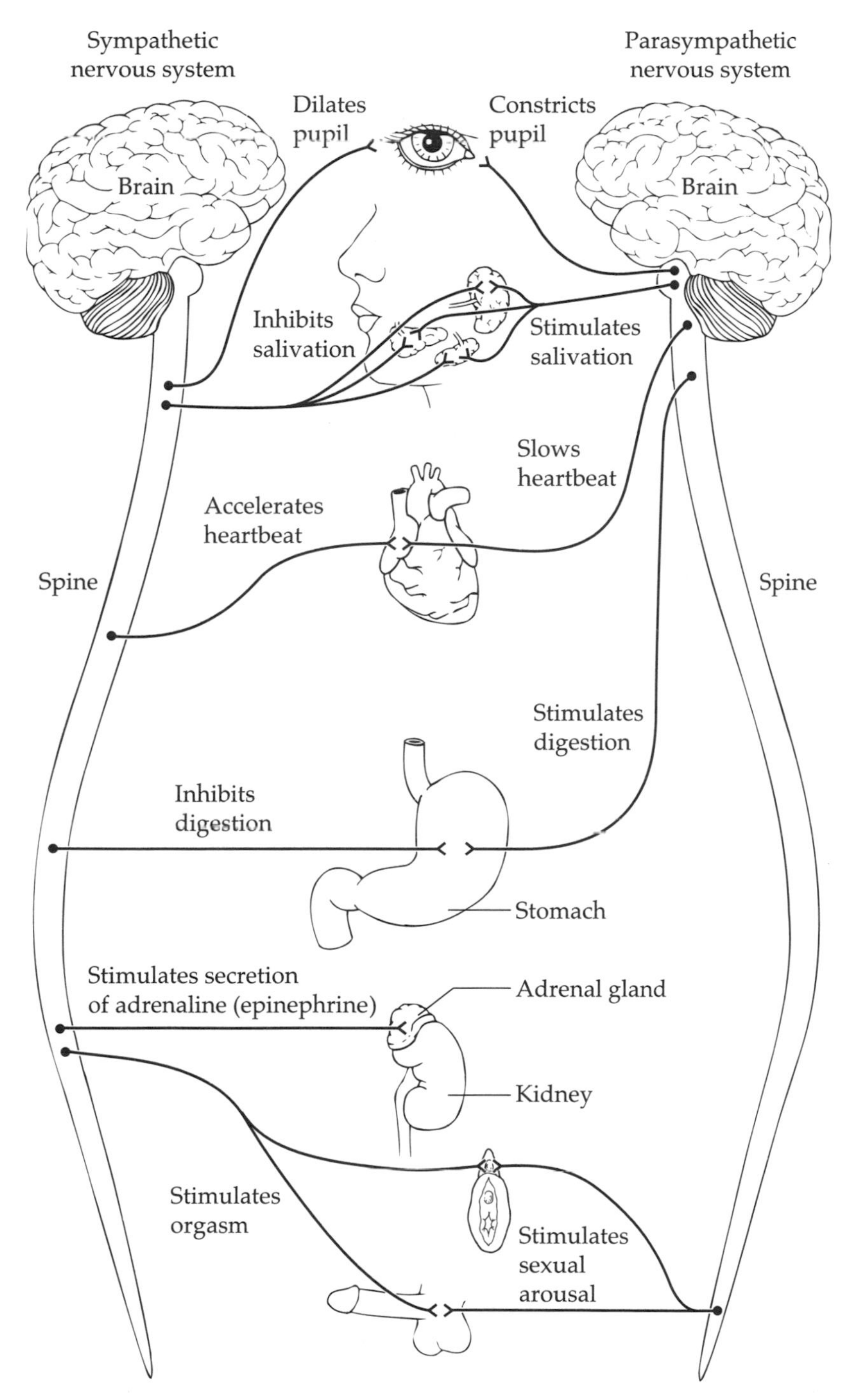

Outline of some of the effects of the sympathetic and parasympathetic nervous systems on various organs and glands.

to our thigh muscles. (The workings of this autonomic nervous system are not entirely out of our control, however; biofeedback, for example, consists of learning to alter autonomic nervous system function consciously. On a more mundane level, we are doing the same thing when we repress a loud burp during a wedding ceremony.) The set of nerve projections to places like sweat glands carry messages that are relatively involuntary and automatic. It is thus termed the *autonomic nervous system,* and it has everything to do with your response to stress. One half of this system is activated in response to stress, one half is suppressed.

The half of the autonomic nervous system that is turned on is called the *sympathetic nervous system.** Originating in the brain, sympathetic projections exit your spine and branch out to nearly every organ, every blood vessel, and every sweat gland in your body. They even project to the scads of tiny little muscles attached to hairs on your body. If you are truly terrified by something and activate those projections, your hair stands on end; gooseflesh results when the parts of your body are activated where those muscles exist but lack hairs attached to them.

The sympathetic nervous system kicks into action during emergencies, or what you think are emergencies. It helps mediate vigilance, arousal, activation, mobilization. To generations of first-year medical students, it is described through the feeble but obligatory joke of mediating the four F's of behavior—flight, fight, fright, and sex. It is the archetypal system that is turned on at times when life gets exciting or alarming—such as during stress. The nerve endings of this system release adrenaline. When someone jumps out from behind a door and startles you, it's your sympathetic nervous system releasing adrenaline that causes your stomach to clutch. Sympathetic nerve endings also

---

* Where'd this name come from? According to the eminent stress physiologist Seymour Levine, this goes back to Galen, who believed that the brain was responsible for rational thought and the peripheral viscera for emotions. Seeing this collection of neural pathways linking the two suggested that it allowed your brain to sympathize with your viscera. Or maybe for your viscera to sympathize with your brain. As well see shortly, the other half of the autonomic nervous system is called the *parasympathetic nervous system. Para,* meaning "alongside," refers to the not very exciting fact that the parasympathetic neural projections sit alongside those of the sympathetic.

"Oh, that's Edward and his fight or flight mechanism."

release the closely related substance noradrenaline. (*Adrenaline* and *noradrenaline* are actually British designations; the American terms, which will be used from now on, are *epinephrine* and *norepinephrine*.) Epinephrine is secreted by the sympathetic nerve endings in your adrenal glands (located just above your kidneys); norepinephrine is secreted by all the other sympathetic nerve endings throughout the body. These are the chemical messengers that kick various organs into gear, within seconds.

The other half of the autonomic nervous system plays an opposing role. This parasympathetic component mediates calm, vegetative activities—everything but the four F's. If you are a growing kid and you have gone to sleep, your parasympathetic system is activated. It promotes growth, energy storage, and other optimistic processes. Have a huge meal, sit there bloated and happily drowsy, and the parasympathetic is going like gangbusters. Sprint for your life across the savanna, gasping and trying to control the panic, and you've turned the parasympathetic component down. Thus, the autonomic system works in opposition: sympathetic and parasympathetic projections from the brain course their way out to a particular organ where, when activated, they bring about opposite results. The sympathetic system speeds up the heart; the parasympathetic system slows it down. The sympathetic system diverts blood flow to your muscles; the parasympathetic does the opposite. It's no surprise that it would be a disaster if both branches were very active at the same time, kind of like putting your foot on the gas and brake simultaneously. Lots of safety features exist to make sure that does not happen. For example, the parts of the brain that activate the sympathetic component during a stressful emergency, or when you are anticipating one, typically inhibit the parasympathetic at the same time.

YOUR BRAIN: THE REAL MASTER GLAND

The neural route represented by the sympathetic system is a first means by which the brain can mobilize waves of activity in response to a stressor. There is another way as well—through the secretion of hormones. If a neuron (a cell of the nervous system) secretes a chemical messenger that travels a thousandth of an inch and causes the next cell in line (typically, another neuron) to do something different, that messenger is called a *neurotransmitter*. Thus, when the sympathetic nerve endings in your heart secrete norepinephrine, which causes heart muscle to work differently, norepinephrine is playing a neurotransmitter role. If a neuron (or any cell) secretes a messenger that, instead, percolates into the bloodstream and affects events far and wide, that messenger is a hormone. All sorts of glands secrete hormones; the secretion of some of them is turned on during stress, and the secretion of others is turned off.

What does the brain have to do with all of these glands secreting hormones? People used to think, "nothing." The assumption was that the peripheral glands of the body—your pancreas, your adrenal, your ovaries, your testes, and so on—in some mysterious way "knew" what they were doing, had "minds of their own." They would "decide" when to secrete their messengers, without directions from any other organ. This erroneous idea gave rise to a rather silly fad during the early part of this century. Scientists noted that men's sexual drive declined with age, and assumed that this occurs because the testicles of aging men secrete less male sex hormone, testosterone. (Actually, no one knew about the hormone "testosterone" at the time; they just referred to mysterious "male factors" in the testes. And in fact, testosterone levels do not plummet with age. Instead, the decline is moderate and highly variable from one male to the next, and even a decline in testosterone to perhaps 10 percent of normal levels does not have much of an effect on sexual behavior.) Making another leap, they then ascribed aging to diminishing sexual drive, to lower levels of male factors. (One may then wonder why females, without testes, manage to grow old, but the female half of the population didn't figure much in these ideas back then.) How, then, to reverse aging? Give the aging males some testicular extracts.

Soon, aged, monied gentlemen were checking into impeccable Swiss sanitariums and getting injected daily in their rears with testicular extracts from dogs, from roosters, from monkeys. You could even go out to the stockyards of the sanitarium and pick out the goat of your choice—just like picking lobsters in a restaurant (and more than one gentlemen arrived for his appointment with his own prized animal in tow). This soon led to an offshoot of such "rejuvenation therapy," namely, "organotherapy"—the grafting of little bits of testes themselves. Thus was born the "monkey gland" craze, the term *gland* being used because journalists were forbidden to print the racy word *testes*. Captains of industry, heads of state, at least one pope—all signed up. And in the aftermath of the carnage of World War I, there was such a shortage of young men and such a surfeit of marriages of younger women to older men, that therapy of this sort seemed pretty important.

Naturally, the problem was that it didn't work. There wasn't any testosterone in the testicular extracts—patients would be injected with a water-based extract, and testosterone does not go into solution in water. And the smidgens of organs that were transplanted would die almost immediately, with the scar tissue

PROFESSOR

BROWN SEQUARD'S

METHOD.

EXTRACTS OF ANIMAL ORGANS.

Testicle Extract,
Grey Matter Extract,
Tyroid Gland Extract, &c., &c.

Concentrated Solutions at 30%.

These preparations, completely aseptic, are mailed to any distance on receipt of a money order. Directions sent with the fluids.

Price for 25 Injections, $2.50.
Syringe Specially Gauged, (3 cubic c.,) $2.50.

Used in the Hospitals of Paris, New York, Boston, etc.
Circular Sent on Application.

New York Biological and Vaccinal Institute,

Laboratory of Bovine Vaccine and of Biological Products.

GEO. G. RAMBAUD, Chemist and Bacteriologist, Superintendent.
PASTEUR INSTITUTE BUILDING, NEW YORK CITY.

Advertisement, New York Therapeutic Review, *1893.*

being mistaken for a healthy graft. And even if they didn't die, they still wouldn't work—if aging testes are secreting less testosterone, it is not because the testes are failing, but because another organ (stay tuned) is no longer telling them to do so. Put in a brand-new set of testes and they should fail also, for lack of a stimulatory signal. But not a problem. Nearly everyone reported wondrous results anyway. If you're paying a fortune for painful daily injections of extracts of a dog's testicles, there's a certain incentive to decide you feel like a young bull. One big placebo effect.

With time, scientists figured out that the testes and other peripheral hormone-secreting glands were not autonomous, but were under the control of something else. Attention turned to the pituitary gland, sitting just underneath the brain. It was known that when the pituitary was damaged or diseased, hormone secretion throughout the body became disordered. In the early part of the century, careful experiments showed that a peripheral gland releases its hormone only if the pituitary first releases a hormone that kicks that gland into action. The pituitary contains a whole array of hormones that run the show throughout the rest of the body; it is the pituitary that actually knows the game plan and regulates what all the other glands do. This realization gave rise to the memorable statement that the pituitary is the master gland of the body.

This understanding was disseminated far and wide, mostly in the *Reader's Digest*, which ran the "I Am Joe's" series of articles ("I Am Joe's Pancreas," "I Am Joe's Shinbone," "I Am Joe's Ovaries," and so on). By the third paragraph of "I Am Joe's Pituitary," out comes that master gland business. By the 1950s, however, scientists were already learning that the pituitary wasn't the master gland after all.

The simplest evidence was that if you removed the pituitary from a body and put it in a small bowl filled with pituitary nutrients, the gland would act abnormally. Various hormones that it would normally secrete were no longer secreted. Sure, you might say, remove any organ and throw it in some nutrient soup and it isn't going to be good for much of anything. But, interestingly, while this "explanted" pituitary stopped secreting certain hormones, it did secrete others at immensely high rates. It wasn't just that the pituitary was traumatized. It was acting erratically because, it turned out, the pituitary didn't really have the whole hormonal game plan. It would normally be following orders from the brain, and there was none on hand in that small bowl to give directions.

The evidence for this was relatively easy to obtain. Destroy the part of the brain right near the pituitary and the pituitary stops secreting some hormones and oversecretes others. This tells you that the brain controls certain pituitary hormones by stimulating their release and controls others by inhibiting them. The problem was to figure out how the brain did this. By all logic, you would look for nerves to project from the brain to the pituitary (like the nerve projections to the heart and elsewhere), and for the brain to release neurotransmitters that called the shots. But no one could find these projections. In 1944, the physiologist Geoffrey Harris proposed that the brain was also a hormonal gland, that it released hormones that traveled to the pituitary and directed the pituitary's actions. In principle, this was not a crazy idea; a quarter-century before, one of the godfathers of the field, Ernst Scharrer, had shown that some other hormones, thought to originate from a peripheral gland, were actually made in the brain. Nevertheless, lots of scientists thought Harris's idea was bonkers. You can get hormones from peripheral glands like ovaries, testes, pancreas—but your *brain* oozing hormones? Preposterous!

Two scientists, Roger Guillemin and Andrew Schally, began looking for these brain hormones. This was a stupendously difficult task. The brain communicates with the pituitary by a minuscule circulatory system, only slightly larger than the period at the end of this sentence. You couldn't search for these hypothetical brain "releasing hormones "and "inhibiting hormones" in the general circulation of blood; if the hormones existed, by the time they reached the voluminous general circulation, they would be diluted beyond detection. Instead, you would have to search in the tiny bits of tissue at the base of the brain containing those blood vessels going from the brain to the pituitary.

Not a trivial task, but these two scientists were up to it. They were highly motivated by the abstract intellectual puzzle of these hormones, by their potential clinical applications, by the acclaim waiting at the end of this scientific rainbow. Plus, the two of them loathed each other, which invigorated the quest. Initially, in the late 1950s, Guillemin and Schally collaborated in the search for these brain hormones. Perhaps one tired evening over the test tube rack, one of them made a snide remark to the other—the actual events have sunk into historical obscurity; in any case a notorious animosity resulted, one enshrined in the annals of science at least on a par with the Greeks versus the Trojans, maybe even with Coke versus Pepsi. Guillemin and Schally went their separate ways, each intent on being the first to isolate the putative brain hormones.

How do you isolate a hormone that may not exist or that, even if it does, occurs in tiny amounts in a minuscule circulation system to which you can't gain access? Both Guillemin and Schally hit on the same strategy. They started collecting animal brains from slaughterhouses. Cut out the part at the base of the brain, near the pituitary. Throw a bunch of those in a blender, pour the resulting brain mash into a giant test tube filled with chemicals that purify the mash, collect the droplets that come out the other end. Then inject those droplets into a rat and see if the rat's pituitary changes its pattern of hormone release. If it does, maybe those brain droplets contain one of those imagined releasing or inhibiting hormones. Try to purify what's in the droplets, figure out their chemical structure, make an artificial version of it, and see if that regulates pituitary function. Pretty straightforward in theory. But it took them years.

One factor in this Augean task was the scale. There was at best a minuscule amount of these hormones in any one brain, so the scientists wound up dealing with thousands of brains at a time. The great Slaughterhouse war was on. Truckloads of pig or sheep brains were collected; chemists poured cauldrons of brain into monumental chemical-separation columns, while others pondered the thimblefuls of liquid that dribbled out the bottom, purifying it further in the next column and the next.... But it wasn't just mindless assembly-line work either. New types of chemistry had to be invented, completely novel ways of testing the effects in the living body of hormones that might or might not actually exist. An enormously difficult scientific problem, made worse by the fact that lots of influential people in the field believed these hormones were fictions and that these two guys were wasting a lot of time and money.

Guillemin and Schally pioneered a whole new corporate approach to doing science. One of our clichés is the lone scientist, sitting there at two in the morning, trying to figure out the meaning of a result. Here there were whole teams of chemists, biochemists, physiologists, and so on, coordinated into isolating these putative hormones. And it worked. A "mere" fourteen years into the venture, the chemical structure of the first releasing hormone was published.* Two years after that, in 1971, Schally got there with the sequence for the next hypothalamic

* "So," asks the breathless sports fan, "who won the race—Guillemin or Schally?" The answer depends on how you define "getting there first." The first hormone isolated indirectly regulates the release of thyroid hormone (that is, it controls the way in which the pituitary regulates the thyroid). Schally and crew were the first to submit a paper for publication saying, in effect, "There really does exist a hormone in the brain

(*continued on following page*)

hormone, and Guillemin published two months later. Guillemin took the next round in 1972, beating Schally to the next hormone by a solid three years. Everyone was delighted, the by-then-deceased Geoffrey Harris was proved correct, and Guillemin and Schally got the Nobel prize in 1976. One, urbane and knowing what would sound right, proclaimed that he was motivated only by science and the impulse to help mankind; he noted how stimulating and productive his interactions with his co-winner had been. The other, less polished but more honest, said the competition was all that drove him for decades and described his relationship with his co-winner as "many years of vicious attacks and bitter retaliation."

So hooray for Guillemin and Schally; the brain turned out to be *the* master gland. It is now recognized that the base of the brain, the hypothalamus, contains a huge array of those releasing and inhibiting hormones, which instruct the pituitary, which in turn regulates the secretions of the peripheral glands. In some cases, the brain triggers the release of pituitary hormone X through the action of a single releasing hormone. Sometimes it

---

that regulates thyroid hormone release, and its chemical structure is X." In a photo finish, Guillemin's team submitted a paper reaching the identical conclusion *five weeks* later. But as a complication, a number of months before, Guillemin and friends had been the first to publish a paper saying, in effect, "If you synthesize a chemical with structure X, it regulates thyroid hormone release and does so in a way similar to the way hypothalamic brain mash does; we don't know yet if whatever it is in the hypothalamus also has structure X, but we wouldn't be one bit surprised if it did." So Guillemin was the first to say, "This structure works like the real thing," and Schally was the first to say, "This structure is the real thing." As I have discovered firsthand, nearly a quarter of a century afterward, the battle-scarred veterans of the Guillemin-Schally prizefight years are still willing to get worked up as to which counts as the knockout.

One might wonder why something obvious wasn't done a few years into this insane competition, like the National Institutes of Health sitting the two down and saying, "Instead of us giving you all of this extra taxpayers' money to work separately, why don't you two work together?" Surprisingly, this wouldn't necessarily be all that great for scientific progress. The competition served an important purpose. Independent replication of results is essential in science. Years into a chase, a scientist triumphs and publishes the structure of a new hormone or brain chemical. Two weeks later the other guy comes forward. He has *every* incentive on earth to prove that the first guy was wrong. Instead, he is forced to say, "I hate that son of a bitch, but I have to admit he's right. We get the identical structure." That is how you know that your evidence is really solid, from independent confirmation by a hostile competitor. When everyone works together, things usually do go faster, but everyone winds up sharing the same assumptions, leaving them vulnerable to small, unexamined mistakes that can grow into big ones.

halts the release of pituitary hormone Y by releasing a single inhibiting hormone. In some cases, a pituitary hormone is controlled by the coordination of both a releasing and an inhibiting hormone from the brain—dual control. To make matters worse, in some cases (for example, the miserably confusing system that I study) there is a whole array of hypothalamic hormones that collectively regulate the pituitary, some as releasers, others as inhibitors.

HORMONES OF THE STRESS-RESPONSE

As the master gland, the brain can experience or think of something stressful and activate components of the stress-response hormonally. Some of the hypothalamus-pituitary-peripheral gland links are activated during stress, some inhibited.

Two hormones vital to the stress-response, as noted already, are epinephrine and norepinephrine, released by the sympathetic nervous system. Another important class of hormones in the response to stress are called *glucocorticoids*. By the end of this book you will be astonishingly informed about glucocorticoid trivia, since I am in love with these hormones. Glucocorticoids are steroid hormones. (*Steroid* is used to describe the general chemical structure of five classes of hormones: androgens—the famed "anabolic" steroids like testosterone that get you thrown out of the Olympics—estrogens, progestins, mineralocorticoids, and glucocorticoids.) Secreted by the adrenal gland, they often act, as we will see, in ways similar to epinephrine. Epinephrine acts within seconds; glucocorticoids back this activity up over the course of minutes or hours.

Because the adrenal gland is basically witless, glucocorticoid release must ultimately be under the control of the hormones of the brain. When something stressful happens or you think a stressful thought, the hypothalamus secretes an array of releasing hormones into the hypothalamic-pituitary circulatory system that gets the ball rolling. The principal such releaser is called *CRF* (corticotropin releasing factor), while a variety of more minor players synergize with CRF. Within fifteen seconds or so, CRF triggers the pituitary to release the hormone ACTH

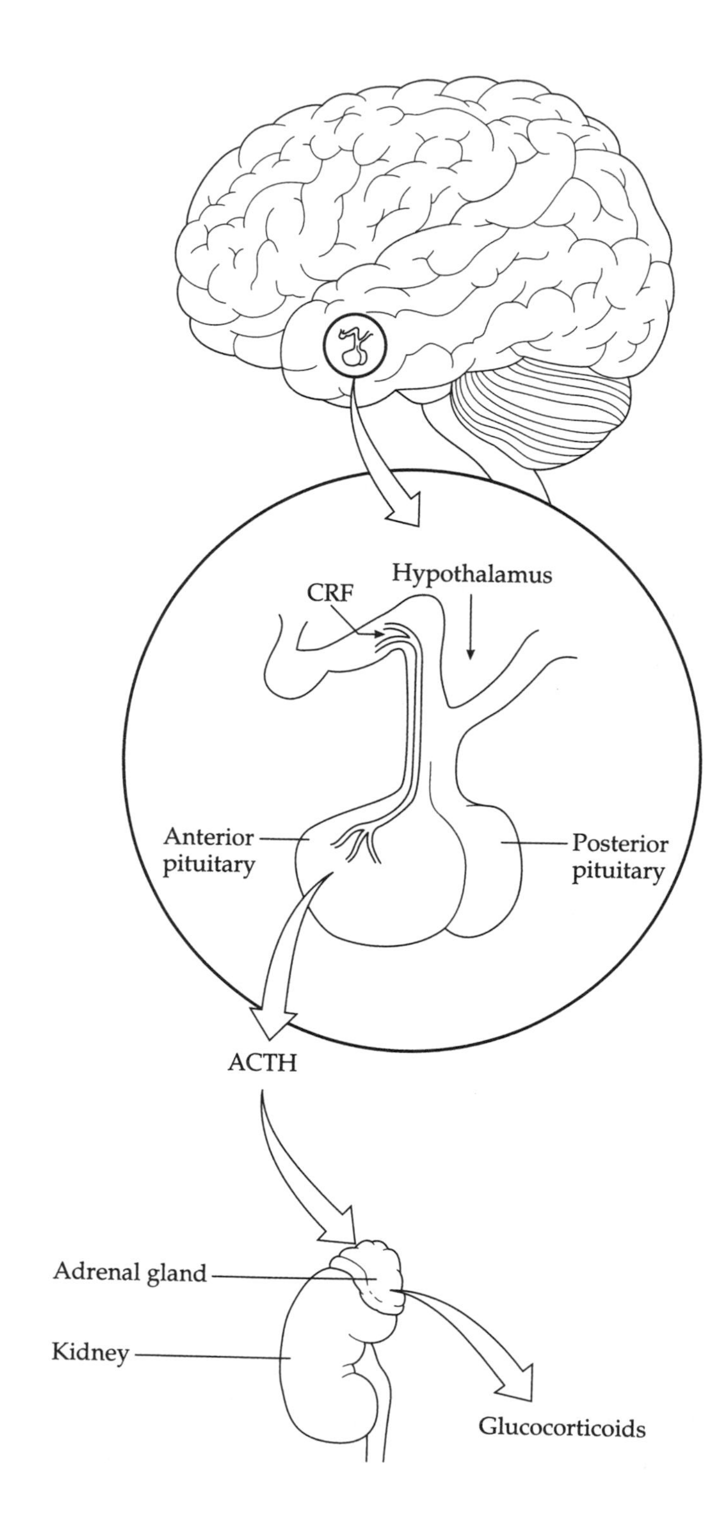
Hypothalamus
CRF
Anterior pituitary
Posterior pituitary
ACTH
Adrenal gland
Kidney
Glucocorticoids

(opposite page) *Outline of the control of glucocorticoid secretion. A stressor is sensed or anticipated in the brain, triggering the release of CRF (and related hormones) by the hypothalamus. These hormones enter the private circulatory system linking the hypothalamus and the anterior pituitary, causing the release of ACTH by the anterior pituitary. ACTH enters the general circulation and triggers the release of glucocorticoids by the adrenal gland.*

(also known as *corticotropin*). After ACTH is released into the bloodstream, it reaches the adrenal gland and, within a few minutes, triggers glucocorticoid release. Together, glucocorticoids and the secretions of the sympathetic nervous system (epinephrine and norepinephrine) account for a large percentage of what happens in your body during stress. These are the workhorses of the stress-response.

In addition, in times of stress your pancreas is stimulated to release a hormone called *glucagon*. Glucocorticoids, glucagon, and the sympathetic nervous system raise circulating levels of the sugar glucose—as we will see, these hormones are essential for mobilizing energy during stress. Other hormones are activated as well. The pituitary secretes prolactin, which, among other effects, plays a role in suppressing reproduction during stress. Both the pituitary and the brain also secrete a class of endogenous morphine-like substances called *endorphins* and *enkephalins*, which help blunt pain perception, among other things. Finally, the pituitary also secretes vasopressin, also known as *antidiuretic hormone*, which plays a role in the cardiovascular stress-response.

Just as some glands are activated in response to stress, various hormonal systems are inhibited during stress. The secretion of various reproductive hormones such as estrogen, progesterone, and testosterone is inhibited. Hormones related to growth (such as growth hormone) are also inhibited, as is the secretion of insulin, a pancreatic hormone that normally tells your body to store energy for later use.

(Are you overwhelmed and intimidated by these terms, wondering if you should have bought some Deepak Chopra self-help book instead? Please, don't even dream of memorizing these names of hormones. The important ones are going to appear so regularly in the coming pages that you will soon be comfortably and accurately slipping them into everyday conversation and birthday cards to favorite cousins. Trust me.)

A FEW COMPLICATIONS

This, then, is an outline of our current understanding of the neural and hormonal messengers that carry the brain's news that something awful is happening. Cannon was the first to recognize the role of epinephrine, norepinephrine, and the sympathetic nervous system. Selye pioneered the glucocorticoid component of the story. Since then the roles of the other hormones and neural systems have been recognized. In just the four years since the previous edition of this book, several new minor hormonal players have been added to the picture, and, undoubtedly, more are yet to be discovered. Collectively, these shifts in secretion and activation form the primary stress-response.

Naturally there are complications in the simple endocrine story outlined in this chapter. One concerns variability among species; not all the features of the stress-response work quite the same way in different species. For example, while stress causes a prompt decline in the secretion of growth hormone in rats, it causes a transient increase in growth hormone secretion in humans (this puzzle and its implication for humans are discussed in the chapter on growth).

Another complication concerns the time course of actions of epinephrine and glucocorticoids. A few paragraphs back, I noted that the former works within seconds, while the latter backs up epinephrine's activity over the course of minutes to hours. That's great—in the face of an invading army, sometimes the defensive response can take the form of handing out guns from an armory (epinephrine working in seconds), and a defense can also take the form of beginning construction of new tanks (glucocorticoids working over hours). But within the framework of lions chasing zebras, how many sprints across the grasslands actually go on for hours? What good are glucocorticoids if some of their actions occur long after your typical dawn-on-the-savanna stressor is over with? This represents an area of ongoing debate. Certain studies suggest that some glucocorticoid actions do not mediate the stress-response but, rather, help mediate the *recovery* from the stress-response. As will be described in Chapter 8, this probably has important implications for a number of autoimmune diseases. In contrast, some glucocorticoid actions appear to prepare you mostly for the *next* stressor, rather than dealing with or recovering from the current one. As will be discussed in Chapter 12, this is critical for understanding the ease with which anticipatory psychological states can trigger glucocorticoid secretion.

Another complication concerns consistency of the stress-response. Central to Selye's conceptualization was the belief that whether you are too hot or too cold, that zebra or that lion, or simply stressed by the repetitiveness of that phrase, you activate the same pattern of secretion of glucocorticoids, epinephrine, growth hormone, estrogen, and so forth for each of those stressors.

It turns out that the pattern of response is not quite that consistent, however. In general, stressors of all kinds, particularly massive physical stressors, involve the hormonal changes outlined in this chapter, with the glucocorticoid and sympathetic components being the most reliable. But the speed and the magnitude with which the secretion of some particular hormone changes may vary according to the stressor, especially for more subtle ones. The orchestration, the patterning of hormone release tends to vary from stressor to stressor, and a hot topic in stress research a few years back was figuring out the hormonal "signature" of a particular stressor.

One example concerns the relative magnitude of the glucocorticoid versus the sympathetic stress-responses. James Henry, who has done pioneering work on the ability of social stressors such as subordinacy to cause heart disease in rodents, has found that the sympathetic nervous system is particularly activated in a socially subordinate rodent that is vigilant and trying to cope with a challenge. In contrast, it is the glucocorticoid system that is relatively more activated in a subordinate rodent that has basically given up on coping. Studies of stressed or depressed humans have shown what may be a human analogue of that dichotomy. Sympathetic arousal is a relative marker of anxiety and vigilance, while heavy secretion of glucocorticoids is more a marker of depression (as glucocorticoid levels are elevated in about half of depressives). Furthermore, all stressors do not cause secretion of both epinephrine and norepinephrine, nor of norepinephrine from all branches of the sympathetic system.

Finally, as will be the topic of Chapter 12, two identical stressors can cause very different stress signatures, depending on the psychological context of the stressors. Thus, every stressor does not generate exactly the same stress-response. This is hardly surprising. Despite the dimensions common to various stressors, it is still a very different physiological challenge to be too hot or too cold, to be extremely anxious or deeply depressed. Despite this, the hormonal changes outlined in this chapter, which occur pretty reliably in the face of impressively different stressors, still constitute the superstructure of the neural and endocrine stress-response. We are now in a position to see how these responses collectively save our skins during acute emergencies but can make us sick in the long run.

3

STROKE, HEART ATTACKS, AND VOODOO DEATH

It's one of those unexpected emergencies: you're walking down the street, on your way to meet a friend for dinner. You're already thinking about what you'd like to eat, savoring your hunger. Come around the corner and—oh, no, a lion! As we now know, activities throughout your body shift immediately to meet the crisis: your digestive tract shuts down and your breathing rate skyrockets. Secretion of sex hormones is inhibited, while epinephrine, norepinephrine, and glucocorticoids pour into the bloodstream. And if your leg muscles are going to save you, one of the most important additional things that better be going on is an increase in your cardiovascular output, in order to deliver oxygen and energy to those exercising muscles.

THE CARDIOVASCULAR STRESS-RESPONSE

Activating your cardiovascular system is relatively easy, so long as you have a sympathetic nervous system plus some glucocorticoids and don't bother with too many details. The first thing you do is shift your heart into higher gear, get it to beat faster and harder. Such effects are accomplished by turning down parasympathetic tone, and turning up sympathetic along with some glucocorticoid secretion. The actual process is extremely complex and beyond the scope of this book—some of the changes in the functioning of the heart, for example, are secondary to changes elsewhere and rely upon some fancy contractile features of heart muscle. The net result, however, is that blood is now moving faster and with more force. In the face of a maximum stressor, this produces about five times the output of the heart during rest.

A second set of changes occurs in the blood vessels. To appreciate this, one needs a detailed familiarity with garden hoses, because the principles are the same. Suppose there is a water tap at one end of your house; you want to attach a hose to it and spray some water on a fire that has just started at the other end of the house. There's a problem, however. You have two hoses of different diameters to choose from, but each is ten feet too short to reach the fire. You want to pick the hose that will allow you to spray the water with the most force across those last ten feet. The first hose, admittedly unlikely for the garden variety, has a diameter of three feet, and its walls are made of a soft expandable material with about as much rigidity as a marshmallow. The second hose has a diameter of one inch and is made of a tough, rigid material. Which hose do you choose?

Obviously if you go for the first one, it's going to take forever to fill, because of its huge diameter; once pressure does begin to build up, it will simply distend the hose outward, because of the soft walls. Eventually, a puddle will form at your feet while your house burns down. If you pick the second hose, however, the water from the tap is going to come streaming out with a lot of force, because of the narrowness and rigidity of the hose. Hoses like that make for increased water pressure. And arteries like that make for increased blood pressure. As the second general step in response to the stressor, your sympathetic nervous system, in conjunction with glucocorticoids, constricts some of your major arteries. Astonishingly, each of these arteries is wrapped in tiny circular muscles, and their sympathetic innervation now causes them to tighten. Up goes blood pressure.

As a result of the constriction of these major vessels, blood is now being delivered with greater speed to exercising muscle. At the same time, there is a dramatic decrease in blood flow to unessential parts of your body, including the mesenteric system, which supplies blood to your digestive tract, along with less blood flow to your kidneys and skin. This was first noted in 1833, in an extended study of a Canadian Native American who had a tube placed in his abdomen after a gunshot wound there. When the man sat quietly, his gut tissues were bright pink, well supplied with blood. Whenever he became anxious or angry, the gut mucosa would blanch, because of decreased blood flow. (Pure speculation perhaps, but one suspects that his transients of anxiety and anger might have been related to those white folks sitting around experimenting on him, instead of doing something useful—like sewing him up.)

There's one final cardiovascular trick in response to stress, involving the kidneys. As that zebra with its belly ripped open, you've lost a lot of blood. Furthermore, you may have to sprint across the veld and go into an hour of evasive maneuvers. It's hot, normally a time you would drink; but that's out of the question now. It makes sense to conserve water. If blood volume goes down because of dehydration or hemorrhage, it doesn't matter what your heart and veins are doing; your ability to deliver glucose and oxygen to your muscles will be impaired. What's the most likely place to be losing water? Urine formation, and the source of the water in urine is the bloodstream. Thus your brain sends a message to the kidneys: stop the process, reabsorb the water into the circulatory system. This is accomplished by the hormone vasopressin (known as *antidiuretic hormone* for its ability to block diuresis, or urine formation), as well as a host of related hormones that regulate water balance.

A question no doubt at the forefront of every reader's mind at this point: if one of the features of the cardiovascular stress-response is to conserve water in the circulation, and this is accomplished by inhibition of urine formation in the kidneys, why is it that when we are *really* terrified, we wet our pants? I congratulate the reader for honing in on one of the remaining unanswered questions of modern science. In trying to answer it, we run into a larger one. Why do we have bladders? They are dandy if you are a hamster or a dog, because species like those fill their bladders up until they are just about to burst and then run around their territories, demarcating the boundaries—odoriferous little "keep out" signs to the neighbors. A bladder is logical for scent-marking species, but I presume that you don't do that sort of thing on a regular basis. For humans, it is a mystery, just a boring storage site. The kidneys, now those are something else. Kidneys are reabsorptive, bidirectional organs, which means you can spend your whole afternoon happily putting water in

THE FAR SIDE By GARY LARSON

"So! Planning on roaming the neighborhood with some of your buddies today?"

from the circulation and getting some back and regulating the whole thing with a collection of hormones. But once the urine leaves the kidneys and heads south to the bladder, you can kiss that stuff good-bye; the bladder is unidirectional. When it comes to a stressful emergency, a bladder means a lot of sloshy dead weight to carry in your sprint across the savanna. The answer is obvious: empty that bladder.*

Everything is great now—you have kept your blood volume up, it is roaring through the body with more force and speed, delivered where it is most needed. This is just what you want when running away from a lion. Interestingly, Marvin Brown of

---

* This, of course, raises that mysterious problem for guys as to why it is so *difficult* to urinate at a urinal when you are stressed by a crowd waiting in line behind you. Operators are standing by for explanations from any card-carrying physiologists out there.

the University of California at San Diego and Laurel Fisher of the University of Arizona have shown that a different picture emerges when one is being vigilant—a gazelle crouching in the grass, absolutely quiet, as a lion passes nearby. The sight of a lion is obviously a stressor, but of a very subtle sort; while having to remain as still as possible, you must also be prepared, physiologically, for a wild sprint across the grasslands with the briefest of warnings. During such vigilance, heart rate and blood flow tend to slow down, and vascular resistance throughout the body increases, including at the muscle. Another example of the complicating point brought up at the end of Chapter 2 about stress signatures—you don't turn on the identical stress-response for every type of stressor.

CHRONIC STRESS AND CARDIOVASCULAR DISEASE

You've escaped the lion, thanks to your cardiovascular system. But if you put your heart, blood vessels, and kidneys to work in this way every time your teenager agitates you, you increase your risk of heart disease. Never is the maladaptiveness of the stress-response during psychological stress clearer than in the case of the cardiovascular system. You sprint through the restaurant district terrified, and you alter cardiovascular functions to divert more blood flow to your thigh muscles. In such cases, there's a wonderful match between blood flow and metabolic demand. In contrast, if you sit and think about the consequences of giving your adolescent the car keys, driving yourself into a hyperventilating panic, you still alter cardiovascular function to divert more blood flow to your limb muscles. Crazy. And, potentially, eventually damaging. But how does stress-induced elevation of blood pressure during chronic psychological stress wind up causing cardiovascular disease?

The short of it is that your heart is just a dumb, simple mechanical pump, and your blood vessels are nothing more exciting than hoses. The cardiovascular stress-response basically consists of making them work harder for a while, and if you do that on a regular basis, they will wear out, just like any pump or hoses you could buy at Sears. The long of it is the rest of this chapter.

Part of the problem is the way that blood vessels wear out. You get in trouble here because of a combination of what is described in this chapter and what you do with your energy stores (described in the next chapter). A general feature of the circulatory system is that, at various points, large blood vessels (your descending aorta, for example) branch into smaller vessels, then into even smaller ones, and so on, down to tiny beds of thousands of capillaries. This process of splitting into smaller and smaller units is called *bifurcation*. (As a measure of how extraordinarily efficient this repeated bifurcation is in the circulatory system, no cell in your body is more than five cells away from a blood vessel—yet the circulatory system takes up only 3 percent of body mass.) One feature of systems that branch in this way is that the points of bifurcation are particularly vulnerable to injury. The branch points in the vessel wall where bifurcation occurs bear the brunt of the fluid pressure slamming into them. Thus, a simple rule: when you increase the force with which the fluid is moving through the system, turbulence increases and those outposts of wall are more likely to wear down (simple fluid mechanics, which applies as readily to bodies as it does to the city's water supply).

With the chronic increase in blood pressure that accompanies repeated stress, damage begins to occur at branch points in arteries throughout the body. The smooth inner lining of the vessel begins to tear, scar, and pit. Once this layer is damaged, the fatty acids and glucose that are mobilized into your bloodstream by the onset of the metabolic stress-response begin to work their way underneath this layer and stick there, thickening the lining. In addition, during stress, the sympathetic nervous system makes your blood more viscous. Specifically, epinephrine makes circulating platelets (a type of blood cell that promotes clotting) more likely to clump together. And if there are tears in the lining of some of your vessels, those clumps will aggregate there, adding to the problem. Moreover, cells full of fatty nutrients, called *foam cells*, begin to form there, too. Before you know it, your vessels are beginning to clog, and blood flow through them decreases.

Thus, chronic stress causes atherosclerosis—the accumulation of plugs called *plaques*, made of fats, starches, foam cells, calcium, and so on underneath the inner lining of blood vessels.

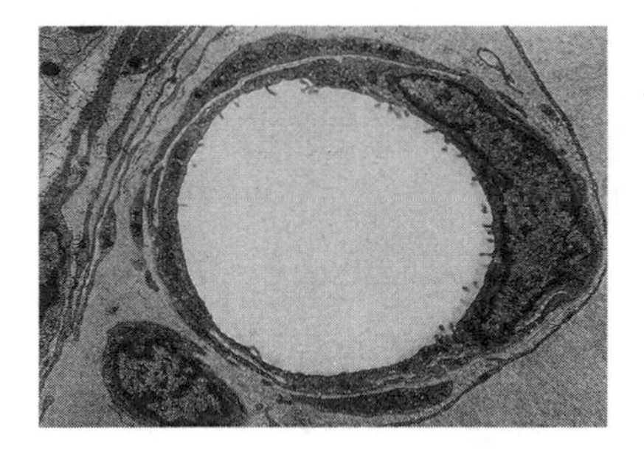
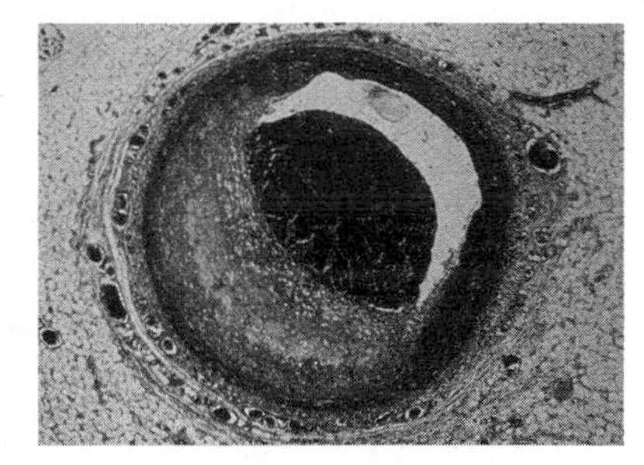

A healthy blood vessel (left), and one with an atherosclerotic plaque (right).

One of the clearest demonstrations of this, with great application to our own lives, is to be found in the work of the physiologist Jay Kaplan at Bowman Grey Medical School. Kaplan built on the landmark work of an earlier physiologist, James P. Henry (who was mentioned in the previous chapter), who showed that purely social stress caused atherosclerosis (as well as high blood pressure) in mice. Kaplan and colleagues have shown a similar phenomenon in primates, bringing the story much closer home to us humans. Establish monkeys in a social group, and over the course of days to months they'll figure out where they stand with respect to one another. Once a stable dominance hierarchy has emerged, the last place you want to be is on the bottom: you have little opportunity to predict what will happen to you, no control over what does, and few outlets when something stressful occurs. Such subordinate animals show a lot of the physiological indices of chronically turning on their stress-responses. Often, these animals wind up with atherosclerotic plaques.

Kaplan showed that another group of animals is also at risk. Suppose you keep the dominance system *un*stable by shifting the monkeys into new groups every month, so that all the animals are perpetually in the tense, uncertain stage of figuring out where they stand with respect to everyone else. Under those circumstances, it is generally the animals precariously holding on to their places at the top of the shifting dominance hierarchy who do the most fighting and who show the most behavioral and hormonal indices of stress. And as it turns out, they have tons of atherosclerosis; some of the monkeys even have heart attacks (abrupt blockages of one or more of the coronary arteries).

In general, the monkeys under the most social stress were most at risk for plaque formation. Kaplan showed that this can even occur with a low-fat diet, which makes sense, since, as will be described in the next chapter, a lot of the fat that forms plaques is being mobilized from stores in the body, rather than coming from the cheeseburger the monkey ate just before the tense conference. But if you couple the social stress with a high-fat diet, the effects synergize, and plaque formation goes through the roof.

The atherosclerotic plaques are probably formed because the sympathetic nervous system is continuously activated in the stressed monkeys. First, the monkeys who developed the most atherosclerosis were what Kaplan termed hot reactors: they responded to stress with the greatest degree of sympathetic nervous system activity. Elevating heart rate and blood pressure, as the sympathetic nervous system does, makes atherosclerosis more likely to occur. Finally, give the monkeys at risk drugs that prevent sympathetic activity (beta-blockers), and they don't form plaques. This makes sense, given what we have already learned about the actions of norepinephrine and epinephrine, as do Kaplan's findings that glucocorticoids make the atherosclerosis worse.

Secretion of large amounts of these stress hormones on a regular basis is a prescription for trouble. Form enough atherosclerotic plaques to seriously obstruct flow to the lower half of the body and you get *claudication*, which means that your legs and chest hurt like hell for lack of oxygen and glucose whenever you walk; you are then a candidate for bypass surgery. If the same thing happens to the arteries going to your heart, you can get coronary heart disease, myocardial ischemia, all sorts of horrible things. Try the same trick with the blood vessels going to your brain and you can get a stroke—a brain hemorrhage.

If chronic stress has made a mess of your blood vessels, each individual new stressor is even more damaging, for a very insidious reason. This has to do with myocardial ischemia—a condition that arises when the arteries feeding your heart have become sufficiently clogged that your heart itself is partially deprived of blood flow and thus of oxygen and glucose. (It may initially seem illogical for the heart to need special arteries feeding it. When the walls of the heart—the heart muscle—require the energy and oxygen stores in the blood, you might imagine that these could simply be absorbed from the vast amounts of

blood passing through the chambers of the heart. But instead it has evolved that heart muscle is fed by arteries coursing from the main aorta. As an analogy, consider people working at a city's water reservoir. Every time they get thirsty, they might go over to the edge of the reservoir with a bucket and pull up some water to drink. Instead, the usual solution is to have a water fountain in the office, fed indirectly by that reservoir just outside.)

Suppose something acutely stressful is happening, and your cardiovascular system is in great shape. You get excited, the sympathetic nervous system kicks into action. Your heart speeds up in a strong, coordinated fashion, and its contractive force increases. As a result of working harder, the heart muscle consumes more energy and oxygen and, conveniently, the arteries going to your heart dilate in order to deliver more nutrients and oxygen to the muscle. Everything is fine.

But if you encounter an acute stressor with a heart that has been suffering from chronic myocardial ischemia, you're in trouble. The coronary arteries, instead of vasodilating in response to the sympathetic nervous system, vaso*constrict*. This is very different from the scenario of vasoconstriction described at the beginning of the chapter, where blood pressure is increased throughout the body by some subtle constriction by your biggest

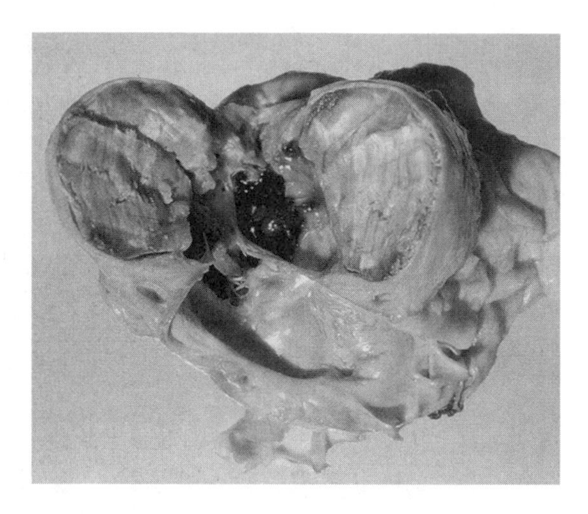

A necrotic heart.

arteries. Instead, these are the small vessels diverting blood right to your heart. Just when your heart needs more oxygen and glucose delivered through these already clogged vessels, acute stress shuts them down even more—exactly the opposite of what you need. Your chest is going to hurt like crazy—angina pectoris. And it turns out that it takes only brief periods of hypertension to cause this vasoconstrictive problem. Therefore, chronic myocardial ischemia from atherosclerosis sets you up for, at the least, terrible chest pain whenever anything physically stressful occurs. This is the perfect demonstration of how stress is particularly good at worsening a preexisting problem.

When cardiology techniques improved in the 1970s, cardiologists were surprised to discover that we are even more vulnerable to trouble in this realm than had been guessed. With the old techniques, you would take someone with myocardial ischemia and wire him (men are more prone to heart disease than are women, despite heart diseases still being the leading cause of death among women in the west) up to some massive ECG machine (same as EKG), focus a huge x-ray camera on his chest, and then send him running on a treadmill until he was ready to collapse. Just as one would expect, blood flow to the heart would decrease and his chest would hurt.

Some engineers invented a miniature ECG machine that can be strapped on while you go about your daily business, and *ambulatory electrocardiography* was invented. Suddenly cardiologists could see how the whole thing worked at a far more subtle level. And everyone got a rude surprise. There were little ischemic crises occurring all over the place in people at risk. Most ischemic episodes turned out to be "silent"—they didn't give a warning signal of pain. Moreover, all sorts of psychological stressors could trigger them—public speaking, pressured interviews, exams. According to the old dogma, if you had heart disease, you had better worry when you were undergoing physical stress and getting chest pains. Now it appears that, for someone at risk, trouble is occurring under all sorts of circumstances of psychological stress in everyday life; one may not even know it. This fits well with studies showing that dogs don't have to be doing some wildly demanding physical activity for them to have uncoordinated heart beats. Instead, it can occur when they are merely angry or fearful. Once the cardiovascular system is damaged, it appears to be immensely sensitive to acute stressors, whether physical or psychological.

SUDDEN CARDIAC DEATH

The preceding sections demonstrate how chronic stress will hammer away at the cardiovascular system, with each succeeding stressor making the system even more vulnerable. Ultimately, this greatly increases the risk of cardiac catastrophe—the scenario of the man sitting there, finishing his breakfast, who is struck down by a fatal heart attack, the number one killer in the United States.

But one of the most striking and best-known features of heart disease is how often that cardiac catastrophe hits during a stressor. A man gets shocking news: his wife has died; he's lost his job; a child long thought to be dead appears at the door; he wins the lottery. The man weeps, rants, exults, staggers about gasping and hyperventilating with the force of the news. Five minutes or five hours later, he suddenly grasps at his chest and he falls over dead from sudden cardiac arrest. We all remember the dramatic example of Bill Hodgman, one of the prosecutors in the O. J. trial, getting chest pains around the twentieth time he jumped up to object to something Johnnie Cochran was saying, and collapsing afterwards (but surviving).

The phenomenon is quite well documented. In one study, a physician collected newspaper clippings on sudden cardiac death in 170 individuals. He identified a number of events that seemed to be associated with such deaths: the collapse, death, or threat of loss of someone close; acute grief; loss of status or self-esteem; mourning, on an anniversary; personal danger; threat of an injury, or recovery from such a threat; triumph or extreme joy. Other studies have shown the same. During the 1991 Persian Gulf war fewer deaths in Israel were due to SCUD missile damage than to sudden cardiac death among frightened elderly people. During the 1992 L. A. earthquake, there was similarly a big jump in heart attacks.*

The actual causes are obviously tough to study (since you can't predict what is going to happen, and you can't interview the people afterward to find out what they were feeling), but the general consensus among cardiologists is that sudden cardiac death is simply an extreme version of acute stress causing ventricular arrhythmia or, even worse, ventricular fibrillation† plus ischemia in the heart. As you would guess, it involves the sympathetic nervous system, and it is more likely to happen in damaged heart tissue than in healthy tissue. People *can* suffer sudden cardiac death without a history of heart disease and despite increased blood flow in the coronary vessels; autopsies have generally shown, however, that these people had a fair amount of atherosclerosis. Mysterious cases still occur, however, of seemingly healthy thirty-year-olds, victims of sudden cardiac death, who show little evidence of atherosclerosis on autopsy.

Fibrillation seems to be the critical event in sudden cardiac death, as judged by animal studies (in which, for example, ten hours of stress for a rat makes its heart more vulnerable to fibril-

* I recently received a letter from the chief medical examiner of Vermont describing his investigation of what he concluded to be a case of stress-induced cardiac arrest: an eighty-eight-year-old man with a history of heart disease, lying dead of a heart attack next to his beloved tractor, while just outside the house, at an angle where she could have seen him prone in the barn, his eighty-seven-year-old wife, more recently dead of a heart attack (but with no history of heart disease and nothing obviously wrong found at autopsy). At her side was the bell she had used to summon him to lunch for who knows how many years.

† Don't panic at the jargon. In ventricular fibrillation: the half of your heart called the *ventricles* begins to contract in a rapid, disorganized way that accomplishes nothing at all in terms of pumping blood. This is in contrast to arrhythmia, in which the overall heartbeat becomes irregular.

lation for days afterward). As one cause, the muscle of a diseased heart becomes more electrically excitable, making it prone to fibrillation. In addition, activation of stimulatory inputs *to* the heart becomes disorganized during a massive stressor. Thc sympathetic nervous system sends two symmetrical nervous projections to the heart; it is theorized that during extreme emotional arousal, the two inputs are activated to such an extent that they become uncoordinated—major fibrillation, clutch your chest, keel over.

FATAL PLEASURES

Embedded in the list of categories of precipitants of sudden cardiac death is a particularly interesting one: triumph or extreme joy. Consider the scenario of the man dying in the aftermath of the news of his winning the lottery, or the proverbial "at least he died happy" instance of someone dying during sex. (When these circumstances apparently claimed the life of an ex-vice president, the medical minutiae of the incident received especially careful examination because he was not with his wife at the time.)

The possibility of being killed by pleasure seems crazy. Isn't stress-related disease supposed to arise from stress? How can joyful experiences kill you in the same way that sudden grief does? Clearly, because they share some similar traits. Extreme anger and extreme joy have different effects on reproductive physiology, on growth, most probably on the immune system as well; but with regard to the cardiovascular system, they have fairly similar effects. Once again, we deal with the central concept of stress physiology in explaining similar responses to being too hot or too cold, a prey or a predator: some parts of our body, including the heart, do not care in which direction we are knocked out of allostatic balance, but rather simply how much. Thus wailing and pounding the walls in grief or leaping about and shouting in ecstasy can place similarly large demands on a diseased heart. Put another way, your sympathetic nervous system probably has roughly the same effect on your coronary arteries whether you are in the middle of a murderous rage or a thrilling orgasm. In that vein, anthropologist Irven DeVore has said that if two people look into each other's eyes for more than about six seconds, they are either preparing to kill each other or to make love. Diametrically opposite emotions then can have surprisingly similar physiological underpinnings (reminding one of the oft-quoted statement from Elie Wiesel, the Nobel laureate writer and Holocaust survivor: "the opposite of love is not hate. The opposite of love is indifference."). When it comes to the cardiovascular system, rage and ecstasy, grief and triumph all represent challenges to allostatic equilibrium.

VOODOO DEATH

The time has come to examine a subject far too rarely discussed in our public schools. Well-documented examples of voodoo death have emerged from all sorts of traditional nonwesternized cultures. Someone eats a forbidden food, insults the chief, sleeps with someone he or she shouldn't have, does something unacceptably violent or blasphemous. The outraged village calls in a shaman who waves some ritualistic hideous gewgaw at the transgressor, makes a voodoo doll, or in some other way puts a hex on the person. Convincingly soon, the hexed one drops dead.

The Harvard team of ethnobotanist Wade Davis and cardiologist Regis DeSilva reviewed the subject.* Davis and DeSilva object to the use of the term voodoo death, since it reeks of western condescension toward nonwestern societies—grass skirts, bones in the nose, and all that. Instead, they prefer the term *psychophysiological death,* noting that in many cases even that term is probably a misnomer. In some instances, the shaman may spot people who are already very sick and, by claiming to have hexed them, gain brownie points when the person kicks off. Or the shaman may simply poison them and gain kudos for his cursing

---

* Wade Davis is the favorite ethnobotanist of horror movie fans far and wide. As detailed in the reference section, his prior research uncovered a possible pharmacological basis of how zombies (people in a deathlike trance with no will of their own) are made in Haiti. Davis's Harvard doctoral dissertation about zombification was first turned into a book, *The Serpent and the Rainbow,* and then a grade B horror movie of the same name—a dream come true for every graduate student whose thesis is destined to be skimmed briefly by a distracted committee member or two.

powers. In the confound (that is, the source of confusion) that I found most amusing, the shaman puts a visible curse on someone, and the community says, in effect, "Voodoo cursing works; this person is a goner, so don't waste good food and water on him." The individual, denied food and water, starves to death; another voodoo curse come true, the shaman's fees go up.

Nevertheless, instances of psychophysiological death do occur, and they have been the focus of interest of some great physiologists in this century. In a great face-off, Walter Cannon (the man who came up with the fight or flight concept) and Curt Richter (a grand old man of psychosomatic medicine) differed in their postulated mechanisms of psychophysiological death. Cannon thought it was due to overactivity of the sympathetic nervous system; in that scheme, the person becomes so nervous at being cursed that the sympathetic system kicks into gear and vasoconstricts blood vessels to the point of rupturing them, causing a fatal drop in blood pressure. Richter thought death was due to too much *para*sympathetic activity. In this surprising formulation, the individual, realizing the gravity of the curse, gives up on some level. The parasympathetic projection to the heart (the vagus nerve) becomes very active, slowing the heart down to the point of stopping—death due to a "vagal storm," as it was called. Both Cannon and Richter kept their theories unsullied by never examining anyone who had died of psychophysiological death, voodoo or otherwise. It turns out that Cannon was probably right. Hearts almost never stop outright in a vagal storm. Instead, Davis and DeSilva suggest that these cases are simply dramatic versions of sudden cardiac death, with too much sympathetic tone driving the heart into ischemia and fibrillation.

All very interesting, in that it explains why psychophysiological death might occur in individuals who already have some degree of cardiac damage. But a puzzling feature about psychophysiological death in traditional societies is that it can also occur in young people who are extremely unlikely to have any latent cardiac disease. This mystery remains unexplained, perhaps implying more silent cardiac risk lurking within us than we ever would have guessed, perhaps testifying to the power of cultural belief. As Davis and DeSilva note, if faith can heal, faith can also kill.

PERSONALITY AND CARDIAC DISEASE: A BRIEF INTRODUCTION

If two people both run like crazy on a treadmill to an equal state of exhaustion, they will not be equally vulnerable to cardiovascular trouble at that point. If those two people experience a more subtle, but still stressful, social situation, they are not equally likely to develop arrhythmia and severe ischemia. Finally, if those two people go through a decade's worth of life's ups and downs, only one of them may have a heart attack.

These individual differences could be due to one persons already having a damaged cardiovascular system—for example, decreased coronary blood flow. They could also be due to genetic factors that influence the mechanics of the system—the elasticity of blood vessels, the numbers of norepinephrine receptors, and so on. They could be the result of differences in how many risk factors each individual experiences—does the person smoke, have high blood pressure, have high circulating fatty acid (triglyceride) levels? (Interestingly, individual differences in these risk factors explain less than half the variability in patterns of heart disease.)

Faced with similar stressors, whether large or small, two people may also differ in their risk for cardiovascular disease as a function of their personalities. A link between certain personality types and certain patterns of increased disease risk has been made in many realms of stress-related disease, and constitutes the subject of Chapter 14. (There, I will review one of the strongest such cases and certainly the best-known one—the famed Type A personality and cardiovascular disease risk. The bad news is that your heart and blood vessels could really do without your being a textbook Type A-er. The good news is that you can probably make at least some changes to reduce Type A-ness, a topic covered in the final chapter.)

This discussion has served as the first example of the style of analysis that will dominate the coming chapters. In the face of a short-term physical emergency, the cardiovascular stress-response is vital. In the face of chronic stress, those same changes are terrible news. These adverse effects are particularly deleterious when they interact with the adverse consequences of too much of a metabolic stress-response, the subject of the next chapter.

STRESS, METABOLISM, AND LIQUIDATING YOUR ASSETS

So you're sprinting down the street with the lion after you. Things looked grim for a moment there, but—your good luck—your cardiovascular system kicked into gear, and now it is delivering oxygen and energy to your exercising muscles. But what energy? There's not enough time to consume a candy bar and derive its benefits as you sprint along; there's not even enough time to digest food already in the gut. Your body must get energy from its places of storage, like fat or liver or nonexercising muscle. To understand how you mobilize energy in this circumstance, and how that mobilization can make you sick at times, we need to learn how the body stores energy in the first place.

PUTTING ENERGY IN THE BANK

The basic process of digestion consists of breaking down chunks of animals and vegetables so that they can then be transformed into chunks of human. We can't make use of the chunks exactly as they are; we can't, for example, make our leg muscles stronger by grafting on the piece of chicken muscle we ate. Instead, complex food matter is broken down into its simplest parts (molecules): amino acids (the building blocks of protein), simple sugars like glucose [the building blocks of more complex sugars and of starches (carbohydrates)], and free fatty acids and glycerol

THE FOUR MAJOR FOOD GROUPS

Regular:

Hamburger, cola, French fries, fruit pie.

Company:

Cracker variety, canapé, "interesting" cheese, mint.

Remorse:

Plain yogurt, soybeans, mineral water, tofu.

Silly:

Space-food sticks, gelatine mold with fruit salad in it, grasshopper pie.

R. Chast

(the constituent parts of fat). This is accomplished in the gastrointestinal tract by enzymes, chemicals that can degrade more complex molecules. The simple building blocks thus produced are absorbed into the bloodstream for delivery to whichever cells in the body need them. Once you've done that, the cells have the ability to use those building blocks to construct the proteins, fats, and carbohydrates needed to stay in business. And just as importantly, those simple building blocks (especially the fatty acids and sugars) can also be burned by the body to provide the energy to do all that construction.

It's Thanksgiving, and you've eaten with porcine abandon. Your bloodstream is teeming with amino acids, fatty acids, glucose. It's far more than you need to power you over to the couch in a postprandial daze. What does your body do with the excess? This is crucial to understand because, basically, the process gets reversed when you're later sprinting for your life.

To answer this question, it's time we talked finances. The works: savings accounts, change for a dollar, stocks and bonds, negative amortization of interest rates, shaking coins out of piggy banks—because the process of transporting energy through the body bears some striking similarities to the movement of money. It is rare today for the grotesquely wealthy to walk around with their fortunes in their pockets, or to hoard their wealth as cash stuffed inside mattresses. Instead, surplus wealth is stored elsewhere, in forms more complex than cash: mutual funds, tax-free government bonds, Swiss bank accounts. In the same way, surplus energy is not kept in the body's form of cash—circulating amino acids, glucose, and fatty acids—but stored in more complex forms. Enzymes in fat cells can combine fatty acids and glycerol to form triglycerides (see the table on the next page). Accumulate enough of these in the fat cells and you grow plump. Meanwhile, enzymes in cells throughout the body can cause a succession of molecules of glucose to stick together. These long chains, sometimes thousands of glucose molecules long, are called *glycogen*. Most glycogen formation occurs in your muscles and liver. Similarly, enzymes in cells throughout the body can combine long strings of amino acids, forming them into proteins.

The hormone that stimulates the transport and storage of these building blocks into target cells is insulin. In a sense, insulin plans for your metabolic future. Eat a huge meal and insulin pours out of the pancreas into the bloodstream, stimulating the transport of fatty acids into fat cells, stimulating glycogen and protein synthesis. It's insulin that's filling out the deposit slips at your fat banks. We even secrete insulin when we are *about* to fill our bloodstream with all those nutritive building blocks: if you eat dinner each day at six o'clock, by five forty-five your parasympathetic nervous system is already stimulating insulin secretion in anticipation.

What you stick in your mouth	How it winds up in your bloodstream	How it gets stored if you have a surplus	How it gets mobilized in a stressful emergency
Proteins →	Amino acids →	Protein →	Amino acids
Starch, sugars, carbohydrates →	Glucose →	Glycogen →	Glucose
Fat →	Fatty acids and glycerol →	Triglycerides →	Fatty acids, glycerol, ketone bodies

ENERGY MOBILIZATION DURING A STRESSOR

This grand strategy of breaking your food down into its simplest parts and reconverting it into complex storage forms is precisely what your body should do when you've eaten plenty. And it is precisely what your body should *not* do in the face of an immediate physical emergency. Then, you want to stop energy storage. Turn up the activity of the sympathetic nervous system, turn down the parasympathetic, and down goes insulin secretion: step one in meeting an emergency accomplished.

The body makes sure that energy storage is stopped in a second way as well. With the onset of the stressful emergency, you secrete glucocorticoids, which block the transport of nutrients into fat cells. This counteracts the effects of any insulin still floating around.

In addition to halting the storage of energy, you want your body to gain access to the energy already stored. You want to dip into your bank account, liquidate some of your assets, turn stored nutrients into your body's equivalent of cash to get you through this crisis. Your body reverses all of the storage steps, through the release of the stress hormones glucocorticoids, glucagon, epinephrine, and norepinephrine. These cause triglycerides to be broken down in the fat cells and, as a result, free fatty acids and glycerol pour into the circulatory system. The same hormones trigger the degradation of glycogen to glucose in cells throughout the body, and the glucose is then flushed into the bloodstream. These hormones also cause protein in nonexercising muscle to be converted back to individual amino acids.

The stored nutrients have now been converted into simpler forms. Your body makes another simplifying move. Amino acids are not a very good source of energy, but glucose is. Your body shunts the circulating amino acids to the liver, where they are converted to glucose. The liver can also generate new glucose, a process called *gluconeogenesis,* and this glucose is now readily available for energy during the disaster.

As a result of these processes, lots of energy is available to your leg muscles. There's a burst of activity; you leave the lion in the dust and arrive at the restaurant only a smidgen late for your five forty-five anticipatory insulin secretion.

The scenario I've been outlining is basically a strategy to shunt energy from storage sites like fat to muscle during an emergency. But it doesn't make adaptive sense to automatically fuel, say, your arm muscles while you're running away from a predator. It turns out that the body has solved this problem. Glucocorticoids and the other hormones of the stress-response also act to block energy uptake into muscles and into fat tissue. Somehow the individual muscles that are exercising during the emergency have a means to override this blockade and to grab all the nutrients floating around in the circulation. No one knows what the local signal is, but the net result is that you shunt energy from fat and from *non*exercising muscle to the exercising ones.

SO WHY DO WE GET SICK?

If the mobilization of energy in response to stress works so wonderfully, why should the process make us sick when we turn on the same stress-response for months on end? For many of the same reasons that constantly running to the bank and drawing on your account is a foolish way to handle your finances.

On the most basic level, it's inefficient. Another financial metaphor helps. Suppose you have some extra money and decide to put it away for a while in a high-interest account. If you agree not to touch the money for a certain period (six months, two years, whatever), the bank agrees to give you a higher-than-normal rate of interest. And typically, if you request the money earlier, you will pay a penalty for the early withdrawal. Suppose, then, that you happily deposit your money on these terms. The next day you develop the financial jitters, withdraw your money, and pay the penalty. The day after, you change your mind again, put the money back in, and sign a new agreement, only to change your mind again that afternoon, withdraw the money, and pay another penalty. Soon you've squandered half your money on penalties.

In the same way, every time you store energy away from the circulation and then return it, you lose a fair chunk of the potential energy. It takes energy to shuttle those nutrients in and out of the bloodstream, to power the enzymes that glue them together (into proteins, triglycerides, and glycogen) and the other enzymes that then break them apart, to fuel the liver during that gluconeogenesis trick. In effect, you are penalized if you activate the stress-response too often: you wind up expending so much energy that, as a first consequence, you tire more readily—just plain, old everyday fatigue.

As a second consequence, your muscles can waste away, although this rarely happens to a significant degree. Muscle is chock-full of proteins. If you are stressed chronically, constantly triggering the breakdown of proteins, your muscles never get the chance to rebuild. While they atrophy ever so slightly each time your body activates this component of the stress-response, it requires a really extraordinary amount of stress for this to happen to a serious extent. However, such myopathy of muscle can occur when massive amounts of glucocorticoids are given to a patient to control any of a variety of diseases; in such cases, the wasting away is typically known as *steroid myopathy*.

Finally, with enough stress, you may begin to have problems with either of two types of diabetes mellitus. This takes some explaining. There are two types of this diabetes. The first is insulin-dependent (or type 1, or juvenile diabetes). For reasons that are just being sorted out, the immune system decides that the cells in the pancreas which secrete insulin are, in fact, foreign invaders and attacks them (such "autoimmune" diseases will be discussed in Chapter 8). This leaves the person with very little ability to secrete insulin and therefore little ability to promote the uptake of glucose (and, indirectly, fatty acids) into target cells. Cells starve—big trouble. In addition, there's now all that glucose and fatty acid circulating in the bloodstream—oleaginous hoodlums with no place to go, and soon there's trouble there as well. They gum up the blood vessels in the kidneys, making it harder for these organs to do their job. They form atherosclerotic plaques in arteries and make it impossible for oxygen and glucose to be delivered to the tissues that depend on those blood vessels, causing little strokes in those tissues and, often, chronic pain. They also link proteins together in the eyes to make cataracts. Bad news.

And what is the best way to manage insulin-dependent diabetes? As we all know, accommodating that dependency with insulin injections. If you're diabetic, you never want your insulin levels to get too low—cells are deprived of energy, circulating glucose levels get too high. But you don't want to take too much insulin—for complex reasons, this deprives the brain of energy, potentially damaging neurons. The better the metabolic control in a diabetic, the fewer the complications and the longer the life expectancy. Thus, there's a major task for any diabetic to keep things just right, to keep food intake and insulin dosages balanced with respect to activity, fatigue, and so on. And this is an area where there has been extraordinary technological progress enabling diabetics to monitor blood glucose levels minute by minute and make minuscule changes in insulin dosages accordingly.

In non-insulin-dependent diabetes (type 2, or adult-onset diabetes), the trouble is not too little insulin, but the failure of the cells to *respond* to insulin. Another name for the disorder is thus insulin-*resistant* diabetes. The problem here arises with the tendency of many people to put on weight as they age. (However, to the extent that people do not put on weight, they show no increased risk of this disease, as is the case among people in nonwesternized populations. This disease is not, therefore, a normal feature of aging; instead, it is a disease of inactivity and fat surplus, conditions that just happen to be more common with age in some societies.) With enough fat stored away, the fat cells essentially get full; once you are an adolescent, the number

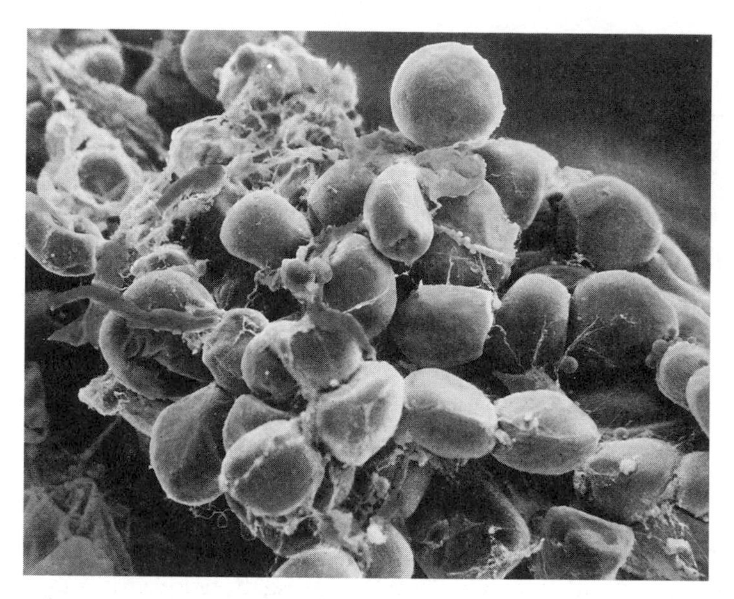

Photomicrograph of bloated fat cells.

of fat cells that you have is fixed, so if you put on weight, the individual fat cells are distended. Yet another heavy meal, a burst of insulin trying to promote more fat storage by the fat cells, and the fat cells refuse—"Tough luck, I don't care if you are insulin; we're completely full." The fat cells become less responsive to insulin trying to promote more fat storage, and less glucose is taken up by these cells.

This decreased sensitivity is mostly due to the cells losing their specialized receptors for insulin, in response to the constant insulin signal—a process called *receptor "down-regulation."**

Do the cells now starve? Of course not; the amount of fat stored in them was the source of the trouble in the first place. They get into trouble because of all that circulating glucose and fatty acid, damaging kidneys and blood vessels and eyes. So in insulin-dependent diabetes, cells starve, there's the potential for insulin overdose, and circulating nutrients accumulate in

---

*The careful reader may be confused at this point—if insulin regulates *glucose* uptake, why does it influence the amount of *fat* being stored in fat cells (since triglyceride storage is a function of the uptake of fatty acids and glycerol, not of glucose)? For immensely complex reasons, the storage of free fatty acids and glycerol as triglycerides requires glucose uptake.

places where they don't belong; in "insulin resistant" diabetes, the glomming is the problem.

How does chronic stress affect this process? First, the hormones of the stress-response cause even more glucose and fatty acids to be mobilized into the bloodstream. For both types of diabetics, this increases the likelihood of the pathologies of glucose and fatty acid gumming up in the wrong places. And chronic elevation of blood sugar levels even increases the likelihood of vascular damage in non-diabetics.

There is another, more subtle problem that occurs with chronic stress as well. When something stressful happens, not only does it make sense for the body to block insulin secretion, but also something else happens. Basically, the brain doesn't quite trust the pancreas not to secrete a little insulin still. So a second step occurs—during stress, glucocorticoids, epinephrine, and norepinephrine act on fat cells throughout the body to make them less sensitive to insulin, just in case there's some still floating around. Stress promotes insulin resistance. [And when people get into this diabetic state because they are taking large amounts of synthetic glucocorticoids (to control any of a variety of diseases that will be discussed later in the book) they have succumbed to "steroid diabetes."]

Why is this bad for someone with insulin-dependent diabetes? They have everything nice and balanced, with a healthy diet, a good sensitivity to their body's signals as to when a little insulin is needed, and so on. But throw in some chronic stress, and suddenly insulin doesn't work quite as well, causing the person to feel terrible until they figure out that they need to inject more of the stuff . . . which can make cells even more resistant to insulin, spiraling the insulin requirements upward . . . until the period of stress is over with, at which point it's not clear when to start getting the insulin dose down . . . because different parts of the body regain their insulin sensitivity at different rates. . . . The perfectly balanced system is completely upended.

Stress, including psychological stress, can thus cause havoc with metabolic control in an insulin-dependent diabetic. In one demonstration of this, diabetics were classified as either having or not having their diabetes well-controlled. Individuals were then stressed by having to speak in public, and it was observed that the uncontrolled diabetics had the far greater glucocorticoid secretion. In other words, having an exaggerated stress-response is associated with less well controlled diabetes.

And what's the relevance of stress-induced insulin resistance for insulin-resistant diabetes? It's obvious. Suppose that you're in your sixties, overweight, and just on the edge of insulin resistance. Along comes a period of chronic stress with those stress hormones repeatedly telling your cells what a great idea it is to be insulin-resistant. Enough of this and you pass the threshold for becoming overtly diabetic; you're set up for more atherosclerotic trouble.

For convenience, physicians usually use an absolute cutoff to decide when someone has insulin-resistant diabetes (that is to say, once you demonstrate a certain level of glucose in the bloodstream during a glucose tolerance test, you get labeled as having the disorder). However, the disease really represents a continuum; there is no hard and fast point of insulin resistance and hyperglycemia (elevated blood glucose levels) at which the body suddenly begins to get into trouble. Instead, for every bit of insulin resistance and hyperglycemia, there is a bit more risk of the types of damage discussed. And people who are genetically susceptible to insulin-resistant diabetes have some sort of metabolic vulnerability, such that stress disrupts their metabolism to an atypical extent long before they have become diabetic.

Insulin-dependent diabetes is devastating—in terms of the management demands, the medical complications, and the impact on life expectancy. And insulin-resistant diabetes is nearly epidemic in the United States, afflicting more than 15 percent of people over sixty-five. The disease more than doubles mortality, and nearly triples the rate of heart disease in men. Furthermore, it is a leading cause of blindness, and the seventh leading cause of death. As such, the impact of stress in this arena is pretty major. This chapter serves as a second example of the double-edged nature of the stress-response and the pathogenic consequences when it is prolonged.

ULCERS, COLITIS, AND THE RUNS

When it comes to your gut, there's no such thing as a free lunch. You've just finished some feast, eaten like a hog—slabs of turkey, somebody's grandma's famous mashed potatoes and gravy, a bare minimum of vegetables to give a semblance of healthiness, and—oh, why not—another drumstick and some corn on the cob, a slice or two of pie for dessert, ad nauseam. You expect your gut to magically convert all that into a filtrate of nutrients in your bloodstream? It takes energy, huge amounts of it. Muscular work. Your stomach not only breaks down food chemically, it does so mechanically as well. It undergoes systolic contractions: the muscle walls contract violently on one side of your stomach, and hunks of food are flung against the far wall, breaking them down in a cauldron of acids and enzymes. Your small intestines do a snake dance of peristalsis (directional contraction), contracting the muscular walls at the top end in order to squeeze the food downstream in time for the next stretch of muscle to contract. After that, your bowels do the same, and you're destined for the bathroom soon. Circular muscles called *sphincters* located at the beginning and end of each organ open and close, serving as locks to make sure that things don't move to the next level in the system until the previous stage of digestion is complete, a process no less complicated than shuttling ships through the locks of the Panama Canal. At your mouth, stomach, and small intestines, water has to be poured into the system to keep everything in solution, to make sure that the sweet potato pie, or what's left of it, doesn't turn into a dry plug. By this time, the action has moved to your large intestines, which have to extract the water and return it to your bloodstream so that you don't inadvertently excrete all that fluid and desiccate like a prune. All this takes energy, and we haven't even considered jaw fatigue. All told, your run-of-the-mill mammals, including us, expend between 10 and 23 percent of their energy on digestion.

So our by-now-familiar drama on the savanna: if you are that zebra being pursued by a lion, you can't waste energy on your stomach walls doing a rumba; there isn't time to get any nutritional benefits from digestion. And if you are that lion, nature dictates that you haven't just roused yourself from a heavy meal.

Digestion is quickly shut down during stress. We all know the first step in that process. If you get nervous, you stop secreting saliva and your mouth gets dry. Your stomach grinds to a halt, contractions stop, enzymes and digestive acids are no longer secreted, your small intestines stop peristalsis, nothing is absorbed. The rest of your body even knows that the digestive tract has been shut down—as we saw two chapters ago, blood flow to your stomach and gut is decreased so that the bloodborne oxygen and glucose can be delivered elsewhere, where they're needed. The parasympathetic nervous system, perfect for all that calm, vegetative physiology, normally mediates the actions of digestion. Along comes stress: turn off the parasympathetic, turn on the sympathetic, and forget about digestion. End of stress; switch gears again, and the digestive process resumes.

As usual, this all makes wonderful sense for the zebra or the lion. And as usual, it is in the face of chronic stress that diseases emerge instead. The remainder of the chapter is concerned with four pressing questions:

- What does stress have to do with ulcers?
- Why, when we're truly terrified, do we soil our pants (and why is it so frequently diarrhea under those charming circumstances)?
- What about colitis and irritable bowel syndrome?
- Why, if the logic of the stress-response is to inhibit digestion during stress, do so many of us overeat during periods of emotional turmoil?

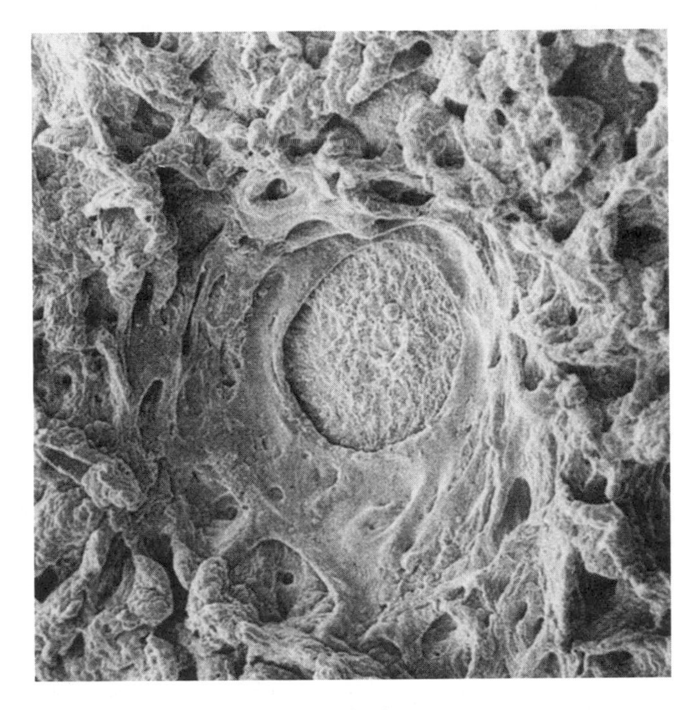

Photomicrograph of a stomach ulcer.

ULCERS

An ulcer is a hole in the wall of an organ, and ulcers originating in the stomach or in the organs immediately bordering it are termed *peptic ulcers*. The ones within the stomach are called *gastric ulcers;* those a bit higher up than the stomach are *esophageal,* and those at the border of the stomach and the intestine are *duodenal* (the most common of peptic ulcers).

As will be recalled, peptic ulcers were among the trio of symptoms Selye noted more than fifty years ago when he exposed his rats to nonspecific unpleasantness. Since then, stomach ulcers have emerged as the disorder most recognized by the lay public as a stress-related disease: in this view, you have upsetting thoughts for a long period of time and holes appear in the walls of your stomach.

Most clinicians agree that there is a subtype of ulcers that forms relatively rapidly (sometimes over the course of days) in humans who are exposed to immensely stressful crises—

hemorrhage, massive infection, trauma due to accident or surgery, burns over large parts of the body, and so on. Such "stress ulcers" can be life-threatening in severe cases.

But where a lot of contention has appeared has been with the issue of gradually emerging ulcers. This used to be a realm where people, including physicians, would immediately think stress. But a revolution over the last decade has dramatically changed thinking about ulcers.

That revolution came with the discovery in 1983 of a bacterium called *Helicobacter pylori*. This obscure microorganism was discovered by an obscure Australian pathologist named Robert Warren. He, in turn, interested an even more obscure younger colleague named Barry Marshall, who documented that this bacteria consistently turned up in biopsies of the stomachs of people with duodenal ulcers and stomach inflammation (gastritis). He theorized that it actually *caused* the inflammation and ulcers, announced this to the (gastroenterological) world at a conference, and was nearly laughed out of the room. Ulcers were caused by diet, genetics, stress—not bacteria. And besides, because the stomach is so incredibly acidic, owing to the hydrochloric acid in stomach juices, no bacteria could survive in there. People had known for years that the stomach was a sterile environment, and that any bacteria that might turn up were just due to contamination by some sloppy pathologist.

Marshall showed that the bacteria caused gastritis and ulcers in mice. That's great, but mice work differently than humans, everyone said. So, in a heroic, soon-to-be-a-movie gesture, he swallowed some *Helicobacter* bilge and caused gastritis in himself. Eventually, some folks in the field got tired of hearing him go on about the damn bacteria at meetings, decided to do some experiments to prove him wrong, and found that he was absolutely right.

Helicobacter pylori turns out to be able to live in the acidic stomach environment, protecting itself by having a structure that is particularly acid-resistant and by wrapping itself in a coat of protective bicarbonate. And, this bacterium probably has a lot to do with 85 to 100 percent of ulcers in western populations (as well as with stomach cancer). Nearly 100 percent of people in the developing world are infected with *Helicobacter*—it is probably the most common chronic bacterial infection in humans. The bacteria infect cells in the lining of the stomach, causing gastritis, which somehow compromises the ability of those cells lining the duodenum to defend themselves against stomach acids. Boom, under the right conditions, you've got a hole in that duodenal wall.

Many of the details remain to be sorted out, but the greatest triumph for Marshall and Warren has been the demonstration that antimicrobial drugs, such as antibiotics, turn out to be the greatest things since sliced bread for dealing with duodenal ulcers—they are as good at getting rid of the ulcers as are antacids or antihistamine drugs (the main prior treatments), and, best of all, unlike the aftermath of other treatments, ulcers now stay away (or at least until the next *Helicobacter* infection).

With the triumph of understanding the role of bacteria has come a view by many in the field that stress can be written out of the equation. In what one pair of investigators has termed the "Helicobacterization" of stress research on ulcers, the number of papers on stress as a component of the ulcer story has plummeted—and why not? Don't bother with this psychosomatic stuff when we finally have gotten some real science here, a bacteria that's got its own Latin name.

The trouble is that one bacteria can't be the whole story. For starters, up to 15 percent of duodenal ulcers form in people who aren't infected with *Helicobacter*, or with any other known bacteria related to it. And more damning, only about 10 percent of the people infected with the bacteria get ulcers. It's got to be *Helicobacter pylori* plus something else. Maybe nonsteroidal inflammatory drugs like aspirin, which can cause ulcers. Maybe a genetic tendency to secrete a lot of acid or to make only minimal amounts of mucus to protect stomach linings from the acid. But one of the additional factors has to be stress.

As will be detailed in Chapter 12, if you expose laboratory animals to psychological stressors, they get duodenal ulcers. And as proof that this is not stress being "ulcerogenic" all on its own, but rather worsening the impact of the bacteria, psychological stress no longer causes ulcers in rats raised in a germ-free environment (and, thus, free of those *Helicobacter* guys). And within the human realm, study after study, even those carried out after the ascendancy of the bacteria, show that duodenal ulceration is more likely to occur in people who are anxious, depressed, or undergoing severe life stressors. And a similar stress-bacteria interaction seems to occur in people as well—

study a zillion ulcer patients, do some fancy statistics that take into account bacterial load, lifestyle risk factors, and stress (something aptly called a *multivariate analysis*), and you observe that only a little bit of stress need be coupled with a ton of bacteria in order for an ulcer to occur, whereas only a smidgen of bacteria are needed when there's a major stressor.

So how does stress exacerbate the process of ulcer formation? Some sixty years after Selye first noticed his rats' ulcers, it is still not quite clear. There are some favorite scenarios, however.

Acid-Rebound To understand this mechanism, we once again have to grapple with the grim reality of what bizarre things we are willing to eat and expect our stomachs to digest. The only way that the stomach is going to be able to handle some of this stuff is if it has powerful degradative weapons. The systolic contractions certainly help, but the main weapon is the hydrochloric acid that pours into your stomach from the cells lining it. Hydrochloric acid is immensely acidic; all well and good, but it raises the obvious question of why your stomach is not itself digested by the digestive acids. Eat somebody else's stomach and your stomach disintegrates it. How do your own stomach walls remain unscathed? Basically, your stomach has to spend a fortune protecting itself. It builds many layers of stomach wall and coats them with thick, soothing mucus that buffers the acid. In addition, bicarbonate is secreted into the stomach to neutralize the acid. This is a wonderful solution, and you happily go about digestion.

Along comes a stressful period that lasts months. Your body cuts down on its acid secretion—there are now frequently times where digestion is being inhibited. During this period, your stomach essentially decides to save itself some energy by cutting corners. It cuts back a bit on the constant thickening of the stomach walls, undersecretes mucus and bicarbonate, and pockets the difference. Why not? There isn't much acid around during this stressful period anyway.

End of stressful period; you decide to celebrate by eating a large chocolate cake inscribed for the occasion, stimulate your parasympathetic nervous system, start secreting hydrochloric acid, and . . . your defenses are down. The walls have thinned, there isn't as thick a protective mucous layer as there used to be, the bicarbonate is overwhelmed. Couple repeated cycles of stress and rebound with a bacterial infection that is already compromising the defenses and you've got an ulcer.

Suppose you are in the middle of a very stressful period, and you worry that you are at risk for an ulcer. What's the solution? You could make sure that you remain under stress every second for the rest of your life. You definitely will avoid ulcers caused by hydrochloric acid secretion, but of course you'll die for a zillion other reasons. The paradox is that, in this scenario, ulcers are not formed so much during the stressor as during the recovery. This idea predicts that several periods of transient stress should be more ulcerative than one long, continuous period, and animal experiments have generally shown this to be the case.

Dramatic Decrease in Blood Flow In an emergency, you want to deliver as much blood as possible to the muscles that are exercising. In response to stress, your sympathetic nervous system diverts blood from the gut to more important places—remember the man with a gunshot wound in the stomach, whose guts would blanch from decreased blood flow every time he became angry or anxious. If your stressor is one that involves a dramatic decrease in blood flow to the gut (for example, following a hemorrhage), it begins to cause little infarcts—small strokes—in your stomach walls, because of lack of oxygen. You develop small lesions of necrotic (dead) tissue, which are the building blocks of ulcers.

This condition probably arises for at least two reasons. First, with decreased blood flow, less of the acid that accumulates is being flushed away. The second reason involves another paradoxical piece of biology. We all obviously need oxygen and would turn an unsightly blue without it. However, running your cells on oxygen can sometimes produce an odd, dangerous class of compounds called *oxygen radicals*. Normally, another group of compounds (free radical quenchers, or scavengers) dispose of these villains. There is some evidence, however, that during periods of chronic stress, when blood flow (and thus oxygen delivery) to the gut decreases, your stomach stops making the scavengers that protect you from the oxygen radicals. Fine for the period of stress (since the oxygen radicals are also in shorter supply); it's a clever way to save energy during a crisis. At the end of stress, however, when blood flow chock-full of oxygen resumes and the normal amount of oxygen radicals is generated, the stomach has its oxidative pants down. Without sufficient scavengers, the oxygen radicals start killing cells in the stomach walls; couple that with cells already in trouble thanks to bacterial infection and you've got an ulcer. Note how similar this scenario is to the acid-rebound mechanism: in both cases, the damage occurs not during the period of stress but in its aftermath, and not so much because stress increases the size of an insult (for example, the amount of acid secreted or the amount of oxygen radicals produced), but because, during the stressful emergency, the gut scrimps on defenses against such insults.

Immune Suppression As you will soon learn in sickening detail (Chapter 8) chronic stress suppresses immunity, and in this scenario, lowered immune defenses equals more *Helicobacters* reproducing happily. A possibility, although it is not clear how much the immune system normally affects the cauldron of the stomach anyway.

Insufficient Amounts of Prostaglandins In this scenario, micro-ulcers begin now and then in your gut, as part of the expected wear and tear on the system. Normally your body can repair the damage by secreting a class of chemicals called *prostaglandins,* thought to aid the healing process by increasing blood flow through the stomach walls. During stress, however, the synthesis of these prostaglandins is inhibited by the actions of glucocorticoids. In this scenario, stress does not so much cause ulcers to form as impair your body's ability to catch them early and repair them. It is not yet established how often this is the route for ulcer formation during stress. (Aspirin also inhibits prostaglandin synthesis, which is why aspirin can aggravate a bleeding ulcer.)

Stomach Contractions For unknown reasons, stress causes the stomach to initiate slow, rhythmic contractions (about one per minute); and for unknown reasons, these seem to add to ulcer risk. One idea is that during the contractions, blood flow to the stomach is disrupted, causing little bursts of ischemia; there's not much evidence for this, however. Another idea is that the contractions mechanically damage the stomach walls. The jury is still out on that mechanism.

Most of these mechanisms are pretty well documented routes by which ulcers can form; of those credible mechanisms, most can occur during at least certain types of stressors. More than one mechanism may occur simultaneously, and people seemingly differ as to how likely each mechanism is to occur in their gut during stress, and how likely it is to interact with bacterial infection. Additional mechanisms for stress's role in ulcer formation will no doubt be discovered, but for the moment, these should be quite sufficient to make anyone sick.

BOWELS IN AN UPROAR

Regardless of how stressful that board meeting or examination is, we're not likely to soil our pants. Nevertheless, we are all aware of the tendency of immensely terrified people—convicts about to be executed, soldiers amid horrifying battle—to defecate spontaneously. (This reaction is consistent enough that in many states, prisoners are clothed in diapers before an execution.)

The logic as to why this occurs is similar to why we lose control of our bladders if we are very frightened, as described in Chapter 3. Most of digestion is a strategy to get your mouth, stomach, bile ducts, and so forth, to work together to break your food down into its constituent parts by the time it reaches the small intestines. The small intestines, in turn, are responsible for absorbing nutrients out of this mess and delivering them to the bloodstream. As is apparent to most of us, not much of what we eat is actually nutritious, and a large percentage of what we consume is left over after the small intestines pick through it. In the large intestines, the leftovers are converted to feces and eventually exit stage left.

Yet again, you sprint across the veld. All that stuff sitting in your large intestines, from which the nutritive potential has already been absorbed, is just dead weight. You have the choice of sprinting for your life with or without a couple of pounds of excess baggage in your bowels. Empty them.

The biology of this is quite well understood. The sympathetic nervous system is responsible. At the same time that it is sending a signal to your stomach to stop its contractions and to your small intestine to stop peristalsis, your sympathetic nervous system is actually stimulating muscular movement in your large intestine. Inject into a rat's brain the chemicals that turn on the sympathetic nervous system, and suddenly the small intestine stops contracting and the large intestine starts contracting like crazy.

But why, to add insult to injury, is it so frequently diarrhea when you are truly frightened? Relatively large amounts of water are needed for digestion, to keep your food in solution as you break it down so that it will be easy to absorb into the circulation when digestion is done. The job of the large intestine is to get that water back, and that's why your bowels have to be so long—the leftovers slowly inch their way through the large intestine, starting as a soupy gruel and ending up, ideally, as reasonably dry stool. Disaster strikes, run for your life, increase that large intestinal motility, and everything gets pushed through too fast for the water to be absorbed optimally. Diarrhea, simple as that.

COLITIS AND IRRITABLE BOWELS

As we have seen, because of the ability of stress to increase large intestinal motility, being executed can often lead to socially embarrassing gastrointestinal consequences. But similar problems, writ small, may occur among people who are less dramatically stressed. Irritable bowel syndromes are the most common gastrointestinal diseases and are probably the most common stress-related disorder, likely to afflict the majority of us at some point in our lives. Personally, all the major rites of passage in my life have been marked by pretty impressive cases of the runs a few days before—my bar mitzvah, going away to college, my doctoral defense, proposing marriage, my wedding, the day my wife gave birth to our son. (Finally, that confessional tone obligatory to successful books these days. Now if I can only name some famous Hollywood starlets with whom I've taken diuretics, this may become a best-seller.)

Irritable bowel syndrome (IBS) is a hodgepodge of disorders in which there is abdominal pain (particularly just after a meal) that is relieved by defecating and which, at least 25 percent of the time, includes symptoms such as diarrhea or constipation, passage of mucus, bloating and abdominal distention. Colitis is inflammation of the colon (also known as the "bowels"); as the most common consequence of colitis, the colon can become irritable, qualifying as a type of IBS. "Spastic colon" is another way of describing irritable colon.

The disorder seems to arise not only from being under a lot of stress, but also from having a gastrointestinal system that is abnormally sensitive to such stress. The first component—being exposed to too many stressors—can probably best be thought of as a problem of different parts of the gastrointestinal tract getting desynchronized in their function. A scenario in a perfect world: the small and large intestines are doing their jobs correctly; something stressful occurs, and both organs respond promptly (with motility decreasing in the small intestines and increasing in the large). End of the stressor, and both organs promptly resume their normal level of activity.

Instead of this ideal, however, with repeated stressors the responses of the small and large intestines may become uncoordinated—one of the organs may have a relatively larger stress-response than the other, or a more rapid recovery of function after the stressor ends. If the relative net result of this uncoordination is small intestinal motility being too inhibited, the result is constipation. If the result is large intestinal motility being too stimulated, the result is diarrhea.

In addition, these are not only disorders of too much stress, but of too much gastrointestinal sensitivity to stress. This can be shown in experimental situations, where a person with IBS is subjected to a stressor (keeping her hand in ice water for a while, trying to make sense of two recorded conversations at once, participating in a pressured interview). These subjects show a greater gastrointestinal response than do healthy controls. Those who tend toward chronic diarrhea are the ones who show the greatest large intestinal responses, while those who tend toward constipation show the greatest small intestinal ones.

The preceding section was pretty much all I had to write on this subject in the first edition of this book. To my great surprise, a number of people with colitis took great exception to it (including one reviewer for a colitis newsletter who wrote that if I was so misinformed on this subject, the rest of the book obviously couldn't be trusted either). There is a great deal of resistance to the link between stress and IBS. The reasons for this appear to be twofold. First, lots of careful studies have failed to show a stress component in these disorders. Second, some personality types have been linked to IBS, and the linkage seems pretty suspect. These studies tended to focus on a lot of psychoanalytic gibberish (there, now I'll get myself into trouble with that crowd)—some hoo-ha about the person being stuck in the anal stage of development, a regression to the period of toilet training where going to the bathroom gained great acclaim and, suddenly, diarrhea was a symbolic reach for parental approval. Or the approval of the doctor as a parental surrogate. Or something or other. I'm not sure how they factored in constipation, but I'm sure they did.

Few gastroenterologists take these ideas seriously anymore, and most other attempts at defining specific personality types that are more prone toward gastrointestinal motility diseases have been discredited in the scientific literature. However, in less scientific circles, some still cling to these views, and it is easy to see how someone suffering from IBS would take umbrage at it being relegated to "blame the victim who's still stuck in the toilet training stage" and thus not want to hear about stress either (especially when the theme is not how much stress the person suffers from but how poorly some people handle everyday stressors).

Nevertheless, stress appears to have something to do with IBS—it worsens colitis in animal models of the disease, the underlying biology of the connection makes sense, experimental stressors trigger gut motility in people with IBS more readily than in controls, and everyday stressors clearly are related to the disease in a subset of patients. If this is true, why have there been careful studies which, nonetheless, fail to find a link between stress and the disease?

First, stress is more likely to play a role in worsening preexisting colitis than in causing it, and instances of worsening are harder to demonstrate than instances of causing. Second, both the severity of IBS symptoms and the intensity of stressors that someone is experiencing tend to wax and wane over time, and detecting a link between two such fluctuating patterns takes some very fancy statistics (typically, a technique called *time-series analysis*, a subject four classes more advanced than the statistics that most biomedical scientists have sort of learned. When my wife had to do a time-series analysis as part of her doctoral dissertation research, it made me nervous just to have a textbook on the subject in the house). Such waxing and waning of stress and of symptoms is particularly difficult to track because most studies are *retrospective* (they look at people who already have IBS and ask them to identify stressors in their past) rather than *prospective* (in which people who do not have a disease are followed to see if stress predicts who is going to get it). The problem here is that people are terribly inaccurate at recalling information about stressors and symptoms that are more than a few months old. Finally, stress is likely to play a role in only a subset of sufferers (probably reflecting the fact, as noted, that gastrointestinal motility diseases reflect an array of causes), and it takes some additional fancy statistics to detect those folks as a meaningful subset of the whole, instead of as just random noise in the data.

At later junctures in this book, we will see other supposed links between stress and some disease, and there will be the same quandary—there definitely is a link in some patients, or clinical impressions strongly support a stress-disease link, yet hard-nosed studies fail to show the same thing. As we will see repeatedly, the trouble is that the supposedly hard-nosed studies are often asking a fairly unsophisticated question: does stress cause the disease in the majority of sufferers? The far more sophisticated questions to ask are whether stress worsens preexisting disease, whether the linkage of symptoms and stressors fluctuates over time, and whether these links occur only in a subset of vulnerable individuals. When asked in those ways, the stress-disease link becomes far more solid.

STRESS AND APPETITE

Clearly, regulation of the gastrointestinal tract during stress involves some component of eating patterns and appetite. Just as clearly, it is pretty confusing what exactly happens. Our everyday experiences are contradictory. You get upset over something, become agitated and anxious, and find yourself mechanically chewing away at some junk food without even tasting it. You get nervous and upset over something, and find you have no

Mark Daughhetee, 1985: The Sin of Gluttony, *oil on silver print.*

appetite for dinner. Dogs lose their beloved masters and pine away, refusing to eat. People go through a traumatic divorce and balloon up. People go through a traumatic divorce and waste away, claiming that they feel queasy every time they eat anything. We could simply conclude that "stress influences appetite," but that wouldn't tell us anything about why it does so in opposite directions in different cases.

Initially, when you study the biology of the effect of stress on appetite, the picture is no clearer. At least one of the critical hormones of the stress-response stimulates appetite, while another inhibits it. You might recall from earlier chapters that a hormone (CRF) is released by the hypothalamus and, by stimulating the pituitary to release ACTH, starts the cascade of events that culminates in adrenal release of glucocorticoids. Evolution has allowed the development of efficient use of the body's chemical messengers, and CRF is no exception. It is also used in parts of the brain to regulate other features of the stress-response. It helps to turn on the sympathetic nervous system, and it plays a role in increasing vigilance and arousal during stress. It also suppresses appetite. This might lead us to the conclusion that stress inhibits food intake. (Unsuccessful dieters should be warned against running to the neighborhood pharmacist for a bottle of CRF. It will probably help you lose weight, but you'll feel awful—as if you were always in the middle of an anxiety-provoking emergency: your heart racing; you, jumpy, hyposexual, irritable. The same scientific renegades who, as outlined in Chapter 2, took on Guillemin and discovered the structure of CRF recently found a new chemical in the brain called *urocortin*. It is structurally very similar to CRF, binds to one of the two types of CRF receptors, and is even better at suppressing appetite but doesn't cause as much of an anxious stress-response. No one knows yet if it is released in the brain during stress. More importantly for those eager to buy urocortin stocks, no one yet knows if it is safe for humans to take synthetic versions of urocortin. Probably just opt for a few more sit-ups.)

On the other side of the picture are glucocorticoids. In addition to the actions already outlined in response to stress, they appear to stimulate appetite. This is typically demonstrated in rats: glucocorticoids make these animals more willing to run mazes looking for food, more willing to press a lever for a food pellet, and so on. To my knowledge, the equivalent has not been tested in humans—stoking people up on adrenal steroids to see how many times they are willing to scurry up and down supermarket aisles in order to find the apricot fruit rolls or the pickled onions that they've been craving. Scientists have a reasonably good idea where in the brain glucocorticoids stimulate appetite, which type of glucocorticoid receptors are involved, and so on.

Thus, the CRF literature leads to the conclusion that stress will decrease food intake, while the glucocorticoid literature predicts that stress will do just the opposite. How do you reconcile these opposite effects of CRF and glucocorticoids? Time may have something to do with it. When a stressful event occurs, there is a burst of CRF secretion within a few seconds. ACTH levels take about fifteen seconds to go up, while it takes many minutes for glucocorticoid levels to surge in the bloodstream (depending on the species). Thus, CRF is the fastest wave of the adrenal cascade, glucocorticoids the slowest. This difference in "timecourse" is also seen in the speed at which these hormones work on various parts of the body. CRF makes its effects felt within seconds, while glucocorticoids take minutes to hours to exert their actions. Finally, when the stressful event is over, it takes mere seconds for CRF and ACTH to be cleared from the bloodstream, while it can take hours for glucocorticoids.

These differences enable us to make some pretty confident predictions. If there are large amounts of CRF and ACTH in your bloodstream, yet almost no glucocorticoids, it is a safe bet that you are in the first few minutes of a stressful event. If there are large amounts of CRF, ACTH, *and* glucocorticoids in the bloodstream, you are probably right in the middle of a sustained stressor. And if there are substantial amounts of glucocorticoids in the circulation but little CRF or ACTH, you have probably started the recovery period.

The conflicting effects of CRF and glucocorticoids on appetite begin to make some sense. At the very beginning of a stress-response, it is logical to shut down digestion, turn off activity in the gut, mobilize energy from sites of storage in your body. If your salivary glands have stopped secreting and your stomach is asleep, this is a reasonable time to lose your appetite, and the burst of CRF release helps bring that about. Then, when the stressful event is finished, digestion starts up again and your body can begin to replenish those stores of energy consumed in that mad dash across the savanna. Appetite is stimulated. In this case, glucocorticoids would not so much serve as the mediator of the stress-response, but as the means of *recovering* from the stress-response.

These processes may even begin to explain why some of us lose our appetites during stress and others get the munchies. It may be related to the pattern, the duration of our stressors. Something truly stressful occurs, and a maximal signal to secrete CRF, ACTH, and glucocorticoids is initiated. If the stressor ends after, for example, ten minutes, there will cumulatively be perhaps a twelve-minute burst of CRF exposure (ten minutes during the stressor, plus the two minutes or so it takes to clear the CRF afterward) and a two-hour burst of exposure to glucocorticoids (the ten minutes of secretion during the stressor plus the much longer time to clear). This may be a case where the net result of the stressor is stimulation of appetite. By contrast, the longer the stressor lasts, the longer the cumulative time of exposure to CRF, causing inhibition of appetite. (This model assumes what is generally the case: that the inhibitory effects of CRF on appetite are stronger than the stimulatory effects of glucocorticoids, if they reach hunger centers in the brain at the same time.) Thus, lots of short stressful events should lead to overeating, while one long, continuous stressor should lead to appetite loss.

I do not think this idea has been tested, but in considering individual responses, one would also have to factor in the precise quantities of CRF and glucocorticoids secreted in a particular person, how long each hormone persists in the bloodstream, just how effectively each influences hunger centers in the brain, and so on. In other words, there are likely to be many complex, individual differences in how people respond. Moreover, there are probably other hormones relevant to the regulation of appetite, and it's not entirely clear what is happening with them during stress. The most exciting example of this is leptin, which people are already buying stock in. This hormone is coded for by the *ob* (for "obesity") gene in fat cells; once it gets into the circulation, it makes its way to the hypothalamus and triggers a satiation signal—food intake and body weight decline, and metabolism increases, burning up fat stores. Not surprisingly, feverish research is now uncovering instances of obesity caused by the secretion of too little leptin, unresponsive of the hypothalamus to leptin, and so on. And where does stress fit in? It's not clear. Glucocorticoids increase the activity of the *ob* gene and increase leptin levels in the bloodstream. Paradoxically, glucocorticoids also *decrease* the efficacy with which leptin blunts appetite in the brain. Part of the appetite-stimulating effects of chronic glucocorticoid exposure might thus be due to its chronically making the hypothalamus resistant to the leptin satiety signal. Stay tuned for this rapidly emerging story.

Eating and digesting are fundamental ways in which a body plans for the future, while sprinting across the veld in a panic represents one of the most fundamental ways in which a body responds to an immediate crisis. The central lesson of this chapter is how incompatible it is to plan for the future and simultaneously deal with a current emergency. This idea will dominate the coming two chapters, which review two of the most optimistic projects your body can undertake in planning ahead—growth and reproduction.

DWARFISM AND THE IMPORTANCE OF MOTHERS

It still surprises me that organisms grow. Maybe I don't believe in biology as much as I should. Eating and digesting a meal seems very real. You put a massive amount of something or other in your mouth, and as a result, all sorts of very tangible things happen—your jaw gets tired, your stomach distends, and eventually something comes out the other end. The results of growth seem pretty tangible, too. Long bones get longer, kids weigh more when you heft them, your squirrelly little nephew starts towering over you. Pretty concrete.

My difficulty is with the intervening steps. I know theoretically what happens in the blood after a meal; my university even allows me to teach impressionable students about it. A lot more glucose, fatty acids, and amino acids wind up in the circulation, and I know that this stuff is nutritive. But you can't look with the naked eye at a vial of blood and tell if it's chock-full of nutrients. Someone ate a mountain of spaghetti, salad, garlic bread, *and* two slices of cake for dessert—and that has been transformed and is now in this test tube of blood? Hard to believe. And somehow it's going to be reconstructed into bone? Just think, your femur is made up of tiny pieces of your mother's chicken pot pie that you ate throughout your youth. Ha! You see, you don't really believe in the process either. Maybe we're too primitive to comprehend the transmogrification of material. Perhaps it would make more sense if we could actually observe globs of bread pudding moving directly from our stomachs and welding themselves onto the ends of our bones to make them longer.

Nevertheless, growth does occur. And in a kid, it's not a trivial process. The brain gets bigger, the shape of the head changes. Cells divide, grow in size, and synthesize new proteins. Long bones lengthen as cartilaginous cells, called *chondrocytes,* at the ends of bones migrate into the shaft and solidify into bone. Baby fat melts away and is replaced by muscle. The larynx thickens and the voice deepens, hair grows in all sorts of unlikely places on the body, breasts develop, testes enlarge—the works.

From the standpoint of understanding the effects of stress on growth, the most important feature of the growth process is that, of course, growth doesn't come cheap. Calcium must be obtained to build bones, amino acids are needed for all that protein synthesis, fatty acids build cell walls—and it's glucose that pays for the building costs. Appetite soars, and nutrients pour in from the intestines. A large part of what various hormones do is to mobilize the energy and the material needed for all these civic expansion projects. Growth hormone dominates the process. Sometimes it works directly on cells in the body—for example, growth hormone helps to break down fat cells, depleting them of their fatty acids so that these stored nutrients can be diverted to the growing cells. Alternatively, sometimes growth hormone must first trigger the release of another class of hormones called *somatomedins,* which actually do the job—for example, promoting cell division. Thyroid hormone plays a role, promoting growth hormone release, making bones more responsive to somatomedins. Insulin does something similar as well. The reproductive hormones come into play around puberty. Estrogen promotes the growth of long bones, both by acting directly on bone and by increasing growth hormone secretion. Testosterone does similar things to long bones and, in addition, enhances muscle growth.

The differential effects of estrogen and of testosterone on growth are the reasons that at puberty girls get taller, whereas boys get taller and more muscular. Testosterone increases muscle mass in adults as well, which is why various football players and weight lifters, as part of their muscle-building programs, abuse "anabolic" (growth-enhancing) steroids, of which testosterone is the prime example. Adolescents stop growing when the ends of the long bones meet and begin to fuse, but for complex reasons, testosterone, by accelerating the growth of the ends of long bones, can actually speed the cessation of growth. Thus, pubescent boys given particularly high concentrations of testosterone will, paradoxically, wind up having their adult stature blunted a bit. Conversely, boys castrated before puberty grow to be quite tall, with lanky bodies and particularly long limbs. Opera history buffs will recognize this morphology: castrati were famed for this body shape.

GROWTH INHIBITION DURING STRESS

Growth is great when you are a ten-year-old lying in bed at night with a full belly. However, it wouldn't make sense to spend much energy on this sort of thing during a stressful event. You're sprinting for your life, trying to evade the lion, that whole scenario. If there is no time to derive any advantages from digesting your meal at that point, there certainly isn't time to get any benefit from growth.

To understand the process by which stress inhibits growth, it helps to begin with extreme cases. A child of, say, eight years, is brought to a doctor because she has stopped growing. There are none of the typical problems—the kid is getting enough food, there is no apparent disease, she has no intestinal parasites that compete for nutrients. No one can identify an *organic* cause of her problem; yet she doesn't grow. In many such cases, there turns out to be something dreadfully stressful in her life—emotional neglect or psychological abuse. In such circumstances, the syndrome is called *stress dwarfism*, or *psychosocial* or *psychogenic dwarfism*.

Perhaps right now a question is running through the minds of the half of you who are below average height. If you are short, yet didn't have any obvious chronic diseases as a kid and can recall an unpleasant period in your childhood, are you a product of mild stress dwarfism? Suppose one of your parents had a job necessitating frequent moves, and every year or two throughout childhood you were uprooted, forced to leave your friends, moved off to a strange school. Is this the sort of situation associated with psychogenic dwarfism? Definitely not. How about something more severe? Suppose that your parents divorced when you were a kid; it was very acrimonious, and at some point it became horrifyingly clear that neither of them particularly wanted you to live with them afterward. Stress dwarfism? Probably not.

The syndrome is extremely rare, and physicians fall over themselves to see the occasional cases. These are the kids who are incessantly harassed and psychologically terrorized by the crazy stepfather. These are the kids who, when the police and the social workers break down the door, are discovered to have been locked in a dark closet for months, fed a tray of food slipped under the door. These are the products of vast, grotesque family psychopathology. And they appear in every endocrinology textbook, standing nude in front of a growth chart. Stunted little kids, years behind their expected growth rate, years behind in mental development, bruised and with distorted, flinching postures, haunted, slack facial expressions, eyes masked by the obligatory rectangles that accompany all naked people in medical texts. And invariably with stories to take your breath away and make you wonder at the potential sickness of the human mind.

And, invariably, on the same page in the text is a surprising second photo—the same child a few years later, after having been placed in a different environment (or, as one pediatric endocrinologist drolly termed it, having undergone a "parentectomy"). No bruises, maybe a tentative smile. And a lot taller. So long as the stressor is removed before the child is far into puberty (when the ends of the long bones fuse together and growth ceases), there is the potential for some degree of "catch-up" growth, although shortness of stature and some degree of stunting of personality and intellect usually persist into adulthood.

Despite the clinical rarity of stress dwarfism, instances pop up throughout history. One possible case arose during the thirteenth century as the result of an experiment by that noted endocrinologist, King Frederick II of Sicily. It seems that his court was engrossed in philosophic disputation over the natural language of humans. In an attempt to resolve the question, Frederick (who was apparently betting on Hebrew, Greek, or Latin) came up with a surprisingly sophisticated idea for an

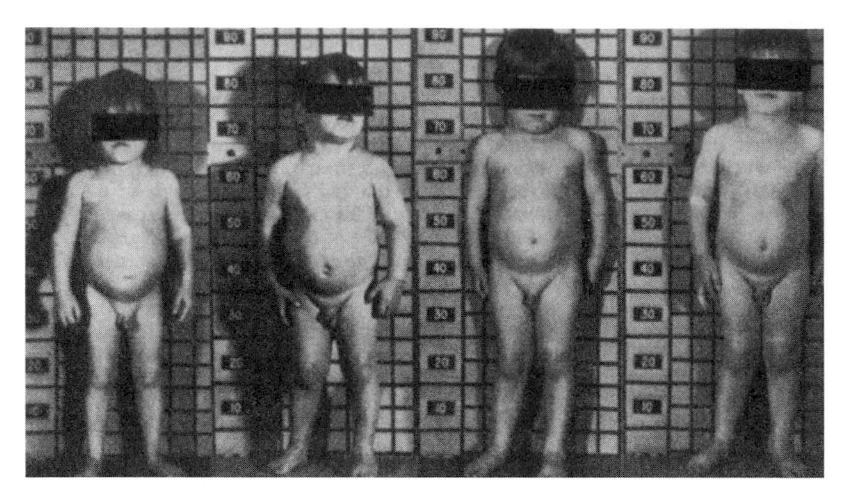

A child suffering from stress dwarfism: changes in appearance during hospitalization (left to right).

experiment. He commandeered a bunch of infants and had each one reared in a room by itself. Every day someone would bring the child food, fresh blankets, and clean clothes, all of the best quality. But they wouldn't stay and play with the infant, or hold it—too much of a chance that the person would speak in the child's presence. The infants would be reared without human language, and everyone would get to see what was actually the natural language of humans.

Of course, the kids did not spontaneously burst out of the door one day reciting poetry in Italian or singing opera. The kids didn't burst out of the door at all. None of them survived. The lesson is obvious to us now—optimal growth and development do not merely depend on being fed the right number of calories and being kept warm. Frederick "laboured in vain, for the children could not live without clappings of hands and gestures and gladness of countenance and blandishments," reported the contemporary historian Salimbene. It seems quite plausible that these kids, all healthy and well fed, died of stress dwarfism.*

---

*Some clinical nomenclature: "maternal deprivation syndrome," "deprivation syndrome," and "nonorganic failure to thrive" usually refer to infants and invariably to the loss of the mother. "Stress dwarfism," "psychogenic dwarfism," and "psychosocial dwarfism" usually refer to children aged three years or older. However, some papers do not follow this age dichotomy; during the nineteenth century, infants dying of failure to thrive in orphanages were said to suffer from "marasmus," Greek for "wasting away."

A recent study makes the same point, if more subtly. This one winds up in half the textbooks, and if it had been planned, it could not have produced a cleaner result. The subjects of the "experiment" were children reared in two different orphanages in Germany just after World War II. Both orphanages were run by the government; thus there were important controls in place—the kids in both had the same general diet, the same frequency of doctors' visits, and so on. The main identifiable difference in their care was the two women who ran the orphanages. The scientists even checked them, and their description sounds like a parable. In one orphanage was Fräulein Grun, the warm, nurturant mother who played with the children, comforted them, and spent all day singing and laughing. In the other was Fräulein Schwarz, a woman who was clearly in the wrong profession. She discharged her professional obligations, but minimized her contact with the children; she frequently criticized and berated them, typically among their assembled peers. The growth rates at the two orphanages were entirely different. Fräulein Schwarz's kids grew in height and weight at a slower pace than the kids in the other orphanage. Then, in an elaboration that couldn't have been more useful if it had been planned by a scientist, Fräulein Grun moved on to greener pastures and, for some bureaucratic reason, Fräulein Schwarz was transferred to the other orphanage. Growth rates in her former orphanage promptly increased; those in her new one decreased.

Until recent decades, being placed in an orphanage was one of the most dangerous events that could happen to a child. In 1915, despite adequate food and health care, the majority of orphanages in the United States had close to 100 percent mortality rates. Even some thirty years later, a third of infants in orphanages were still dying, "in spite of good food and meticulous medical care." It should be noted that this was a period when child-rearing practices were intensely Spartan, and the leading expert, Dr. Luther Holt of Columbia University, warned parents of the adverse effects of the "vicious practice" of using a cradle, picking up the child when it cried, or handling it too often. Ironically, in the less wealthy orphanages, where the staff was often not up to date on the advice of this savant, mortality rates were lower than in the "better" orphanages.

A final and truly disturbing example comes to mind. If you ever find yourself reading chapter after chapter about growth

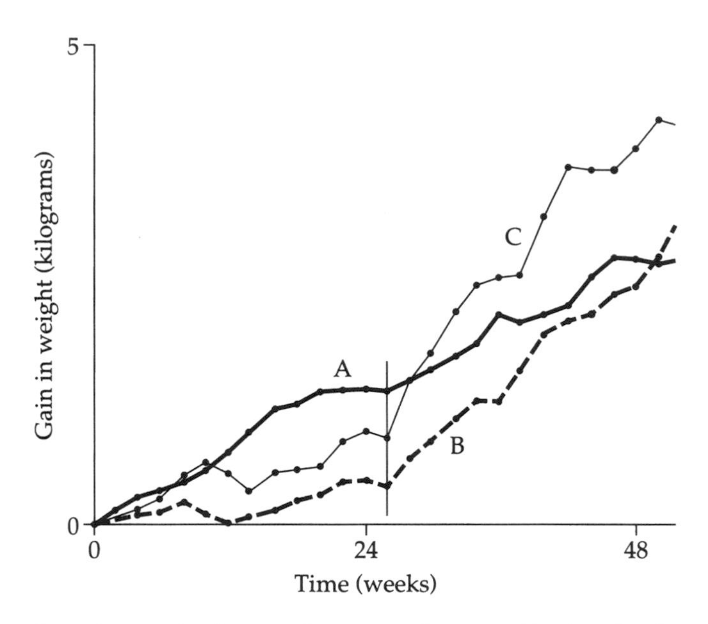

Growth rates in the two German orphanages. During the first 26 weeks of the study, growth rates in Orphanage A, under the administration of the warm Fräulein Grun, were much greater than those in Orphanage B, with the stern Fräulein Schwarz. At 26 weeks (vertical line), Fräulein Grun left Orphange A and was replaced by Fräulein Schwarz. The rate of growth in that orphanage promptly slowed; growth in Orphanage B, now minus the stern Fräulein Schwarz, accelerated and soon surpassed that of Orphanage A. A fascinating elaboration emerges from the fact that Schwarz was not completely heartless, but had a subset of children who were her favorites (Curve C), whom she had transferred with her.

endocrinology (which I don't recommend), you will note an occasional odd reference to Peter Pan—perhaps a quotation from the play, or a snide comment about Tinker Bell. For a long time, I couldn't make sense of this phenomenon. Finally, buried in a chapter in one textbook, I found the explanation.

The chapter reviewed the regulation of growth in children and the capacity for severe psychological stress to trigger psychogenic dwarfism. It gave an example that occurred in a British Victorian family. A son, age thirteen, the beloved favorite of the mother, is killed in a skating accident. The mother, despairing and bereaved, takes to her bed in grief for years afterward, utterly ignoring her other, six-year-old son. Horrible scenes ensue. The boy, on one occasion, enters her darkened room; the mother, in her delusional state, briefly believes it is the dead son—"David, is that you? Could that be you?"—before realizing: "Oh, it is only you." On the rare instances when the mother interacts with the younger son, she repeatedly expresses the same obsessive thought: the only solace that she feels is that David died when he was still perfect, still a boy, never to be ruined by growing up and growing away from his mother.

The younger boy, ignored (the rather stern and distant father appears to have had no interactions with any of the children), seizes upon this idea; by remaining a boy forever, by not growing up, he will at least have some chance of pleasing his mother, winning her love. Although there is no evidence of disease or malnutrition in his well-to-do family, he ceases growing. As an adult, he is just barely five feet in height, and his marriage is unconsummated.

The forlorn boy became the author of the much-beloved children's classic *Peter Pan*. J. M. Barrie's plays and novels are filled with children who didn't grow up, who were fortunate enough to die in childhood, who came back as ghosts to visit their mothers.

THE MECHANISMS UNDERLYING STRESS DWARFISM

Despite the rarity of stress dwarfism, it is known that the affected children tend to have extremely low growth hormone levels in their circulation, which would certainly account for the dwarfism. Why do growth hormone levels decline in these kids? Growth hormone is secreted by the pituitary gland, which in turn is regulated by the hypothalamus in the brain (see Chapter 2). The hypothalamus controls growth secretion through the release of two hormones—one that stimulates growth hormone release (*growth hormone releasing hormone,* or GHRH) and one that inhibits release (*growth hormone inhibiting hormone,* GHIH, or, as it is more commonly termed, *somatostatin*). The normal fluctuation in growth hormone levels represents the integration of the brain's stimulatory GHRH signal and inhibitory GHIH signal.

If growth hormone levels are suppressed in kids with stress dwarfism, it could be because of a problem at the level of the hypothalamus, perhaps secretion of too little GHRH, or too much GHIH, or both. This has been tough to study in humans, but animal studies indicate that it is probably too much GHIH that drives down growth hormone levels. In addition, the pituitary might be responding strangely to the hypothalamic hormones. If it becomes overly sensitive to GHIH, or insensitive to GHRH, that could account for the declining growth hormone levels as well.

Once growth hormone is secreted, it binds, along with a related class of other hormones called *somatomedins*, to receptors in target cells and causes these cells to begin growing and dividing. There is some indication that psychogenic dwarfism results not only from too little growth hormones being secreted, but also from target cells becoming insensitive to growth hormone and somatomedins. Cynthia Kuhn and Saul Schanberg at Duke University have studied an enzyme called *ornithine decarboxylase* (ODC), which is critical for cell division during growth. If infant rats are deprived of their mothers, growth hormone levels and ODC activity both rapidly decline. That may suggest that ODC activity declines because there is no longer growth hormone around to stimulate it. However, if you inject those infant rats with growth hormone, ODC activity still does not return to normal—the cells no longer respond to the growth hormone signal.

There are hints that other hormones may also mediate the decline in growth hormone concentrations associated with dwarfism. The sympathetic nervous system may be overactive in these children, which would block growth hormone secretion. As evidence for the sympathetic overactivity hypothesis, scientists have found that giving a drug that blocked one branch of the sympathetic system to a child with stress dwarfism caused his growth hormone levels to return to normal. This certainly would fit a picture of kids under stress and the lesson of the four Fs from Chapter 2.

Glucocorticoids may also be involved. While in most studies kids with stress dwarfism are found to have normal to low levels of these stress hormones, a few report elevated levels; moreover, when infant rats are separated from their mothers, their glucocorticoid levels soar. Glucocorticoids disrupt growth in many ways. They block the secretion of growth hormone, the sensitivity of target cells to growth hormone, and the synthesis of new proteins and of new DNA in dividing cells. In human children, levels of glucocorticoids need to be maintained at only two to three times higher than normal to arrest growth (with major stressors being capable of causing up to tenfold increases).

In addition to hormonal problems, kids with stress dwarfism appear to have gastrointestinal problems as well. With perfectly adequate diets, they fail to absorb the nutrients out of their guts. This is probably because of the enhanced activity of their sympathetic nervous systems, which are producing excessive levels of epinephrine and norepinephrine. As discussed in Chapter 5, these sympathetic hormones will halt the release of various digestive enzymes, stop the muscular contractions of the stomach and intestinal walls, and block nutrient absorption. (Stress, or psychogenic, dwarfism, the kind that hits kids aged three years and older, seems to involve growth hormone shortage more than gastrointestinal malabsorption problems, whereas the failure-to-thrive syndrome of infants seems to involve gastrointestinal problems more than hormonal ones.)

One study confirms some of the lessons of the German orphanage report and, in addition, shows how dramatically the hormones involved in growth respond to environmental subtleties. The paper follows a single child with stress dwarfism. When brought to the hospital, he was assigned to a special nurse who spent a great deal of time with him and to whom he became emotionally attached. Row A in the table on the next page shows his physiological profile upon entering the hospital: extremely low growth hormone levels and a low rate of growth. Row B shows his profile a few months later, while still in the hospital: his growth hormone levels have more than doubled (without his having received any synthetic hormones), and there has been a vast increase in his growth rate. These data graphically demonstrate that stress dwarfism is not a problem of insufficient food—the boy was eating more at the time he entered the hospital than a few months later, when his growth resumed.

Row C in the table profiles the period during which the nurse, to whom the boy was already greatly attached, went on a three-week vacation. Despite the same food intake, growth hormone levels and growth rates plummeted back to the minimal

A demonstration of the sensitivity of growth to the emotional state of a child

Condition	Growth hormone	Growth	Food intake
A. Entry into hospital	5.9	0.5	1663
B. 100 days later	13.0	1.7	1514
C. Favorite nurse on vacation	6.9	0.6	1504
D. Nurse returns	15.0	1.5	1521

Source: From Saenger and colleagues, 1977. Growth hormone is measured in nanograms of the hormone per milliliter of blood following insulin stimulation; growth is expressed as centimeters per 20 days. Food intake is expressed in calories consumed per day .

values seen when the boy first came into the hospital. Finally, row D shows the boy's profile after the nurse returned from vacation.

What is the critical element missing for a child reared in pathological isolation, for a rat separated from its mother? There are obviously all sorts of possibilities. Kuhn and Schanberg—and in separate studies, Myron Hofer of the New York State Psychiatric Institute—have studied that question in infant rats. Is it the absence of the smell of mom? Is it something in her milk that stimulates growth? Do the rats get chilly without her? Is it the rat lullabies that she sings? You can imagine the various ways scientists test for these possibilities—playing recordings of mom's vocalizations, pumping her odor into the cage, seeing what substitutes for the real thing.

It turns out to be touch, and it has to be active touching. Separate a baby rat from its mother and its growth hormone levels plummet; growth stops. Allow it contact with its mother while she is anesthetized, and growth hormone is still low. Mimic active licking by the mother by stroking the rat pup in the proper pattern, and growth normalizes. In a similar set of findings, other investigators have observed that handling neonatal rats causes them to grow faster and larger.

The same seems to apply in humans, as was demonstrated in an extremely important study. Tiffany Field of the University of Miami School of Medicine, along with Schanberg, Kuhn, and

Pigtailed macaque mother and infant.

others, performed an incredibly simple experiment, inspired by the rat work just described. Studying premature infants in neonatology wards, they noted that the premature kids, while pampered and fretted over and maintained in near-sterile conditions, were hardly ever touched. So Field and crew went in and started touching them: fifteen-minute periods, three times a day, stroking their bodies, moving their limbs. It worked wonders. The kids grew nearly 50 percent faster, were more active and alert, matured faster behaviorally, and were released from the hospital nearly a week earlier than the premature infants who weren't touched. Months later, they were still doing better than infants who hadn't been touched. If these studies turn out to be generally replicable, their implications are enormous. Given the cost of hospitalization for these infants and the number who wind up in neonatology wards, by reducing the length of the hospital stay, daily touching would save approximately 1 billion dollars a year—and perhaps make for healthier kids for years afterward. It's rare that the highest technology of medical instrumentation—MRI machines, artificial organs, pacemakers—has the potential for as much impact as this simple intervention.

Touch is one of the central experiences of an infant, whether rodent, primate, or human. We readily think of stressors as consisting of various unpleasant things that can be done to an organism. Sometimes a stressor can be the *failure* to provide something for an organism, and the absence of touch is seemingly one of the most marked of developmental stressors that we can suffer.

GROWTH AND GROWTH HORMONE IN ADULTS

Personally I don't grow much anymore, except wider. According to the textbooks, another ten Groundhog Days or so and I'm going to start shrinking. Yet I, like other adults, still secrete growth hormone into my circulation (although much less frequently than when I was an adolescent). What good is it in a nongrowing adult?

Like the Red Queen in *Alice in Wonderland,* the bodies of adults have to work harder and harder just to keep standing in the same place. Once the growth period of youth is finished and the edifice is complete, the hormones of growth mostly work at rebuilding and remodeling—carrying out the paint job, plastering the cracks that appear here and there.

Much of this repair work takes place in bone. Most of us probably view our bones as pretty boring and phlegmatic—they just sit there, inert. In reality, they are dynamic outposts of activity. They are filled with blood vessels, with little fluid-filled canals, with all sorts of cell types that are actively growing and dividing. New bone is constantly being formed, in much the same way as in a teenager. Old bone is being broken down, disintegrated by ravenous enzymes (a process called *resorption*).

New calcium is shuttled in from the bloodstream; old calcium is flushed away. Growth hormone, somatomedins, parathyroid hormone, and vitamin D stand around in hard hats, supervising the project.

Why all the tumult? Some of this bustle is because bones serve as the Federal Reserve for the body's calcium, constantly giving and collecting loans of calcium to and from other organs. And part is for the sake of bone itself, allowing it to gradually rebuild and change its shape in response to need. How else do cowboys' legs get bowed from too much time on a horse? The process has to be kept well balanced. If the bones sequester too much of the body's calcium, much of the rest of the body shuts down; if the bones dump too much of their calcium into the bloodstream, they become fragile and prone to fracture, and that excess circulating calcium can start forming calcified kidney stones.

Predictably, the hormones of stress wreak havoc with the trafficking of calcium, biasing bone toward disintegration, rather than growth. The main culprits are glucocorticoids. They inhibit the growth of new bone by disrupting the division of the bone-precursor cells in the ends of bones. Furthermore, they reduce the calcium supply to bone. Glucocorticoids block the uptake of dietary calcium in the intestines (uptake normally stimulated by vitamin D), increase the excretion of calcium by the kidney, and accelerate the resorption of bone.

If your body secretes excessively large amounts of glucocorticoids, your bones are likely to give you problems. This is seen in people with Cushing's syndrome (in which glucocorticoids are secreted at immensely high levels because of a tumor), and in people being treated with high doses of glucocorticoids to control some disease. In those cases, bone mass decreases markedly, and patients are at greater risk for osteoporosis (softening and weakening of bone). Any situation that greatly elevates glucocorticoid concentrations in the bloodstream is a particular problem for older people, in whom bone resorption is already predominant (in contrast to adolescents, in whom bone growth predominates, or young adults, in which the two processes are balanced). This is especially a problem in older women. Tremendous attention is now being paid to the need for calcium supplements to prevent osteoporosis in postmenopausal women.

Estrogen potently inhibits bone resorption, and as estrogen levels drop after menopause, the bones suddenly begin to degenerate. A hefty regimen of glucocorticoids on top of that is the last thing you need.

These findings suggest that chronic stress can increase the risk of osteoporosis and cause skeletal atrophy. Most clinicians would probably say that the glucocorticoid effects on bone are "pharmacological" rather than "physiological." This means that normal (physiological) levels of glucocorticoids in the bloodstream, even those in response to normal stressful events, are not enough to damage bone. Instead, it takes pharmacological levels of the hormone (far higher than the body can normally generate), due to a tumor or to ingestion of prescription glucocorticoids, to cause these effects. However, work from Jay Kaplan's group has shown that chronic social stress leads to loss of bone mass in female monkeys. While these are among the first reports of this kind, they raise the possibility that stress can damage bone.

STRESS AND GROWTH HORMONE SECRETION IN HUMANS

The pattern of growth hormone secretion during stress differs in humans from rodents, and the implications can be fascinating. But the subject is a tough one, not meant for the fainthearted.

When a rat is first stressed, growth hormone levels begin to go down in the circulation almost immediately. If the stressor continues, growth hormone levels remain depressed. And as we have seen, in humans major and prolonged stressors cause a decrease in growth hormone levels as well. The weird thing is that during the period immediately following the onset of stress, growth hormone levels actually go up in humans and some other species. In these species, in other words, short-term stress actually *stimulates* growth hormone secretion for a time.

Why? As we have seen, growth hormone stimulates the secretion of somatomedins that, in turn, stimulate bone growth, cell division, and other processes. But growth hormone does something else as well to promote growth—it helps supply the energy needed to fuel physical development. During growth, nutrients are taken out of fat cells and other storage sites and shipped to the growing tissue. Makes sense—this is why kids suddenly lose their baby fat when they start their pubescent growth spurt. Growth hormone is partially responsible for this. It works directly at fat cells to break down stored fats (triglycerides); the building blocks of fat, as we saw in Chapter 4, are then dumped into the bloodstream in the form of fatty acids and glycerol, where they can be utilized by the growing muscle. In effect, growth hormone not only runs the construction site for the new building, but arranges financing for the work as well.

Thus, during development, growth hormone causes the breakdown of nutrients tucked away in storage sites and diverts them to *growing tissue*. As we saw in Chapter 4, the main thing that your body does metabolically with glucocorticoids, epinephrine, norepinephrine, and glucagon during stress is to break down nutrients tucked away in storage sites and divert them to *exercising muscle*. Similar first steps, very dissimilar second steps. During stress, therefore, it is adaptive to secrete growth hormone insofar as it helps to mobilize energy, but a bad move to secrete growth hormone insofar as it stimulates an expensive, long-term project like growth.

As discussed, the bulk of growth hormone's actions on tissue growth and division are mediated by somatomedins. If you block somatomedin release or action, growth hormone will not have its growth-promoting effects, while still free to have its energy-mobilizing actions. And as we have seen, somatomedin levels and tissue sensitivity to somatomedin indeed decline during stress. This winds up being a clever mechanism for taking advantage of some of growth hormone's skills while blocking others. To extend the metaphor used earlier, growth hormone has just taken out cash from the bank, aiming to fund the next six months of construction; instead, the cash is used to solve the body's immediate emergency.

So long as somatomedin release or action is blocked, your body can secrete growth hormone forever, enjoying the advantages of energy mobilization without causing undue growth during a stressor. Why, then, do growth hormone levels decline at all during stress (whether immediately, as in the rat, or after a while, as in humans)? It is probably because the system does not work perfectly—somatomedin action is not completely shut down during stress. In practice, your body can use the energy-mobilizing effects of growth hormone for only so long before causing growth. Perhaps the timing of the decline of growth hormone levels represents a compromise between the trait triggered by the hormone that is good news during stress and the trait that is undesirable.

What impresses me is how careful and calculating the body has to be during stress in order to coordinate hormonal activities just right. It must perfectly balance the costs and benefits, knowing exactly when to stop secreting the hormone. If the body miscalculates in one direction and growth hormone secretion is blocked too early, there is relatively less mobilization of energy for dealing with the stressor. If it miscalculates in the other direction and growth hormone secretion goes on too long, stress may actually enhance growth. One oft-quoted study suggests that the second type of error occurs during some stressors.

In the early 1960s, Thomas Landauer of Dartmouth and John Whiting of Harvard methodically studied the rites of passages found in various nonwesternized societies around the world; they wanted to know whether the stressfulness of the ritual was related to how tall the kids wound up being as adults. Landauer and Whiting classified cultures according to whether and when they subjected their children to physically stressful development rites. Stressful rites included piercing the nose, lips, or ears; circumcision, inoculation, scarification, or cauterization; stretching or binding of limbs, or shaping the head; exposure to hot baths, fire, or intense sunlight; exposure to cold baths, snow, or cold air; emetics, irritants, and enemas; rubbing with sand, or scraping with a shell or other sharp object. (And you thought having to play the piano at age ten for your grandmother's friends was stressful.)

Reflecting some of the anthropological tunnel vision of the time, Landauer and Whiting only studied males. They examined 80 cultures around the world and controlled their study in an important way—they collected examples from cultures from the same gene pools, with and without those stressful rituals. For example, they compared the West African tribes of Yoruba (stressful rituals) and Ashanti (nonstressful), and similarly matched Native American tribes. With this approach, they attempted to control for genetic contributions to stature (as well as nutrition, since related ethnic groups were likely to have similar diets) and to examine cultural differences instead.

Given the effects of stress on growth, it was not surprising that among cultures where kids of ages six to fifteen went through stressful maturational rituals, growth was inhibited (relative to cultures without such rituals, the difference was about 1.5 inches). Surprisingly, going through such rituals at ages two to six had no effect on growth. And most surprising, in cultures in which those rituals took place with kids under two years of age, growth was stimulated—adults were about 2.5 inches taller than in cultures without stressful rituals.

There are some possible confounds that could explain the results. One is fairly silly—maybe tall tribes like to put their young children through stressful rituals. One is more plausible—maybe putting very young children through these stressful rituals kills a certain percentage of them, and you inadvertently select survivors who are more robust and likely to wind up as tall adults. Landauer and Whiting noted that possibility and could not rule it out. In addition, even though they attempted to pair similar groups, there may have been differences other than just the stressfulness of the rites of passage—perhaps in diet or child-rearing practices. Not surprisingly, no one has ever measured levels of growth hormone, somatomedins, and so on, in Shilluk or Hausa kids while they are undergoing some grueling ritual, so there is no direct endocrine evidence that such stressors actually stimulate growth hormone secretion in a way that increases growth. Despite these problems, these cross-cultural studies have been interpreted by many biological anthropologists as evidence that some types of stressors in humans can actually stimulate growth.

THE "L" WORD

In looking at research on how stress and understimulation can disrupt growth, a theme pops up repeatedly: an infant human or animal can be well-fed, maintained at an adequate temperature, peered at nervously, and ministered to by the best of neonatologists, yet still not thrive. Something is still missing. Perhaps we can even risk scientific credibility and detachment and mention the word *love* here, because that most ephemeral of phenomena lurks between the lines of this chapter; something roughly akin to love is needed for proper biological development, and its absence is among the most aching, distorting stressors that we can suffer. Scientists and physicians and other caregivers have often been dim at recognizing its importance in the mundane biological processes by which organs and tissues grow and develop. Yet young organisms were able to teach this fact to surprised scientists in a classic set of studies conducted in the 1950s through the 1970s—studies that are, in my opinion, among the most haunting and troubling of all the pages of science.

The work was carried out by Harry Harlow of the University of Wisconsin, a renowned and controversial scientist. Harlow helped to answer a seemingly obvious question in a nonobvious way. Why do infants become attached to their mothers? Psychology at that time was dominated by a rather extreme school of thought called *behaviorism,* in which behavior (of an animal or a human) was thought to operate according to rather simple rules: an organism does something more frequently because it has been rewarded for it in the past; an organism does something less frequently because it has failed to be rewarded, or has even been punished for that behavior. In this view, there are just a few basic things that lie at the basis of reinforcement—primary drives like hunger, pain, sex. Look at the behaviors, view organisms as machines responding to stimuli, and develop a predictive mathematics built around the idea of rewards and punishments. Naturally, behaviorists had a theory as to why babies get attached to their mothers. Mothers are good at relieving the state of hunger; and that's why kids like their moms.

Harlow smelled a rat and decided to test what everyone considered obvious. He raised infant rhesus monkeys without mothers. Instead, he gave them a choice of two types of artificial "surrogate" mothers. One pseudo-mother had a monkey head constructed of wood and a wire-mesh tube resembling a torso. In the middle of the torso was a bottle of milk. This surrogate mother gave nutrition. The other surrogate mother had a similar head and wire-mesh torso. But instead of containing a milk bottle, this one's torso was wrapped in terry cloth. We know exactly

Infant monkey and cloth mother, in a Harlow study.

what the behaviorists would say. "Aha, the first surrogate mother supplies food and thus is more rewarding!" But the baby monkeys chose the terry-cloth mothers. Kids don't love their mothers because mom balances their nutritive intake, these results suggested. They love them because, usually, mom loves them back, or at least is someone soft to cling to. "Man cannot live by milk alone. Love is an emotion that does not need to be bottle- or spoon-fed, and we may be sure that there is nothing to be gained by giving lip service to love," wrote Harlow.

Harlow's work remains controversial because of the nature of these experiments and variations on them (for example, raising monkeys in complete social isolation, in which they never see another living animal). These were brutal studies, and they are often among the primary ones cited by those opposed to animal experimentation. Moreover, Harlow's scientific writing displayed a striking callousness to the suffering of these animals, and that is often cited by animal rights activists as well. "Why were these experiments necessary?" they ask; "everyone knows that love is important. Why torture baby monkeys to prove the obvious?" The other side retorts, "Oh yeah? Everyone predicted just the opposite back then; it wasn't obvious at all." Then the rejoinder, "But wasn't the point made? Why has this experimentation gone on for decades with endless elaborations?" And the answer typically given is that such studies have enabled us to generate animal models of childhood trauma and abuse, in order to understand why, for example, humans and monkeys abused in childhood are far more likely than average to be abusive parents someday; other variants on Harlow's work have taught us how repeated separations of infants from their mothers can predispose those individuals to depression when they are adults.

Personally, I am torn on this one. To animal rights activists who would ban all animal experimentation, I unapologetically say that I am in favor of the use of animals in research, and that much good has come out of this particular type of research. To the scientist who would deny the brutality of some types of animal research, I unapologetically say that things can go too far, and the Harlow school of research has constituted a case of this at times. It is sad and pathetic when we must experiment on infant animals in order to be taught the importance of love. But it is sadder and more pathetic to consider that we have learned about love so poorly and still have to be reminded of its importance at every opportunity.

7

SEX AND REPRODUCTION

Sure, we all care deeply about what our kidneys do during stress, and heart disease is a drag, but what we really want to know is why, when we are being stressed, our menstrual cycles become irregular, erections are more difficult to achieve, and we lose our interest in sex. As it turns out, there are an astonishing number of ways in which reproductive mechanisms may go awry when we are upset.

MALES: TESTOSTERONE AND LOSS OF ERECTIONS

It makes sense to start simple, so let's initially consider the easier reproductive system, that of males. In the male, the brain releases the hormone LHRH (luteinizing hormone releasing hormone), which stimulates the pituitary to release LH (luteinizing hormone) and FSH (follicle-stimulating hormone). LH, in turn, stimulates the testes to release testosterone. Since men don't have follicles to be stimulated by follicle-stimulating hormone, FSH instead stimulates sperm production. This is the reproductive system of your basic off-the-rack male.

With the onset of a stressor, the whole system is inhibited. LHRH concentrations decline, followed shortly thereafter by declines in LH and FSH, and then the testes close for lunch. The

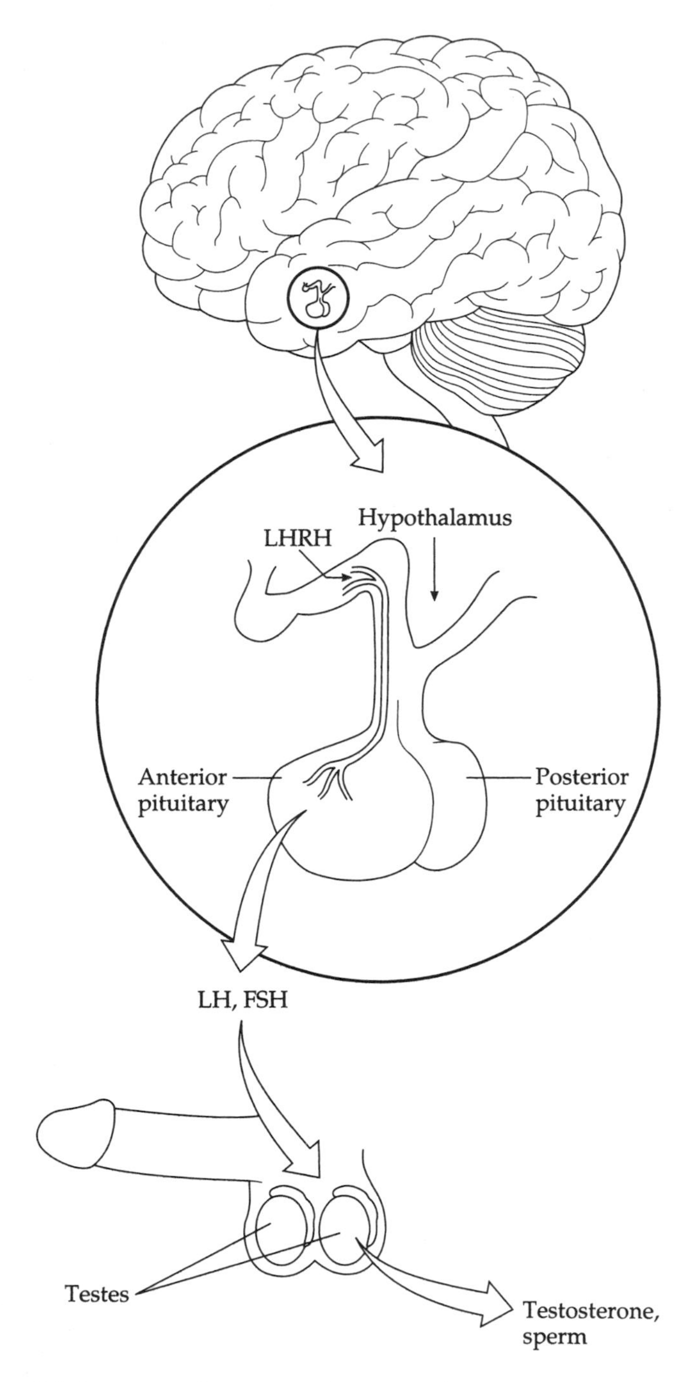

A simplified version of male reproductive endocrinology. The hypothalamus releases LHRH into the private circulatory system that it shares with the anterior pituitary. LHRH triggers the release by the pituitary of LH and FSH, which work at the testes to cause testosterone secretion and sperm production.

result is a decline in circulating testosterone levels. The most vivid demonstrations of this occur during physical stress. If a male goes through surgery, within seconds of the first slice through his skin, the reproductive axis begins to shut down. Injury, illness, starvation, surgery—all of these drive down testosterone levels. Anthropologists have even shown that in human societies in which there is constant energetic stress (for example, those of rural Nepalese villagers), there are significantly lower testosterone levels than among sedentary Bostonian controls.

But subtle psychological stressors are just as disruptive. Lower the dominance rank of a social primate and down go his testosterone levels. Put a person or a monkey through a stressful learning task and the same occurs. In a celebrated study several decades ago, U.S. Officer Candidate School trainees who underwent an enormous amount of physical and psychological stress were subjected to the further indignity of having to pee into Dixie cups so that military psychiatrists could measure their hormone levels. Lo and behold, testosterone levels were down; maybe not to the levels found in cherubic babies, but still it's worth keeping in mind the next time you see some leatherneck at a bar bragging about his circulating androgen concentrations.

Why do testosterone concentrations plunge with the onset of a stressor? For a variety of reasons. The first occurs at the brain. With the onset of stress, two important classes of hormones, the endorphins and enkephalins (mostly the former), act to block the release of LHRH from the hypothalamus. As will be discussed in Chapter 9, endorphins play a role in blocking pain perception and are secreted in response to exercise (helping to account for the famed "runner's high" or "endorphin high" that hits many hardy joggers around the thirty-minute mark). If males secrete endorphins when they are experiencing runner's high, and these compounds inhibit testosterone release, will exercise suppress male reproduction? Sometimes. Males who do extreme amounts of exercise (for example, professional soccer players and runners who cover more than 40 or 50 miles a week) have less LHRH, LH, and testosterone in their circulation, smaller testes, less functional sperm. (A similar decline in reproductive function is found in men who are addicted to opiate drugs.) To jump ahead to the female section, reproductive dysfunction is also seen in women athletes, and this is at least partially due to endorphin release as well. Serious runners often stop having menstrual cycles, and highly athletic girls reach puberty later than usual. For example, in one study of 14-year-olds, approximately 95 percent of control subjects had started menstruating, whereas only 20 percent of gymnasts and 40 percent of runners had.

This brings up a broader issue important to our era of lookin' good. Obviously if you don't exercise at all, it is not good for you. A little exercise helps all sorts of physiological systems to function well. More exercise helps more. But at some point, too much begins to damage various physiological systems. Everything in physiology follows this rule—just because more of something is better, a lot more of something isn't necessarily a lot better. Too much can be as bad as too little. There are optimal points of allostatic balance. In our age of jogging mania, thirty-year-old athletes who run more than 40 or 50 miles a week are showing up with dramatically decalcified bones, decreased bone mass, increased risk of stress fractures and scoliosis (sideways curvature of the spine)—their skeletons look like those of seventy-year-olds. In general, a moderate amount of exercise leads to an increase in bone mass, particularly in the bones that are bearing the greatest brunt of the exercise (for example, the ankle bones of runners). But too much causes trouble, and it appears that these very serious athletes have passed the optimal point in the exercise–bone-mass curve and are beginning to damage themselves.

To put exercise in perspective, imagine this: sit with a group of hunter-gatherers from the African grasslands and explain to them that in our world we have so much food and so much free time that some of us run 26 miles a day, simply for the sheer pleasure of it. They are likely to say, "Are you crazy? That's stressful." Throughout history, hominids who run 26 miles in a day have generally been pretty intent on finding food, or about to be eaten by someone. Not all that normal.

Thus, we have a first step. With the onset of stress, LHRH secretion declines. In addition, prolactin, another pituitary hormone that is released during major stressors, decreases the sensitivity of the pituitary to LHRH. A double whammy—less of the hormone dribbling out of the brain, and the pituitary no longer responding as effectively to it. Finally, glucocorticoids

Overexercise can have a variety of deleterious effects. Max Ernst, 1920: (left) Health Through Sport, *photographic enlargement of a photomontage mounted on wood; (right)* Above the Clouds Midnight Passes, *collage with fragments of photographs and pencil.*

block the response of the testes to LH, just in case any of that hormone manages to reach them during the stressor (and serious athletes tend to have pretty dramatic elevations of glucocorticoids in their circulation, no doubt adding to the reproductive problems just discussed).

Decline in testosterone secretion is only half the story of what goes wrong with male reproduction during stress. The other half concerns the nervous system and erections. Getting an erection to work properly is so incredibly complicated physiologically that if men ever actually had to understand it, none of us would be here. Fortunately, it runs automatically. In order for a male to have an erection, his parasympathetic nervous system must be turned on. In some species, including humans, the parasympathetic causes hemodynamic erections—blood flow into the penis is increased, the exit route for the blood via the veins is blocked, and the penis fills with blood and stiffens. In other species, such as rodents, erections are muscular—a particular muscle contracts, yanks on a penile bone, and up goes the penis. Hemodynamic erections take longer to occur, but last longer. In either case, erections are mediated by the parasympathetic system. Calm, vegetative, relaxed.

What happens next for you, if you are male? You are having a terrific time with someone. Maybe you are breathing faster, your heart rate has increased. Your body is taking on a sympathetic tone—remember the four F's of sympathetic function introduced in Chapter 2. Your autonomic nervous system is functioning in an oddly compartmentalized state. After awhile, most of your body is screaming sympathetic while, heroically, you are trying to hold on to parasympathetic tone in that one lone outpost as long as possible. Finally, when you can't take it any more, the parasympathetic shuts off at the penis, the sympathetic comes roaring on, and you ejaculate. (Incredibly complicated choreography of these two systems; don't try this un-supervised.) This new understanding generates tricks that sexual therapists advise—if you are close to ejaculating and don't want to yet, take a deep breath. Expanding the chest muscles briefly triggers a parasympathetic volley that defers the shift from parasympathetic to sympathetic.

What, then, changes during stress? First, it becomes difficult to establish parasympathetic activity if you are nervous or anxious. You have trouble having an erection. Impotency. And if you already have the erection, you get in trouble as well. You're rolling along, parasympathetic to your penis, having a wonderful time. Suddenly, you find yourself worrying about the seriousness of our trade deficit with Japan and—shazaam—you switch from parasympathetic to sympathetic far faster than you wanted. Premature ejaculation.

It is extremely common for problems with impotency and premature ejaculation to arise during stressful times. A number of studies have shown that more than half the visits to doctors by males complaining of reproductive dysfunction turn out to be due to "psychogenic" impotency rather than organic impotency (there's no disease there, just too much stress). How do you tell if it is organic or psychogenic impotency? This is actually diagnosed with surprising ease, because of a quirky thing about human males. As soon as they go to sleep and enter REM (rapid eye movement) dream sleep, they get erections. No one has a clue why, but that's how they work. So a man comes in complaining that he hasn't been able to have an erection in six months. Is he just under stress? Does he have some neurological disease? Take a handy little penile cuff with an electronic pressure transducer attached to it. Have him put it on just before he goes to sleep. By the next morning you may have your answer—if this guy gets an erection when he goes into REM sleep, his problem is likely to be psychogenic. (I've been told about an advance on this technology. Instead of having to use one of these fancy electronic cuffs, you simply tape a few postage stamps to the penis just before going to sleep. By the morning, if the stamps have been pulled loose on one side or torn, you had a REM-stage erection during the night—fabulous, a lab result for 96 cents.)

Thus, stress will knock out male sexual responses quite readily. In general, the problems with erections are more disruptive than problems with testosterone secretion. Testosterone and sperm production have to shut down almost entirely to affect performance. A little testosterone and a couple of sperm wandering around and most males can muddle through. But no erection, and forget about it. The erectile component is exquisitely sensitive to stress in an incredible array of species. To put this in context, it might be worth describing here the one species that constitutes an exception, a species that breaks all the rules in terms of the effects of stress on erectile function. It is time we had a little talk about hyenas.

OUR FRIEND, THE HYENA

The spotted hyena is a vastly unappreciated, misrepresented beast. I know this because over the years, in my work in East Africa, I have shared my campsite with the hyena biologist Laurence Frank of the University of California at Berkeley. For lack of distracting televisions, books, or telephones, he has devoted his time there to singing the hyena's praises to me. They are wondrous animals who have gotten a bad rap from the press.

We all know the scenario. It's dawn on the savanna. Marlin Perkins of Mutual of Omaha's *Wild Kingdom* is there filming lions eating something dead. We are delighted, craning to get a good view of the blood and guts. Suddenly, on the edge of our field of vision, we spot them—skulky, filthy, untrustworthy hyenas looking to dart in and steal some of the food. Scavengers! We are invited to heap our contempt on them (a surprising bias, given how few of the carnivorous among us ever wrestle down our meals with our canines). It wasn't until the Pentagon purchased a new line of infrared night viewing scopes and decided to unload its old ones on various university zoologists that, suddenly, researchers could watch hyenas at night (important, given that hyenas mostly sleep during the day). Turns out that they are fabulous hunters. And you know what happens? Lions, who are not particularly effective hunters, because they are big and slow and conspicuous, spend most of their time keying in on hyenas and ripping off their kills. No wonder when it's dawn on the savanna the hyenas on the periphery are looking cranky, with circles under their eyes. They stayed up all night hunting that thing, and who's having breakfast now?

Having established a thread of sympathy for these beasts, let me explain what is really strange about them. Among hyenas, females are socially dominant. They are more muscular and more aggressive, and have more of a male sex hormone (a close relative of testosterone called *androstenedione*) in their bloodstreams than males. It's also almost impossible to tell the sex of a hyena by looking at its external genitals.

More than two thousand years ago, Aristotle, for reasons obscure to even the most learned, dissected some dead hyenas, discussing them in his treatise *Historia Animalium*, VI, XXX. The conclusion among hyena savants at the time was that these animals were hermaphrodites—animals that possess all the machinery of both sexes. Hyenas are actually what gynecologists would call pseudohermaphrodites (they just look that way). The female has a fake scrotal sac made of compacted fat cells; she doesn't really have a penis but, instead, an enlarged clitoris that can become erect. The same clitoris, I might add, with which she has sex and through which she gives birth. It's pretty wild. Laurence Frank, who is one of earth's experts on hyena genitals, will dart some animal and haul it, anesthetized, into camp. Excitement; we go to check it out, and maybe twenty minutes into examining it, he kind of thinks he knows what sex this particular one is. (Yes, the hyenas themselves know exactly who is which sex, most probably by smell.)

Behold, the female hyena.

Perhaps the most interesting thing about hyenas is that there is a fairly plausible theory as to why they evolved this way, a theory complicated enough for me mercifully to relegate it to the endnotes. For our purposes here, what is important is that hyenas have evolved not only genitals that look unique, but also unique ways to use these organs for social communication. This is where stress comes into play.

Among many social mammals, males have erections during competitive situations as a sign of dominance. If you are having a dominance display with another male, you get an erection and wave it around in his face to show what a tough guy you are. Social primates do this all the time. In hyenas, an erection is, instead, a sign of social *subordinance*. When a male is menaced by a terrifying female, he gets an erection—"Look, I'm just some poor no-account male; don't hit me, I was just leaving." Low-ranking females do the same thing; if a low-ranking female is about to get trounced by a high-ranking one, she gets a conspicuous clitoral erection—"Look, I'm just like one of those males; don't attack me; you know you are dominant, why bother?" If you're a hyena, you get an erection *when you are stressed*. Among male hyenas, the autonomic wiring has got to be completely reversed in order to account for the fact that stress causes erections. This hasn't yet been demonstrated, but as I write, Berkeley scientists are slaving away at the study of issues like this, squandering tax dollars that could otherwise be spent on Cruise missiles.

Thus the hyena stands as the exception to the rule about erectile functions being adversely affected by stress, a broader demonstration of the importance of looking at a zoological oddity as a means of better seeing the context of our own normative physiology, and a friendly word of warning before you date a hyena.

FEMALES: LENGTHENED CYCLES AND AMENORRHEA

We now turn to female reproduction. Its basic outline is similar to that of the male. LHRH is released by the brain, which releases LH and FSH from the pituitary. The latter stimulates the ovaries to release eggs; the former stimulates ovaries to synthesize estrogen. During the first half of the menstrual cycle, the "follicular" stage, levels of LHRH, LH, FSH, and estrogen build up, heading toward the climax of ovulation. This ushers in the second half of the cycle, the "luteal" phase. Progesterone, made in the corpus luteum of the ovary, now becomes the dominant hormone on the scene, stimulating the uterine walls to mature so that an egg, if fertilized just after ovulation, can implant there and develop into an embryo. Because the release of hormones has the fancy quality of fluctuating rhythmically over the menstrual cycle, the part of the hypothalamus that regulates the release of these hormones is generally more structurally complicated in females than in males.

The first way in which stress disrupts female reproduction concerns a surprising facet of the system. There is a small amount of male sex hormone in the bloodstream of females, even nonhyena females. In human beings, this doesn't come from the ovaries (as in the hyenas), but from the adrenals. The amount of these "adrenal androgens" is only about 5 percent of that in males, but enough to cause trouble. (Point of information: the adrenal androgens are usually not testosterone, but androstenedione.) An enzyme in the fat cells of females usually

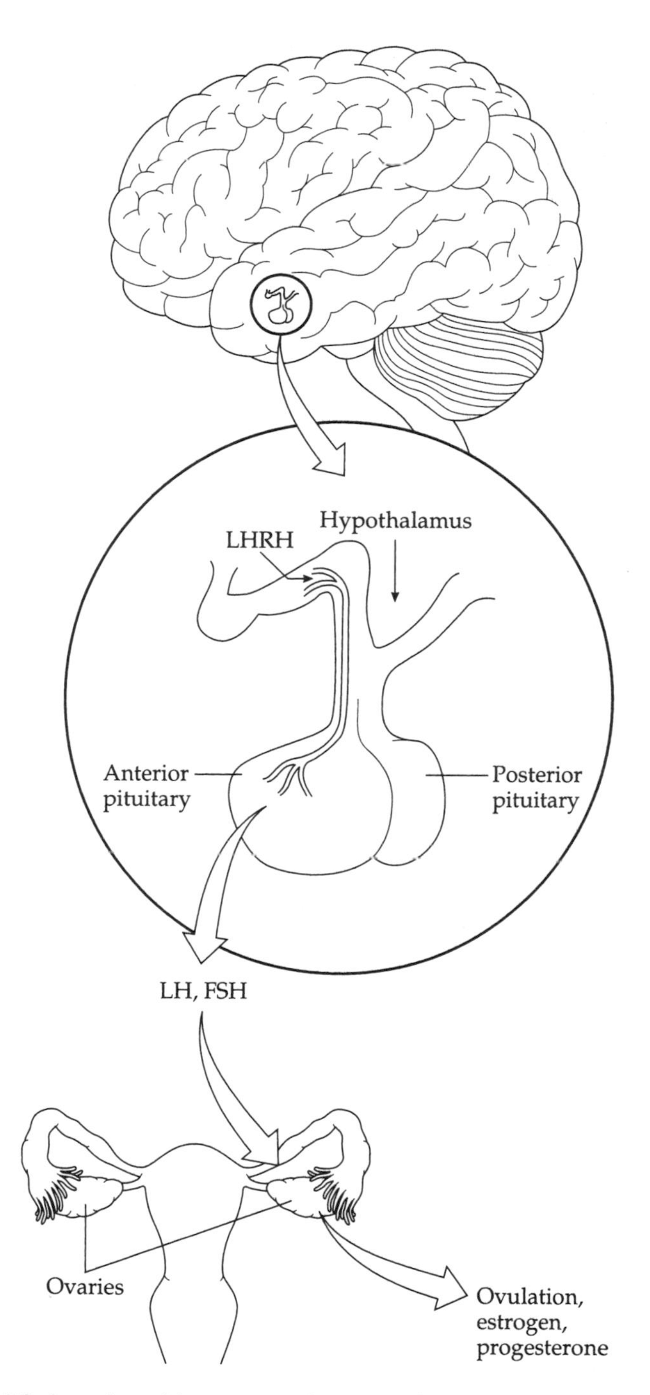

A simplified version of female reproductive endocrinology. The hypothalamus releases LHRH into the private circulatory system that it shares with the anterior pituitary. LHRH triggers the release by the pituitary of LH and FSH, which in turn bring about ovulation and hormone release from the ovaries.

eliminates these androgens by converting them to estrogens. Problem solved. But what if you are starving because the crops failed this year, or there is a drought? Body weight drops, fat stores are depleted, and suddenly there isn't enough fat around to convert all the androgen to estrogen. Less estrogen, therefore, is produced. More importantly, androgen concentrations build up, which inhibits numerous steps in the reproductive system.

Reproduction is similarly inhibited if you starve voluntarily. One of the hallmarks of anorexia nervosa is disruption of reproduction in the (typically) young women who are starving themselves. And loss of body fat leading to androgen buildup is one of the mechanisms by which reproduction is impaired in females who are extremely active physically. As noted above, this has been best documented in young girls who are very serious dancers or runners, in whom puberty can be delayed for years, and in women who exercise enormous amounts, in whom cycles can become irregular or cease entirely. Overall, this is a very logical mechanism. In the human, an average pregnancy costs approximately 50,000 calories, and nursing costs about a thousand calories a day; neither is something that should be gone into without a reasonable amount of fat tucked away.

Stress also can inhibit reproduction in ways other than shrinkage of fat cells. Many of the same mechanisms apply as in the male. Endorphins and enkephalins will inhibit LHRH release (as discussed, this occurs in female athletes as readily as in males); prolactin and the glucocorticoids will block pituitary sensitivity to LHRH; and glucocorticoids will also affect the ovaries, making them less responsive to LH. The net result is lowered secretion of LH, FSH, and estrogen, making the likelihood of ovulating decrease. As a result, the follicular stage is extended, making the entire cycle longer and less regular. At an extreme, the entire ovulatory machinery is not merely delayed, but shut down, a condition termed *anovulatory amenorrhea*.

Stress can also cause other reproductive problems. Progesterone levels are often inhibited, which disrupts maturation of the uterine walls. The release of prolactin during stress adds to this effect, interfering with the activity of progesterone. Thus even if there is still enough hormonal action during the follicular period to cause ovulation, and the egg has become fertilized, it is now much less likely to implant normally.

The loss of estrogen with sustained stress has some consequences beyond the reproductive realm. As noted earlier, athletes who exercise very heavily can begin to decalcify their bones, decreasing bone mass and increasing the risk of fractures and osteoporosis. Studies of these effects are predominantly of women, in whom the consequences are most severe, and it is amenorrheic women in whom the problems are most likely to occur. As noted in the chapter on growth, estrogen helps recalcify bone (a reason osteoporosis becomes so serious in postmenopausal women). Furthermore, as noted in Chapter 3, estrogen helps protect the heart against atherosclerosis, and stress can endanger the cardiovascular system by lowering estrogen levels. For example, Jay Kaplan and colleagues have shown that social subordinance in monkeys suppresses estrogen to levels as low as those seen in females with their ovaries removed, and that such subordinance increases the risk for atherosclerosis.

Of all the hormones that inhibit the reproductive system during stress, prolactin is probably the most interesting. It is extremely powerful and versatile; if you don't want to ovulate, this is the hormone to have lots of in your bloodstream. It not only plays a major role in the suppression of reproduction during stress and exercise, but also is the main reason that breast feeding is such an effective form of contraception.

Oh, you are shaking your head smugly; everyone knows that's an old wives' tale; nursing isn't an effective contraceptive. On the contrary; nursing works fabulously. It probably prevents more pregnancies than any other type of contraception. All you have to do is do it right.

Breast feeding causes prolactin secretion. There is a reflex loop that goes straight from the nipples to the hypothalamus. If there is nipple stimulation for any reason (in males as well as females), the hypothalamus signals the pituitary to secrete prolactin. And as we now know, prolactin in sufficient quantities causes reproduction to cease.

The problem with nursing as a contraceptive is how it is done in western societies. During the six months or so that she breastfeeds, the average mother in the west allows perhaps half a dozen periods of nursing a day, each for 30 to 60 minutes. Each time she nurses, prolactin levels go up in the bloodstream within seconds, and at the end of the feeding, prolactin settles back to prenursing levels fairly quickly. This most likely produces a scalloping sort of pattern in prolactin release.

This is not how most women on earth nurse. A prime example emerged a few years ago in a study of hunter-gatherer Bushmen in the Kalahari desert of southern Africa (the folks depicted in the movie *The Gods Must Be Crazy*). Bushman males and females have plenty of intercourse, and no one uses contraceptives, but the women have a child only about every four years. Initially, this seemed easy to explain. Western scientists looked at this pattern and said, "They're hunter-gatherers: life for them must be short, nasty, and brutish; they must all be starving." Malnutrition-induced cessation of ovulation.

However, when anthropologists looked more closely, they found that the Bushmen were anything but suffering. If you are going to be nonwesternized, choose being a hunter-gatherer over being a nomadic pastoralist or an agriculturist. The Bushmen hunt and gather only a few hours a day, and spend much of the rest of their time sitting around chewing the fat. Scientists have called them the original affluent society. Out goes the idea that the four-year birth interval is due to malnutrition.

Instead, the lengthy interval is probably due to their nursing pattern. This was discovered by a pair of scientists, Melvin Konner and Carol Worthman. When a hunter-gatherer woman gives birth, she begins to breast-feed her child for a minute or two approximately every fifteen minutes. Around the clock. For the next three years. (Suddenly this doesn't seem like such a hot idea after all, does it?) The young child is carried in a sling on the mother's hip so he can nurse easily and frequently. At night, he sleeps near his mother and will nurse every so often without even waking her (as Konner and Worthman, no doubt with their infrared night viewing goggles and stopwatches, scribble away on their clipboards at two in the morning). Once the kid can walk, he'll come running in from play every hour or so to nurse for a minute.

When you breast-feed in this way, the endocrine story is going to be very different. At the first nursing period, prolactin levels rise. And with the frequency and timing of the thousands of subsequent nursings, prolactin stays high for years. Estrogen and progesterone levels are suppressed, and you don't ovulate.

This pattern has a fascinating implication. Consider the life history of a hunter-gatherer woman. She reaches puberty at

A Kalahari Bushman mother with her child in a hip sling.

about age 13 or 14 (a bit later than in our society). Soon she is pregnant. She nurses for three years, weans her child, has a few menstrual cycles, becomes pregnant again, and repeats the pattern until she reaches menopause, or dies (an event unlikely to wait for her eighth decade of life). Think about it: over the course of her life span, she has perhaps two dozen periods. Contrast that with modern western women, who average perhaps 500 periods over their lifetime. Huge difference. The hunter-gatherer pattern, the one that has occurred throughout most of human history, is what you see in nonhuman primates. Perhaps some of the gynecological diseases that plague modern westernized women have something to do with this activation of a major piece of physiological machinery 500 times when it may have evolved to be used only 20 times; an example of this is probably endometriosis (having uterine lining thickening and sloughing off in places in the pelvis and abdominal wall where it doesn't belong), which is more common among women with fewer pregnancies and who start at a later age.

FEMALES: DISRUPTION OF LIBIDO

The preceding section describes how stress disrupts the nuts and bolts of female reproduction—uterine walls, eggs, ovarian hormones, and so on. But what about its effects upon sexual behavior? Just as stress does not do wonders for erections or for the desire of a male to do something with his erections, stress disrupts female libido. This is a commonplace experience among women stressed by any of a number of circumstances, as well as among laboratory animals undergoing stress.

It is relatively easy to document a loss of sexual desire among women when they are stressed—just hand out a questionnaire on the subject and hope it is answered honestly. But how is sexual drive studied in a laboratory animal? How can one possibly infer a libidinous itch on the part of a female rat, for example, as she gazes into the next cage at the male with the limpid eyes and cute incisors? The answer is surprisingly simple—how often would she be willing to press a lever in order to gain access to that male? This is science's quantitative way of measuring rodent desire (or, to use the jargon of the trade, "proceptivity").* A similar experimental design can be used to measure proceptive behavior in primates. Proceptive and receptive behaviors

---

* Quick primer on how to describe animal sex the way professionals do: *Attractivity* refers to how much the subject animal interests another animal. This can be operationally defined as how many times the other animal is willing to press a lever, for example, to gain access to the subject. *Receptivity* describes how readily the subject responds to the entreaties of the other animal. Among rats, this can be defined by the occurrence of the "lordosis" reflex, a receptive stance by the female in which she arches her back, making it easier for the male to mount. Female primates show a variety of receptive reflexes that facilitate male mounting, depending on the species. *Proceptivity* refers to how actively the subject pursues the other animal.

fluctuate among animals as a function of factors like the point in the reproductive cycle (both of these measures of sexual behavior generally peak around ovulation), the recency of sex, the time of year, or vagaries of the heart (who is the male in question). In general, stress suppresses both proceptive and receptive behaviors.

This effect of stress is probably rooted in its suppression of the secretion of various sex hormones. Among rodents, both proceptive and receptive behaviors disappear when a female's ovaries are removed, and the absence of estrogen after the ovariectomy is responsible; as evidence, injection of ovariectomized females with estrogen reinstates these sexual behaviors. Moreover, the peak in estrogen levels around ovulation explains why sexual behavior is almost entirely restricted to that period. A similar pattern holds in primates, but it is not as dramatic as in rodents. A decline in sexual behavior, although to a lesser extent, follows ovariectomy in a primate. For humans, estrogen plays a role in sexuality, but a still weaker one—social and interpersonal factors are far more important.

Estrogen exerts these effects both in the brain and peripheral tissue. Genitals and other parts of the body contain ample amounts of estrogen receptors and are made more sensitive to tactile stimulation by the hormone. Within the brain, estrogen receptors occur in areas that play a role in sexual behavior; through one of the more poorly understood mechanisms of neuroendocrinology, when estrogen floods those parts of the brain, salacious thoughts follow.

Surprisingly, adrenal androgens also play a role in proceptive and receptive behaviors; as evidence, sex drive goes down following removal of the adrenals and can be reinstated by administration of synthetic androgens. This appears to be more of a factor in primates and humans than in rodents. While the subject has not been studied in great detail, there are some reports that stress suppresses the levels of adrenal androgens in the bloodstream. And stress certainly suppresses estrogen secretion. As noted in Chapter 3, Jay Kaplan has shown that the stressor of social subordinance in a monkey can suppress estrogen levels as effectively as removing her ovaries. Given these findings, it is relatively easy to see how stress disrupts sexual behavior in a female.

STRESS AND THE SUCCESS OF HIGH-TECH FERTILIZATION

In terms of psychological distress, the strain placed on a relationship with a significant other, the disruption of daily activities and ability to concentrate at work, the estrangement from friends and family,* and the rates of depression, few medical maladies have as much of an adverse impact as infertility. Thus, circumventing infertility with recent high-tech advances has been a wonderful medical advance.

There is now a brave new world of assisted fertilization: *artificial insemination. In vitro fertilization (IVF),* in which sperm and egg meet in a petri dish, and fertilized eggs are then implanted in the woman. *Preimplantation screening,* carried out when one of the couple has a serious genetic disorder; after eggs are fertilized, their DNA is analyzed, and only those eggs that do not carry the genetic disorder are implanted. *Donor eggs, donor sperm.* Injection of an individual sperm into an egg, when the problem is an inability of the sperm to penetrate the egg's membrane on its own.

Some forms of infertility are solved with some relatively simple procedures, but some involve extraordinary, innovative technology.

There are two problems with that technology, however. The first is that it is an astonishingly stressful experience for the individuals who go through it. Furthermore, it's expensive as hell, and is often not paid for by insurance, especially when some of the fancier, new experimental techniques are being tried. How many young couples can afford to spend ten to fifteen thousand dollars out of pocket each cycle they attempt to get pregnant? Next, most IVF clinics are located only near major medical centers, meaning that many participants have to spend weeks in a motel room in some strange city, far from friends and family. For some genetic screening techniques, only a handful of places in

* Two of the most common subjects discussed in infertility support groups: (1) how to handle the damage done to friendships and family relations when you can no longer attend baby showers, can no longer join the family at holidays because of all those nieces and nephews just learning to walk, no longer see the old friend who is pregnant and (2) what happens to a relationship with one's significant other when sex has been turned into a medical procedure, especially an unsuccessful one.

the world are available, thus adding a long waiting list to the other stress factors.

But those stress-induced factors pale compared with the stress generated by the actual process. Weeks of numerous, painful daily shots with synthetic hormones and hormone suppressors that can do some pretty dramatic things to mood and mental state. Daily blood draws, daily sonograms, the constant emotional roller-coaster of whether the day's news is good or bad: how many follicles, how big are they, what circulating hormone levels have been achieved? A surgical procedure and then the final wait to see whether you have to try the whole thing again.

The second problem is that it rarely works. It is very hard to figure out how often natural attempts at fertilization actually succeed in humans. And it is hard to find out what the success rates are for the high-tech procedures, as clinics often fudge the numbers in their brochures—"We don't like to publish our success rates, because we take on only the most difficult, challenging cases, and thus our numbers must superficially seem worse than those of other clinics that are wimps and take only the easy ones"—and thus it is hard to gauge just how bad the odds are for a couple with an infertility problem going this route. Nevertheless, going through one of those grueling IVF cycles has something on the order of a 10 to 20 percent chance of working.

All that has preceded in this chapter would suggest that the first problem, the stressfulness of IVF procedures, contributes to the second problem, the low success rate. A number of researchers have specifically examined whether women who are more stressed during IVF cycles are the ones less likely to have successful outcomes. And the answer is a resounding maybe. The majority of studies do show that the more stressed women (as determined by glucocorticoid levels, cardiovascular reactivity to an experimental stressor, or self-report on a questionnaire) are indeed less likely to have successful IVFs. Why, then, the ambiguity? For one thing, some of the studies were carried out many days or weeks into the long process, where women have already gotten plenty of feedback as to whether things are going well; in those cases, an emerging unsuccessful outcome might cause the elevated stress-response, rather than the other way around. And even in studies in which stress measures are taken at the beginning of the process, the number of previous cycles must be controlled for. In other words, a stressed women may indeed be less likely to have a successful outcome, but both traits may be due to the fact that she is an especially poor candidate who has already gone through eight unsuccessful prior attempts and is a wreck.

In other words, more research is needed. And if the correlation does turn out to be for real, one hopes that the outcome of that will be something more constructive than clinicians saying, "And try not to be stressed, because studies have shown it cuts down the chances of IVF will succeed." It would be kind of nice if progress in this area actually resulted in eliminating the stressor that initiated all these complexities in the first place, namely, the infertility.

MISCARRIAGE AND PSYCHOGENIC ABORTIONS*

The link between stress and spontaneous abortion in humans prompted Hippocrates to caution pregnant women to avoid unnecessary emotional disturbances. Since then, it is a thread that runs through some of our most florid and romantic interpretations of the biology of pregnancy, whether it is Anne Boleyn attributing her miscarriage to the shock of seeing Jane Seymour sitting on King Henry's lap, or Rosamond Vincy losing her baby when frightened by a horse in *Middlemarch*. In the movie *Pacific Heights* (which took the Reagan-Bush era to its logical extreme, encouraging us to root for the poor landlords being menaced by a predatory tenant), the homeowner, played by Melanie Griffith, has a miscarriage in response to psychological harassment by the Machiavellian renter.

Stress can cause miscarriages in other animals as well. This may occur, for example, when pregnant animals in the wild or in a corral have to be captured for some reason (a veterinary exam) or are stressed by being transported.

---

* *Miscarriage* and *abortion* are used interchangeably in medical texts and will be so used throughout this section. In everyday clinical usage, however, spontaneous termination of a pregnancy when the fetus is close to being viable is more likely to be termed *miscarriage* than *abortion*.

Studies of social hierarchies among animals in the wild have revealed many instances of miscarriage and, not infrequently, infanticide. In many social species, not all males do equivalent amounts of reproducing. Sometimes the group contains only a single male (typically called a *harem male*) who does all the mating; sometimes there are a number of males, but only one or a few dominant males who reproduce. Suppose the harem male is killed or driven out by an intruding male, or a new male migrates into the multimale group and moves to the top of the dominance hierarchy. Typically, the now-dominant male goes about trying to increase his own reproductive success, at the expense of the prior male. What does the new guy do? In some species, males will systematically try to kill the infants in the group (a pattern called *competitive infanticide* and observed in a number of species, including lions and some monkeys), thus reducing the reproductive success of the preceding male. Following the killing, moreover, the female ceases to nurse and, as a result, is soon ovulating and ready for mating, to the convenient advantage of the newly resident male. Grim stuff, and a pretty strong demonstration of something well recognized by most evolutionists these days; contrary to what Marlin Perkins taught us, animals very rarely behave "for the good of the species." Instead, they typically act for the good of their own genetic legacy and that of their close relatives. Among some species—wild horses, for example, and baboons—the male will also systematically harass any pregnant females to the point of miscarriage, by the same logic.

This pattern is seen in a particularly subtle way among rodents. A group of females resides with a single harem male. If he is driven out by an intruder male who takes up residence, within days, females who have recently become pregnant fail to implant the fertilized egg. Remarkably, this termination of pregnancy does not require physical harassment on the part of the male. It is his new, strange odor that causes the failed pregnancies by triggering a disruptive rise in prolactin levels. As proof of this, researchers can trigger this phenomenon (called the *Bruce-Parkes effect*) with merely the odor of a novel male. Why is it adaptive for females to terminate pregnancy just because a new male has arrived on the scene? If the female completes her pregnancy, the kids will promptly be killed by this new guy. So, making the best of a bad situation, it is seemingly more adaptive to at least save the further calories that would be devoted to the futile pregnancy, terminate it, and ovulate a few days later. [Not surprisingly, females have evolved many strategies of their own to salvage reproductive success from these battling males. One is to go into a fake heat (in primates, called *pseudo-estrus*) to sucker the new guy into thinking he's the father of the offspring she is already carrying. And given the appalling lack of knowledge about obstetrics among most male rodents and primates, it usually works. Touché.]

Despite the drama of the Bruce-Parkes effect, stress-induced miscarriages are relatively rare among animals, particularly among humans. It is not uncommon to decide retrospectively that when something bad happens (such as a miscarriage), there was significant stress beforehand. To add to the confusion, there is a tendency to attribute miscarriages to stressful events occurring for the *day* or so preceding them. In actuality, most miscarriages involve the expelling of a dead fetus, which has typically died quite a while before. If there was a stressful cause, it is likely to have come days or even weeks before the miscarriage, not immediately preceding it.

When a stress-induced miscarriage does occur, however, there is a fairly plausible explanation of how it happens. The delivery of blood to the fetus is exquisitely sensitive to blood flow in the mother, and anything that decreases uterine blood flow will be disruptive to the fetal blood supply. Moreover, fetal heart rate closely tracks that of the mother, and various psychological stimuli that stimulate or slow down the heart rate of the mother will cause a similar change a minute or so later in the fetus. This has been shown in a number of studies of both humans and primates.

Trouble seems to occur during stress as a result of repeated powerful activation of the sympathetic nervous system, causing increased secretion of norepinephrine and epinephrine. Studies of a large number of different species show that these two hormones will decrease blood flow through the uterus—dramatically, in some cases. Exposing animals to something psychologically stressful (for example, a loud noise in the case of pregnant sheep, or the entrance of a strange person into the room in which a pregnant rhesus monkey is housed) will cause a similar reduction in blood flow. As a result, fetuses rapidly become "hypoxic" and "bradycardic" (their blood pressure drops and their heart rate slows down). The general assumption in the field is that a few of these events cause little problem, but that repeated episodes of fetal hypoxia will eventually cause asphyxiation.

HOW DETRIMENTAL TO FEMALE REPRODUCTION IS STRESS?

As we have seen, there is an extraordinary array of mechanisms by which reproduction can be disrupted in stressed females—fat depletion; secretion of endorphins, prolactin, and glucocorticoids acting on the brain, pituitary, and ovaries; lack of progesterone; excessive prolactin acting on the uterus. Moreover, possible blockage of implantation of the fertilized egg and changes in blood flow to the fetus generate numerous ways in which stress can make it less likely that a pregnancy will be carried to term. With all these different mechanisms implicated, it seems as if even the mildest of stressors would shut down the reproductive system completely. Surprisingly, however, this is not the case; collectively, these mechanisms are not all that effective.

One way of appreciating this is to examine the effects of chronic low-grade stress on reproduction. Consider traditional nonwesternized agriculturists with a fair amount of background disease (seasonal malaria, for example), a high incidence of parasites, and some seasonal malnutrition thrown in—farmers in Kenya, for example. Before family planning came into vogue, the average number of children born to a Kenyan woman was about eight. Compare this with the Hutterites, non-mechanized farmers who live a life similar to that of the Amish. Hutterites experience none of the chronic stressors of the Kenyan farmers, use no contraceptives, and have essentially the identical reproductive rate—an average of nine children per woman. (It is difficult to make a close quantitative comparison of these two populations. The Hutterites, for example, delay marriage, decreasing their reproductive rate, whereas Kenyan agriculturists traditionally do not. Conversely, Kenyan agriculturists typically breast-feed for at least a year, decreasing their reproductive rate, in contrast to the Hutterites, who typically nurse far less. The main point, however, is that even with such different lifestyles, the two reproductive rates are nearly equal.)

How about reproduction during extreme stress? This has been studied in a literature that always poses problems for those discussing it: how to cite a scientific finding without crediting the monsters who did the research? These are the studies of women in the Third Reich's concentration camps, conducted by Nazi doctors. (The convention has evolved never to credit the names of the doctors, and always to note the criminality of their collaboration.) In a study of the women in the Theresienstadt concentration camp, 54 percent of the reproductive-age women were found to have stopped menstruating. This is hardly surprising; starvation, slave labor, unspeakable psychological terror are going to disrupt reproduction. The point typically made is that, of the women who stopped menstruating, the majority stopped within their first month in the camps—before starvation and labor had pushed fat levels down to the decisive point. Many researchers cite this as a demonstration of how disruptive even psychological stress can be to reproduction.

To me, the surprising fact is just the opposite. Despite starvation, exhausting labor, and the daily terror that each day would be their last, only 54 percent of those women ceased menstruating. Reproductive mechanisms were still working in nearly half the women (although a certain number may have been having anovulatory cycles). And I would wager that despite the horrors of their situation, there were still many men who were reproductively intact. That reproductive physiology still operated in any individual to any extent, under those circumstances, strikes me as extraordinary.

Reproduction represents a vast hierarchy of behavioral and physiological events that differ considerably in subtlety. Some steps are basic and massive—the eruption of an egg, the diverting of rivers of blood to a penis. Others are as delicate as whether the line of a poem awakens your heart or the whiff of a person's scent awakens your loins. Not all the steps are equally sensitive to stress. The basic machinery of reproduction can be astoundingly resistant to stress in a subset of individuals, as evidence from the Holocaust shows. Reproduction is one of the strongest of biological reflexes—just ask a salmon leaping upstream to spawn, or males of various species risking life and limb for access to females, or any adolescent with that steroid-crazed look. But when it comes to the pirouettes and filigrees of sexuality, stress can wreak havoc with subtleties. That may not be of enormous consequence to a starving refugee or a wildebeest in the middle of a drought. But it matters to us, with our culture of multiple orgasms and minuscule refractory periods and oceans of libido. And while it is easy to make fun of those obsessions of ours, the *Cosmos* and *GQ*s and other indices of our indulged lives, those nuances of sexuality matter to us. They provide us with some of our greatest, if also our most fragile and evanescent, joys.

8

IMMUNITY, STRESS, AND DISEASE

The halls of academe are filling with a newly evolved species of scientist—the psychoneuroimmunologist—who makes a living studying the extraordinary fact that what goes on in your head can affect how well your immune system functions. Those two realms were once thought to be fairly separate—your immune system kills bacteria, makes antibodies, hunts for tumors; your brain thinks up poetry, invents the wheel, has favorite TV shows. Yet the dogma of the separation of the immune and nervous systems has fallen by the wayside. The autonomic nervous system sends nerves into the tissues that form or store the cells of the immune system that wind up in the circulation. Furthermore, tissue of the immune system turns out to be sensitive to (that is, it has receptors for) all the interesting hormones released by the pituitary under the control of the brain. The result is that the brain has a vast potential for sticking its nose into the immune system's business.

The evidence for the brain's influence on the immune system goes back at least a century, dating to the first demonstrations that an artificial rose could trigger an allergic response in a patient. Here's a charming and more recent demonstration of the same: take some professional actors and have them spend a day doing either a depressing negative scene, or an uplifting euphoric one. Those in the former state show decreased immune responsiveness, while those in the latter manifest an increase. (And where was such a study carried out? In Los Angeles, of course, at UCLA.) But the study that probably most solidified the link between the brain and the immune system was carried out a few years ago, using a paradigm called *conditioned immunosuppression*.

Give an animal a drug that suppresses the immune system. Along with it, provide, à la Pavlov's experiments, a "conditioned stimulus"—for example, an artificially flavored drink, something that the animal will associate with the suppressive drug. A few days later, present the conditioned stimulus by itself—and down goes immune function. In 1982 the report of an experiment using a variant of this paradigm carried out by two pioneers in this field, Robert Ader and Nicholas Cohen, stunned scientists. The two researchers experimented with a strain of mice that spontaneously develop disease because of overactivity of their immune systems. Normally, the disease is controlled by treating the mice with an immunosuppressive drug. Ader and Cohen showed that by using their conditioning techniques, they could substitute the conditioned stimulus for the actual drug—and sufficiently alter immunity in these animals to extend their life spans.

Studies such as these convinced scientists that there is a strong link between the nervous system and the immune system. It should come as no surprise that if the sight of an artificial rose, or the taste of an artificially flavored drink, can alter immune function, then stress can too. In the first half of this chapter, I discuss what stress does to immunity, and how this might be useful during a stressful emergency. In the second half, I'll examine whether sustained stress, by way of chronic suppression of immunity, can impair the ability of a body to fight off infectious disease. This is a fascinating question, which can be answered only with a great deal of caution and many caveats. Although evidence is emerging that stress-induced immunosuppression can indeed increase the risk and severity of disease, the connection is probably relatively weak and its importance often exaggerated.

In order to evaluate the results of this confusing but important field, we need to start with a primer about how the immune system works.

IMMUNE SYSTEM BASICS

The primary job of the immune system is to defend the body against infectious agents such as viruses, bacteria, fungi, and parasites. The process is dauntingly complex. For one thing, the immune system must tell the difference between cells that are normal parts of the body and cells that are invaders—in immunologic jargon, distinguishing between "self" and "non-self." Somehow, the immune system can remember what every cell in your body looks like, and any cells (for example, bacteria) that lack your distinctive cellular signature are attacked. Moreover, when your immune system does encounter a novel invader, it can even form an immunologic memory of what the infectious agent looks like, to better prepare for the next invasion—a process which is exploited when you are vaccinated with a mild version of an infectious agent in order to prime your immune system for a real attack.

Such immune defenses are brought about by a complex array of circulating cells called *lymphocytes* and *monocytes* (which are collectively known as *white blood cells; cyte* is a term for cells). There are two classes of lymphocytes: T cells and B cells. Both originate in the bone marrow, but T cells migrate to mature in the thymus (hence the "T"), while B cells mature in the bone marrow. B cells principally produce antibodies, but there are several kinds of T cells (T helper and T suppressor cells, cytotoxic killer cells, and so on).

The T and B cells attack infectious agents in very different ways. T cells bring about cell-mediated immunity (see the illustration on the facing page). When an infectious agent invades the body, it is recognized by a type of monocyte called a *macrophage,* which presents the foreign particle to a T helper cell. A metaphorical alarm is now sounded, and T cells begin to proliferate in response to the invasion. This alarm system ultimately results in the activation and proliferation of cytotoxic killer cells, which, as their name implies, attack and destroy the infectious agent. It is this, the T-cell component of the immune system, that is knocked out by the AIDS virus.

By contrast, B cells cause antibody-mediated immunity (see the illustration on the facing page). Once the macrophage–T helper cell collaboration has occurred, the T helper cells then

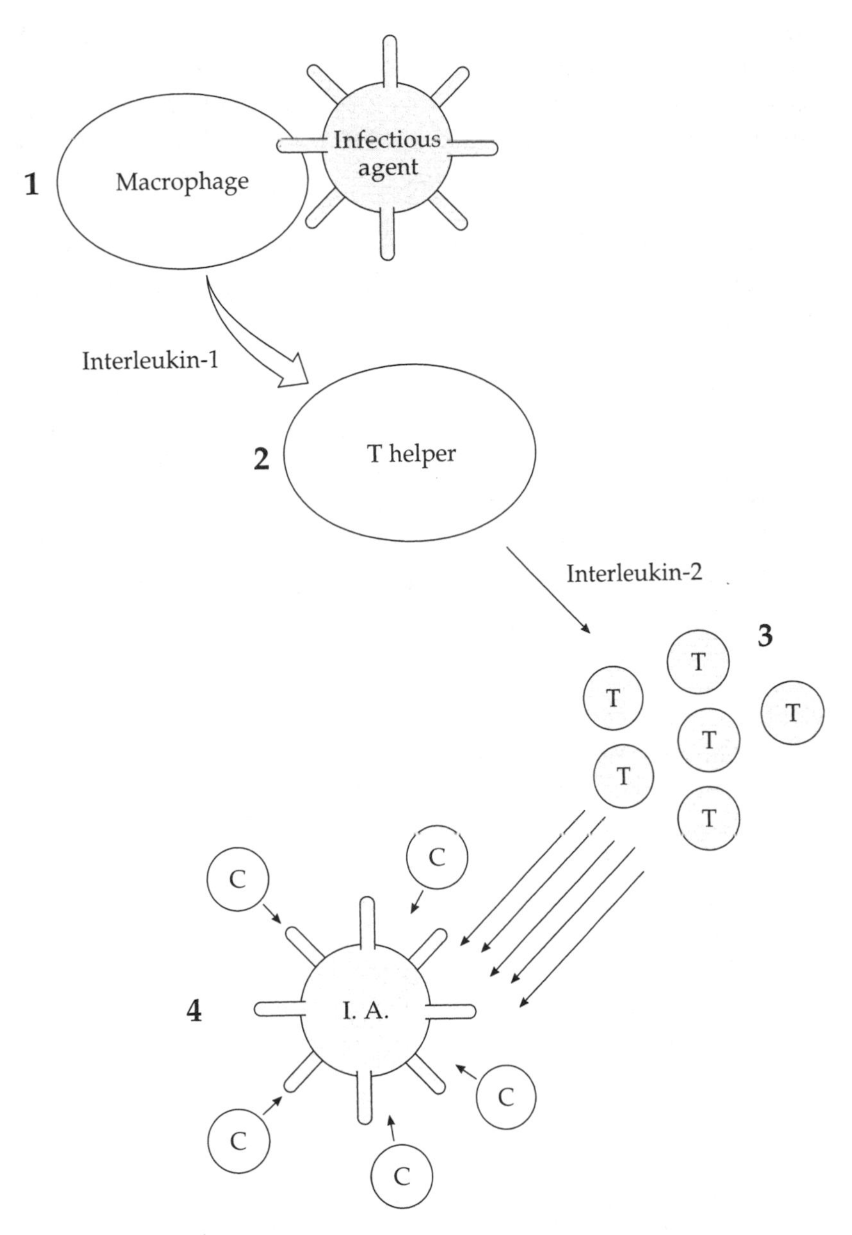

The cascade of cell-mediated immunity. (1) An infectious agent is encountered by a type of monocyte called a macrophage. (2) This stimulates the macrophage to present the infectious agent to a T helper cell (a type of white blood cell) and to release interleukin-1 (IL-1), which stimulates T helper cell activity. (3) The T helper cell, as a result, releases interleukin-2 (IL-2), which triggers T-cell proliferation. (4) This eventually causes another type of white blood cell, cytotoxic killer cells, to proliferate and destroy the infectious agent.

stimulate B-cell proliferation. The main task of the B cells is to differentiate and generate antibodies, large proteins that will recognize and bind to some specific feature of the invading infectious agent (typically, a distinctive surface protein). This specificity is critical—the antibody formed has a fairly unique shape, which will conform perfectly to the shape of the distinctive feature of the invader, like the fit between a lock and key. In binding to the specific feature, antibodies immobilize the infectious agent and target it for destruction.

There is an additional twist to the immune system. If different parts of the liver, for example, need to coordinate some activity, they have the advantage of sitting adjacent to each other. But the immune system is distributed throughout the circulation. In order to sound immune alarms throughout this farflung system, there must be bloodborne chemical messengers that communicate between different cell types. For example, when macrophages first recognize an infectious agent, they release a messenger called *interleukin-1*. This triggers the T helper cell to release interleukin-2, which stimulates T-cell growth (to make life complicated, there are at least half a dozen additional interleukins with more specialized roles). On the antibody front, T cells also secrete B-cell growth factor. Other classes of messengers, such as interferons, activate broad classes of lymphocytes.

The process of the immune system sorting self and nonself usually works well (although truly insidious tropical parasites like those that cause schistosomiasis have evolved to evade your immune system by pirating the signature of your own cells). Your immune system happily spends its time sorting out self from nonself: Red blood cells. Part of us. Eyebrows. Our side. Virus. No good, attack. Muscle cell. Good guy. . . .

What if something goes wrong with the immune system's sorting? One obvious kind of error could be that the immune system misses an infectious invader; clearly, bad news. Equally bad is the sort of error in which the immune system mistakes a normal part of the body for an infectious agent and attacks it. When the immune system erroneously attacks a normal part of the body, a variety of horrendous "autoimmune" diseases may result. In multiple sclerosis, for example, part of your nervous system is attacked; in juvenile diabetes, it's the cells in the pancreas that normally secrete insulin. As we'll see shortly, stress has some rather confusing effects on autoimmune diseases.

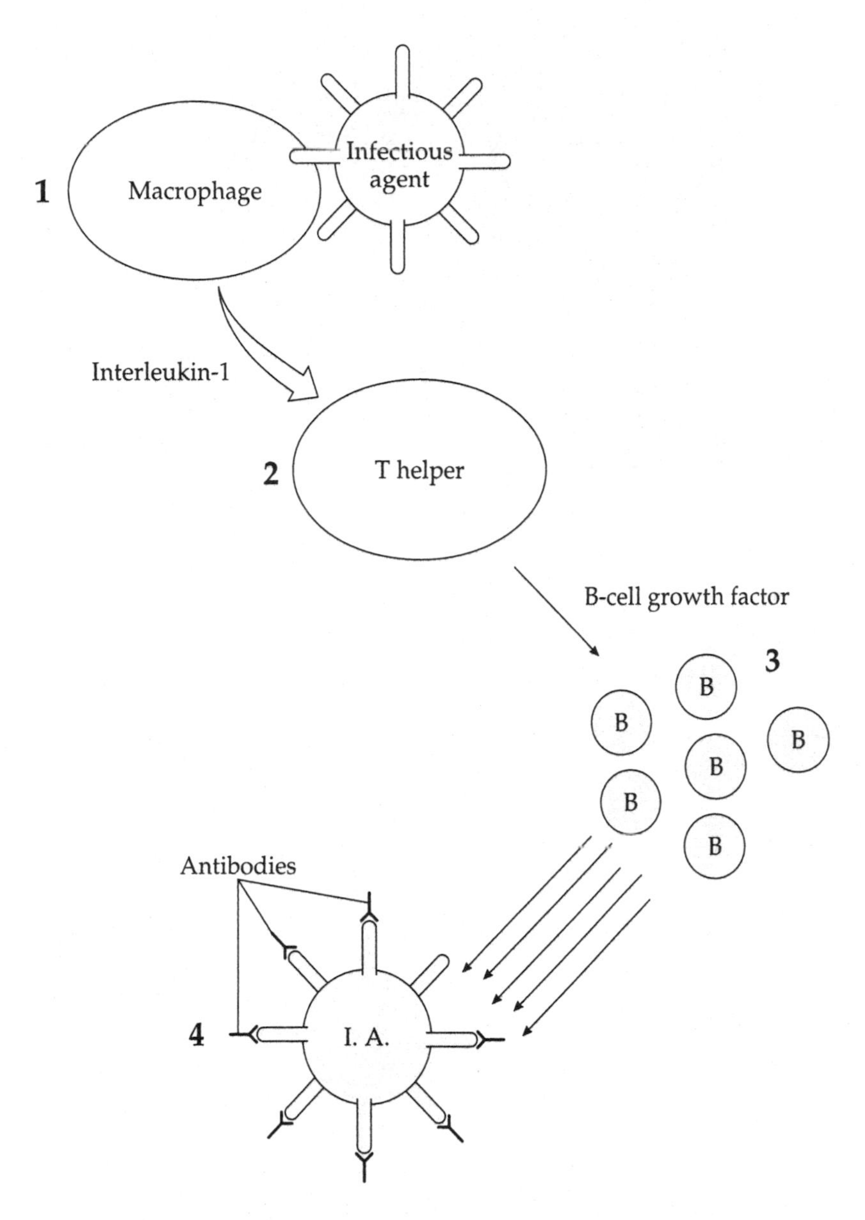

The cascade of antibody-mediated immunity. (1) An infectious agent is encountered by a macrophage. (2) This encounter stimulates it to present the infectious agent to a T helper cell and to release interleukin-1 (IL-1), which stimulates T helper cell activity. (3) The T helper cell then secretes B-cell growth factor, triggering differentiation and proliferation of another white blood cell, B cells. (4) The B cells make and release specific antibodies that bind to surface proteins on the infectious agent, targeting it for destruction by a large group of circulating proteins known as complement.

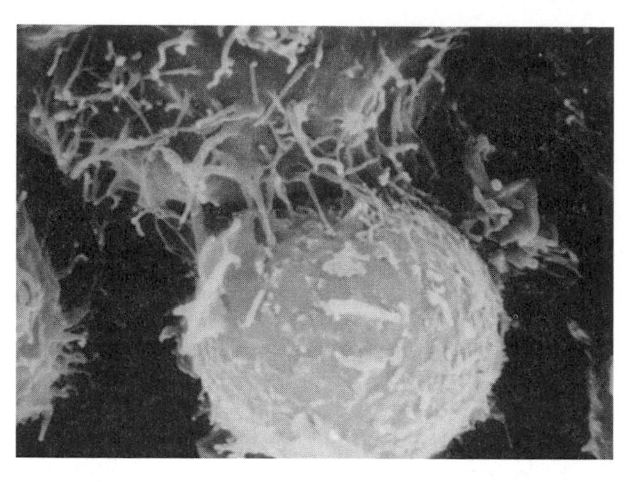

Photomicrograph of a killer T cell attacking a tumor cell.

HOW DOES STRESS INHIBIT IMMUNE FUNCTION?

It's been almost sixty years since Selye discovered the first evidence of stress-induced immunosuppression, noting that immune tissues like the thymus gland atrophied among rats subjected to nonspecific unpleasantness. Scientists have learned more about the subtleties of the immune system since then, and it turns out that a period of stress will disrupt a wide variety of immune functions—the formation of new lymphocytes and their release into the circulation, the time preexisting lymphocytes stay in the circulation, the manufacture of antibodies in response to an infectious agent, and communication among lymphocytes through the release of relevant messengers, to name just a few of these actions. And this is true in rats, in primates, and, very clearly, in humans as well.

The best-documented way in which such immune suppression occurs is via glucocorticoids. Glucocorticoids, for example, can cause shrinking of the thymus gland; this is such a reliable effect that in the olden days (circa 1960), before it was possible to measure directly the amount of glucocorticoids in the bloodstream, one indirect way of doing so was to see how much the thymus gland in an animal had shrunk. The smaller the thymus, the more glucocorticoids in the circulation. Glucocorticoids halt the formation of new lymphocytes in the thymus, and most of thymic tissue is made up of these new cells, ready to be secreted into the bloodstream. Because glucocorticoids inhibit the release of messengers like interleukins and interferons, they also make circulating lymphocytes less responsive to an infectious alarm. Glucocorticoids, moreover, cause lymphocytes to be yanked out of the circulation. And most impressively, glucocorticoids can actually kill lymphocytes. This occurs through a mechanism that cell biologists are wild about: glucocorticoids can enter a lymphocyte and cause it to synthesize a suicide protein that chops the DNA in the lymphocyte into thousands of tiny pieces. (Then the lymphocyte is dead as a doorknob, since you need DNA to generate the proteins that keep you in business.)

Sympathetic nervous system hormones, beta-endorphin, and corticotropin releasing factor (CRF) within the brain also play a role in suppressing immunity during stress. The precise mechanisms by which this happens are nowhere near as well understood as with glucocorticoid-induced immune suppression, and these other hormones have traditionally been viewed as less important than the glucocorticoid part of the story. However, a number of experiments have shown that stressors can suppress immunity independently of glucocorticoid secretion, strongly implicating these other routes.

WHY IS IMMUNITY SUPPRESSED DURING STRESS?

Figuring out exactly how glucocorticoids and the other stress hormones suppress immunity is a very hot topic these days in cell and molecular biology, especially the part about killing lymphocytes. It would be perfectly reasonable to ask somewhere in that excitement *why* you should want your immune system suppressed during stress. In the first chapter, I offered an explanation for this; now that the process of stress-induced immunosuppression has been explained in a little more detail, it should be obvious that my early explanation makes no sense. I suggested that during stress it is logical for the body to shut down long-term building projects in order to divert energy for more immediate needs—this inhibition includes the immune system, which, while fabulous at spotting a tumor that will kill you in six months or making antibodies that will help you in a week, is not vital in the next few moments' emergency. That explanation would make sense only if stress froze the immune system right where it was—no more immune expenditures until the emergency is finished. However, that is not what happens. Instead, stress causes the active expenditure of energy in order to *disassemble* the preexisting immune system—tissues are shrunk, cells are destroyed. This cannot be explained by a mere halt to expenditures—so out goes this extension of the long-term versus short-term theory.

Why should evolution set us up to do something as apparently stupid as disassembling our immune system during stress? Maybe there isn't a good reason. This actually isn't as crazy of a response as you might think. Some scientists, for example the paleontologist Stephen Jay Gould, believe that not everything in the body has an explanation in terms of evolutionary adaptiveness. Maybe stress-induced immunosuppression is simply a by-product of something else that is adaptive; it just came along for the ride.

This is probably not the case. During infections, the immune system releases the chemical messenger interleukin-1, which among other activities stimulates the hypothalamus to release CRF. As noted in Chapter 2, CRF stimulates the pituitary to release ACTH, which then causes adrenal release of glucocorticoids. These in turn suppress the immune system. In other words, under some circumstances, the immune system will *ask* the body to secrete hormones that will ultimately suppress the immune system. For whatever reason the immunosuppression occurs, the immune system sometimes encourages it. It is probably not just an accident.*

Various ideas have floated around over the years to explain why you actively disassemble immunity during stress with the willing cooperation of the immune system. Some seemed fairly plausible until people learned a bit more about immunity and

---

* My tiny footnote in science: I was *the* discoverer of the fact that interleukin-1 stimulates CRF release. Or at least I thought I was. It was a few years ago. The idea made some sense, and the lab I was in jumped on it under my prompting. We worked like maniacs, and at two o'clock one morning I had one of those moments of euphoria that scientists die for: looking at the printout from one of the machines and realizing, "Aha, I was right, it does work that way. Interleukin-1 released CRF." We wrote up the findings, they were accepted by the prestigious journal *Science*, everyone was very excited,

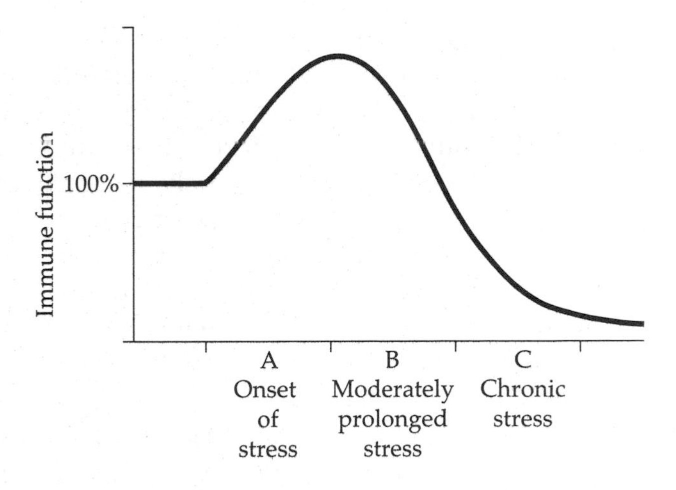

Stress turns out to transiently stimulate the immune system.

could rule them out. Others were quite nutty, and I happily advocated a few of these in the previous edition. But in the last few years, an answer has begun to emerge, and it really turns this field on its head.

When you use very sensitive assays of immune function that can pick up small differences over short periods of time, it becomes clear that during the first few minutes (say, up to about thirty) after the onset of a stressor, you don't suppress immunity, you *enhance* it (phase A on the accompanying graph). More of certain immune cell types are dumped into the circulation, and different classes of immune cells are more responsive. And this doesn't occur only after some infectious challenge. Physical stressors, psychological stressors all appear to cause an early stage of immune activation. Even more surprisingly, the archvillains of immunosuppression, glucocorticoids, appear to play a major role in this (as does the sympathetic nervous system).

I called my parents, and so on. Paper gets published, and right next to it was an *identical* study from a group in Switzerland, sent in to the journal the same exact week. So I became *a* discoverer of this obscure fact. (To hark back to a theme of Chapter 2, if you are a mature, confident individual—which unfortunately I am only rarely—you take pleasure in this sort of thing: two labs, working independently on opposite sides of the globe, come up with the same novel observation. It must be true. Scientific progress lurches forward an inch.)

By about the one-hour mark, more sustained glucocorticoid and sympathetic activation begins to have the opposite effect, namely, suppressing immunity. And for most stressors of moderate intensity and duration, the net effect of those mean, damaging, suppressive glucocorticoids is merely to bring things back to where they started (phase B). Glucocorticoids don't suppress the immune system during most stressors (which is to say they don't cause the system to end up worse off than before the stressor started). They initially stimulate the system and then help it to return to baseline. It is only with major stressors of longer duration, or with really major exposure to glucocorticoids, that the immune system does not just return to baseline, but plummets below into a range that really does qualify as immunosuppressing (phase C).

The idea of temporarily perking up your immune system with the onset of a stressor makes a fair amount of sense (certainly at least as much as some of the convoluted theories as to why suppressing it makes sense). As does the notion that what goes up must come down. And as does the frequent theme of this book, namely, that if you have a stressor that goes on for too long, an adaptive decline back to baseline can overshoot and you get into trouble.

Why did it take people so long to figure this out? Because most scientists in this field study major, unsubtle stressors, or administer major unsubtle amounts of glucocorticoids. This represents a reasonable bias in how experiments are done—start with an enormous experimental manipulation. If nothing happens, pick a new field to study. If something does happen and it's been replicated enough times that you are confident about it, only then begin to think about more subtle elaborations. Which has taken people years to do in this field, because many of the techniques for measuring what's happening in the immune system have only recently become sensitive enough to pick up small, rapid differences, the thing needed to catch phase A, that fast immunostimulatory blip at the beginning of a stressor.

This reorientation of the field represents a triumph for Allan Munck of Dartmouth University, an endocrinologist who predicted most of these new findings in the mid-1980s. He also predicted what turns out to be the answer to a question that pops up after a while. Why would you want to bring immune function back down to the prestress level (phase B in the diagram)?

Why not just let it remain at the enhanced, improved level achieved in the first thirty minutes, get the benefits of an activated immune system all the time? Besides the fact that this would be too costly, there's another problem, which Munck predicted: immune systems that are chronically in that activated state become even more active, and they start generating autoimmune diseases. It has now been shown that if you block the compensatory recovery to baseline during stressors, the immune system spirals higher and higher, until it overshoots into autoimmunity.

Recently, another subtle wrinkle in this story has been uncovered as well. Back to phase B, just when the inhibitory return to baseline begins. It turns out that in the very start of that phase, what appears to be immune suppression is actually still an indirect version of enhancing immunity.

This is seen in two ways. Give someone massive amounts of glucocorticoids, or a huge stressor that has gone on for many hours, and the hormones will be killing lymphocytes indiscriminately, just mowing them down. Have a subtle rise in glucocorticoid levels for a short time (like what is going on at the start of B), and the hormones kill only a particular subset of lymphocytes—older ones, ones that don't work as well. Glucocorticoids, at that stage, are helping to *sculpt* the immune response, getting rid of lymphocytes that aren't ideal for the immediate emergency.

A second subtlety reflects reinterpretation of something people have known for decades. As noted, glucocorticoids not only kill lymphocytes, but also yank some remaining lymphocytes out of the circulation, and stick them back into storage in immune glands. The interpretation has always correctly been that this results in fewer lymphocytes that are available for defensive duty. Recent work by Firdhaus Dhabhar, in Bruce McEwen's laboratory at Rockefeller University, suggests that around phase B, glucocorticoids are indeed yanking lymphocytes out of the circulation. But at that point, the lymphocytes, instead of being stuck in mothballs, are rushed to a specific site of infection, such as the skin. They're not being deactivated, but are instead being transferred to the front lines.

Thus, during stress, glucocorticoids and other stress-responsive hormones cause a transient activation of the immune system, enhancing immune defenses, sharpening them, redistributing immune cells to the scenes of infectious battle. Because of the dangers of the systems overshooting into autoimmunity, more prolonged glucocorticoid exposure begins to reverse these effects, bringing the system back to baseline. And with the pathological scenario of truly major, sustained stressors, immunity is suppressed below baseline.

These new findings help to explain one of the persistent paradoxes in this field, one that I approached only gingerly in the previous edition, as I was obviously utterly confused by the subject. The paradox concerns people with autoimmune diseases.

Two facts about autoimmunity:

1. Insofar as autoimmune diseases involve overactivation of the immune system (to the point of considering a healthy constituent of your body actually to be something invasive), the most time-honored treatment for such diseases is to put people "on steroids"—to give them massive amounts of glucocorticoids. The logic here is obvious; by dramatically suppressing the immune system, it can no longer attack your pancreas or nervous system, or whatever is the inappropriate target of the autoimmunity (and, as an obvious side effect to this approach, your immune system will also not be very effective at defending you against real pathogens). Thus, administration of large amounts of these stress hormones make autoimmune diseases less damaging.
2. At the same time, it appears that stress can *worsen* autoimmune diseases. This has often been reported anecdotally by patients, and is typically roundly ignored by clinicians who know that stress hormones help reduce autoimmunity, not worsen it. But some objective studies also support this view, for autoimmune diseases such as multiple sclerosis, rheumatoid arthritis, Grave's disease (an attack on the thyroid gland), and juvenile diabetes. There have been only a handful of such reports, and they suffer from the weakness of reliance on self (patient)-reported retrospective data, rather than on prospective data. Nevertheless, their findings are relatively consistent—there is a subset of patients whose initial onset of an autoimmune disease and, to an even greater extent, their intermittent flare-ups of bad symptoms, are yoked to stress.

So, do glucocorticoids and stress worsen or decrease the symptoms of autoimmunity? The graph on page 135 gives an obvious answer that wasn't clear in earlier years. Massive stressors, or administration of the big hefty doses of glucocorticoids used by physicians, put the system in phase C—dramatic immune suppression, which decreases the symptoms of autoimmunity. But repetition of the everyday array of stressors that are clinically associated with worsening of autoimmune symptoms puts you in the range of phases A and B—and something about the repeated ups and downs ratchet the system upward, biasing toward autoimmunity. While this process is not understood, it is reminiscent of the problems of repeated shifting motility in the small and large intestines with repeated stress. The system apparently did not evolve for dealing with great repetitions of coordinating the various on and off switches, and ultimately, something uncoordinated occurs. In the case of the bowels, a result can be a propensity toward irritable bowel syndrome. And in the immune systems of people who are in some way at risk, the danger is that of spiraling into autoimmunity.

CHRONIC STRESS AND DISEASE RISK

A repeated theme in this book is how some physiological response to your average, run-of-the-mill mammalian stressor, if too long or too frequent, gets you into trouble. The ability of major stressors to suppress immunity below baseline certainly seems like a candidate for this category. How damaging is stress-induced immunosuppression when it actually occurs? As the AIDS virus has taught us, if you suppress the immune system sufficiently, a 30-year-old will fester with cancers and pneumonias that doctors used to see once in fifty-year careers. But can chronic stress suppress the immune system to the point of making you more susceptible to diseases you wouldn't otherwise get? Once you have a disease, are you less capable of fighting it off?

Evidence pouring in from many quarters suggests that stress may indeed impair our immune systems and increase the risk of illness. Examples of them will be reviewed shortly, and some are quite striking. But despite these fascinating findings, it remains far from clear just how much chronic stress makes you more vulnerable to diseases that would normally be fought off by the immune system. In order to appreciate the current disarray of the research, let us try to break down the findings above into their component parts.

Essentially, all these studies show a link between something that increases or decreases stress and some disease or mortality outcome. The approach of many psychoneuroimmunologists is based on the assumption that this link is established through the following steps:

1. We differ as to the pattern and frequency of stressors to which we are exposed.
2. These variations will determine the magnitude and frequency with which we turn on the stress-response (glucocorticoids, epinephrine, and so on).
3. The magnitude and frequency of the stress-response regulates immune competence.
4. Our level of immune competence, in turn, determines what diseases we get and how readily we resist them.

Now, let's begin to analyze the separate steps.

First, how much stress are you exposed to? In studies of nonhuman animals, the general consensus is that with enough stress, you are going to get to steps 2 through 4. But a problem in extrapolating to humans is that the experimental stressors used in animal studies are usually more awful than would be considered ethical to subject humans to. Not only that, but we differ tremendously among ourselves as to what we experience as truly stressful—that whole realm of individual differences that will be the focus of the last chapter of this book. Therefore, if you try to study the effects of everyday, natural stressors on people's immune systems, you must wrestle with the problem of whether these things actually seem stressful to a given individual or not.

There is another problem with step 1: it's often not clear whether humans are really exposed to the stressors to which they claim they're exposed. We tend to be notoriously bad reporters of what goes on in our lives. An imaginary experiment: take 100 lucky people and slip them a drug that will give them a bad stomachache for a few days. Then send them to a doctor participating in this experiment, who tells them that they have developed stomach ulcers. The doctor asks innocently, "Have things been particularly stressful for you recently?" Perhaps 90 of those subjects will come up with something or other putatively stressful to which they will now attribute the ulcer. In retrospective studies, people confronted with an illness are extremely likely to decide there were stressful events going on. When you rely heavily on retrospective studies with humans, you are likely to get a falsely strong link between stress and disease; and the trouble is, most studies in this field are retrospective (a problem that popped up in the chapter on digestive disorders, as well). The very expensive and lengthy prospective studies are just becoming more common—pick a bunch of healthy people and follow them for decades to come, recording as an objective outsider when they are being exposed to stressors and whether they are becoming sick.

We move to the next step: from the stressor to the stress-response (step 1 to step 2). Again, if you give an organism a massive stressor, it will reliably have a strong stress-response. With more subtle stressors, we have more subtle stress-responses.

The same thing holds for the move from step 2 to step 3. In experimental animal studies, large amounts of glucocorticoids will cause the immune system to hit the floor. The same occurs if a human has a tumor that causes massive amounts of glucocorticoids to be secreted (Cushing's syndrome), or if a person is taking huge doses of synthetic glucocorticoids to control some other disease. But as we now know, the moderate rises in glucocorticoid levels seen in response to many more typical stressors stimulate the immune system, rather than suppress it.

Moving from step 3 to step 4, how much does a change in immune profile alter patterns of disease? The odd thing is that immunologists are not sure about this. If your immune system is massively suppressed, you are more likely to get sick, no doubt about that. People taking high doses of glucocorticoids as medication, who are thus highly immunocompromised, are vulnerable to all sorts of infectious diseases, as are people with Cushing's syndrome. Or AIDS.

The more subtle fluctuations in immunity are less clear in their implications, however. Few immunologists would be likely to assert that "for every tiny decrease in some measure of immune function, there is a tiny increase in disease risk." Their hesitancy is because the relationship between immune competence and disease may be nonlinear. In other words, once you pass a certain threshold of immunosuppression, you are up the creek without a paddle; but before that, immune fluctuations may not really matter much. The immune system is so complex that being able to measure a change in one little piece of it in response to stress may mean nothing about the system as a whole. Thus, the link between relatively minor immune fluctuation and patterns of disease in humans winds up being relatively weak.

There is another reason why it may be difficult to generalize from findings in the laboratory to the real world. In the laboratory, you might be studying the effects of steps 1, 2, and 3 on disease outcome 4. It is inconvenient for most scientists to manipulate a rat's levels of stress, glucocorticoids, or immunity and then wait for the rest of the rat's lifetime to see if it is more likely to become ill than is a control rat. That's slow and expensive. Typically, instead, scientists study induced diseases. Manipulate step 1, 2, or 3 in a rat that has been exposed to a certain virus; then see what happens. When you do that, you get information about the links to 4 when dealing with severe, artificially induced disease challenges—but that approach unfortunately misses the point that we don't get sick because some scientist deliberately exposes us to disease. Instead, we spend our lives passing through a world filled with scattered carcinogenic substances, occasional epidemics, someone sneezing from across the room. Relatively few experimental animal studies have looked at spontaneous diseases, rather than induced ones.

Finally, many scientists assume that if there is a link between stress (step 1) and disease (step 4)—it occurs by way of a stress-response (step 2) causing suppression of immunity (step 3). But, as we will see shortly, there may be other ways to get from step 1 to step 4, having to do with lifestyle, risk factors, and so on.

These are a lot of caveats. Let's consider some areas where there are links between stress and diseases associated with immune dysfunction. This will let us evaluate to what extent these links are a function of progressing from steps 1 through 4, versus some alternative route.

Social Isolation The fewer social relationships a person has, the shorter the life expectancy, and the worse the impact of various infectious diseases. Medically, protective relationships can take the form of marriage, contact with friends and extended family, church membership, or other group affiliations. Moreover, the process of becoming isolated from your significant other—going through a divorce or having severe marital problems—is also associated with worse immune functioning. To continue these themes, people who score high on loneliness scales have been found to have relatively depressed immune function. Moreover, social stressors tend to be more immunosuppressive than noninterpersonal stressors.

This is a fairly consistent pattern that cuts across a lot of different settings. Moreover, these general findings are based on some careful prospective studies and are seen in both sexes and in different races, in American and European populations living in both urban and rural areas. Most importantly, this effect is big. The impact of social relationships on life expectancy appears to be at least as large as that of variables such as cigarette smoking, hypertension, obesity, and level of physical activity. People with the fewest social connections have approximately two-and-a-half times as much chance of dying as those with the most connections, after controlling for such variables as age, gender, and health status.

This is very exciting, and we can readily apply steps 1 through 4 to these studies—socially isolated people are more stressed, for lack of social outlets and support (step 1); this leads to chronic activation of stress-responses (step 2), leading to immune suppression (step 3) and more infectious diseases (step 4). Nice, but there are lots of other ways to get from social isolation to poor health. What if the problem is that socially isolated people lack that special someone to remind them to take their daily medication? It is known that isolated people are less likely to comply with a medical regime. What if they're more likely to subsist on reheated fast food instead of something nutritious? Or more likely to indulge in some foolish risk-taking behavior, like smoking, because there's no one to try to convince them to stop? There are many lifestyle patterns that could link social isolation with step 4 and bypass 1 through 3 entirely. Or what if the causality is reversed—what if the linkage occurs because sickly people are less likely to be able to maintain stable social relationships?

Despite those counterinterpretations, it looks as if the linkage between social isolation and poor health is at least partially mediated through the psychoneuroimmune cascade of steps 1 through 4 that we've outlined. For one thing, the careful studies have controlled for variables such as smoking, diet, or medication compliance. Even stronger support comes from primate studies: social isolation impairs the immunity and health of monkeys, and animals put into stressful environments that would normally suppress their immune systems are buffered from that happening if they are in the company of friends. Just as in humans—but without the human confounds of medication, Big Macs, alcohol, and smoking.

Bereavement An extreme version of social isolation is, of course, the loss of a loved one, and a common scenario in literature is of the one left behind—the grieving spouse, the bereft parent, even the masterless pet—now pining away to an early death. A number of studies suggest that bereavement does indeed increase the risk of dying. This does not appear to be the case among all grieving individuals, however. Instead, it appears that the person has to have an additional physiological or psychological risk factor coupled with the bereavement. In one carefully controlled prospective study, the parents of all the Israeli men who died in the Lebanese war were followed for ten years afterward. Loss of a son did not affect mortality rate in the population of grieving parents in general; however, significantly higher mortality rates occurred among parents who were already widowed or divorced. In other words, this stressor is associated with increased mortality in the subset of parents with the added risk factor of minimal social support.

Sometimes the additional risk factor can be immunological. This is seen among HIV-positive people. When comparing individuals matched for severity of the disease, those who are in the process of grieving for a lost loved one show a more dramatic decline in immune function in the aftermath than those not grieving.

Insofar as there is a link between bereavement and increased mortality, we once again have to consider whether this is due to steps 1 through 4. Once again, the psychoneuroimmune sequence: an elderly man loses his wife and has an increased risk of dying in the next year. An obvious interpretation would be that the stressor of bereavement (step 1) activates his stress-response (step 2), causing enough immunosuppression (step 3) to make him sick (step 4). However, there are the obvious alternative routes—he doesn't bother to eat healthfully, takes to drinking, doesn't take his medication. Perhaps he drives more recklessly than he would have otherwise. Sometimes the confound is more subtle. People tend to marry people who are ethnically and genetically quite similar to themselves. Intrinsic in this trend toward "homogamy" is a tendency of married couples to have higher-than-random chances of sharing certain genetic disease tendencies, which makes it more likely that they will get sick around the same time. Thus, a variety of alternative routes might explain the bereavement literature, though it still remains unclear how much these plausible confounds come into play.

The Common Cold Everybody knows that being stressed increases your chances of getting a cold. Just think back to being run down, frazzled, and sleep-deprived during final exams, and, sure enough, there's that cough and runny nose. Examine the records at university health services and you'll see the same thing—students succumbing to colds left and right around exam period. And many of us continue to see the same pattern decades later—burn the candle at both ends for a few days and, suddenly, there's that scratchy throat.

Looks like a pretty straightforward transition through steps 1 through 4. But maybe there's some really subtle confound floating around. Maybe when we are under stress, we are more likely to consort with people who sneeze. Or maybe the disruptive effects of stress on memory (stay tuned for Chapter 10) cause us to forget to button up our overcoats. Still, 1 through 4 seem pretty plausible here.

A few years ago, a celebrated study provided some strong support for this view. Some cheerfully compliant volunteers were pulled together under conditions where some major lifestyle confounds were controlled for, and they filled out questionnaires regarding how stressed they were. Subjects were then exposed to rhinovirus, the bug that causes the common cold, and (fanfare) more stress equaled greater likelihood of a cold. To rule out lifestyle confounds even further, the same authors next followed up by showing the same stress–nose cold connection in a group of primates.

Cancer For my money, there's a pretty convincing link between stress and the common cold, and I think it involves our sequence of steps 1 through 4. Now, to the other end of the psychoimmune spectrum—how about stress and cancer? I am going to devote quite a few careful words to this, as this is one of the most contentious subjects of this book, and one of the most important.

The first piece of evidence in support of a stress-cancer link comes from animal studies. There is, by now, a reasonably convincing animal-experimentation literature showing that stress affects the course of some types of cancer. For example, the rate at which some tumors grow in mice can be affected merely by what sort of cages the animals are housed in—the more noisy and stressful, the faster the tumors grow. Other studies show that if you expose rats to electric shocks from which they can eventually escape, they reject transplanted tumors at a normal rate. Take away the capacity to escape, yet give the same total number of shocks, and the rats lose their capacity to reject tumors. Stress mice by putting their cages on a rotating platform (basically, a record player), and there is a tight relationship between the number of rotations and the rate of tumor growth. Substitute glucocorticoids for the rotation stressor, and tumor growth is accelerated, as well. These are the results of very careful studies performed by some of the best scientists in the field.

By now, we should recognize the limitations of these approaches, in that these were studies of induced tumors, limiting how much they tell us about spontaneous cancer in humans. No animal studies to my knowledge have shown that stress increases the incidence of spontaneous tumors. Furthermore, most of these studies have relied on tumors that are caused by viruses. In such cases, viruses take over the replicative machinery of a cell and cause it to start dividing and growing out of control. In humans most cancers are currently thought to arise from genetic factors or exposure to environmental carcinogens, rather than from viruses, and those have not been the subject of study with laboratory animals. So a cautionary note from the animal studies: stress can accelerate the growth of a number of tumors, but of a kind with only limited relevance to human cancer.

The second type of evidence linking stress and cancer comes from the laboratory, as well. It makes sense that stress and glucocorticoids have the effects reported, given their biological actions. How might stress have made tumors grow faster in those laboratory animals? Probably through a number of mechanisms. The first is probably immunologic. The immune system contains a specialized class of cells (most notably, natural killer cells) that prevent the spread of tumors. Stress suppresses the numbers of circulating natural killer cells. A second route is probably nonimmunologic. Once a tumor starts growing, it needs enormous amounts of energy, and one of the first things that tumors do is send a signal to the nearest blood vessel to grow a bush of capillaries into the tumor. Such *angiogenesis* allows for the delivery of blood and nutrients to the hungry tumor. Glucocorticoids, at the concentration generated during stress, aid angiogenesis. A final route may involve glucose delivery. Tumor cells are very good at absorbing glucose from the bloodstream. Recall the zebra sprinting away from the lion: Energy storage has stopped, in order to increase concentrations of circulating glucose to be used by the muscles. But, as my own lab recently discovered, when circulating glucose concentrations are elevated in rats during stress, at least one kind of experimental tumor can grab the glucose before the muscle does. Your storehouses of energy, intended for your muscles, are being emptied and inadvertently transferred to the ravenous tumor instead.

So we have some stress-cancer links in animals, and some biology to explain those links. What about humans? Our first, simplest question: is a history of major stressors associated with an increased risk of having cancer somewhere down the line?

A number of studies seemed to show this, but they all suffered from the same problem, namely, that they were retrospective. Again, someone with a cancer diagnosis is more likely to remember stressful events than someone with a bunion. A number of studies, especially of breast cancer patients, have had a "quasi-prospective" design, assessing stress histories of women at the time that they are having a biopsy for a breast lump, comparing those who get a cancer diagnosis with those who don't. Some of these studies have shown a stress-cancer link, and this should be solid—after all, there can't be a retrospective bias, if the women don't know yet if they have cancer. What's the problem here? Apparently, people can guess whether it turns out to be cancer at a better than chance rate, possibly reflecting knowledge of a family history of the disease, or personal exposure to risk factors. Thus, such quasi-prospective studies are already quasi-retrospective as well.

When you rely on the rare prospective studies, there turns out not to be good evidence for a stress-cancer link. For example, as we will see in the chapter on depression, having a major depression is closely linked to both stress and excessive glucocorticoid secretion, and one famous study of two thousand men at a Western Electric plant showed that depression was associated with doubling of the risk of cancer, even up to decades later. But a careful re-examination of those data showed that the depression-cancer link was attributable to a subset of men who were depressed as hell because they were stuck working with some major carcinogens. Subsequent prospective studies of other populations have shown there to be either no depression-cancer link, or a tiny one that is biologically meaningless. Moreover, these studies have not ruled out the confound that depressed people smoke and drink more, two routes to increase risk of cancer. And similar findings emerge from the careful prospective studies of bereavement as a stressor—no link with subsequent cancer. And likewise for the majority of prospective studies examining life stressors in general and cancer incidence.

By now, we should be trained to think of the other side of the coin. Okay, there appears to be no link between getting cancer and having been exposed to a lot of stressors. Maybe it's an issue of coping—maybe certain folks, exposed to the normal array of stressors, deal with them in a certain maladaptive way that predisposes to cancer. We already saw how, in Chapter 5, the notion of there being a colitis-prone personality has raised a lot of hackles, in part because there is often the suggestion between the lines that "disease-prone personality" means blaming the victim. Is there a cancer prone personality, and can it be interpreted in the context of coping poorly with stress?

Some scientists think so. Much of the work in this area has been done with breast cancer, in part because of the prevalence and seriousness of the disease. However, the same pattern has been reported since for other cancers as well. The cancer-prone personality, we're told, is one of repression—emotions held inside, particularly those of anger. This is a picture of an introverted, respectful individual with a strong desire to please—

conforming and compliant. Hold those emotions inside and out comes cancer, according to this view.

Most of these studies have been retrospective or quasi-prospective, and we have seen the problems endemic to such studies. Nonetheless, the prospective studies have shown there to be some link, though a small one.

Are we in the realm of steps 1 through 4? No one has shown that yet, in my opinion. First, there is no biology I know of that links having a repressed, compliant personality with the sort of immune changes associated with increased cancer risk. And, in addition, none of the good prospective studies have ruled out alternative routes, whereby the "cancer-prone" personality leads to more risk factors for cancer (such as smoking, drinking or, in the case of breast cancer, more fat consumption). So the jury remains out on this one.

And what about once cancer has developed? Does stress accelerate the growth of a tumor in a person? In general, there is no clear evidence for that. And the flip side—can stress reduction techniques or a certain personality style slow down the progression of cancer? There are some intriguing findings in that realm. When you compare patients who respond to their cancer with a "fighting spirit" (that is, they are optimistic and assertive) with those who collapse into depression, the former live longer, after controlling for cancer severity. Moreover, some (but not all) studies have shown that group therapy that emphasizes the same fighting spirit extends the life expectancy of cancer patients as well. The best known of these was carried out by David Spiegel and colleagues of Stanford Medical School, whose group therapy regime extended life span an average of eighteen months. Finally, social support is associated with longer survival as well. While immensely exciting, it's not clear whether these findings have anything to do with steps 1 through 4—does a fighting spirit, social support, or the right group psychotherapy lead to less of a stress-response and thus a more combative immune system? No one has shown this yet. Moreover, these studies have not adequately controlled for variation in adherence to treatment regimes. About a quarter of cancer patients don't take medication as often as prescribed, or miss chemotherapy appointments. Maybe naturally having an optimistic, fighting spirit, or being in a supportive group where fellow patients cheer on those fighting traits results in life-extending adherence.

What we have here are some extremely interesting but murky waters. There appears to be no link between a history of a lot of stress and a greater incidence or faster progression of cancer. There seems to be a link between a certain personality type and a somewhat greater cancer risk, but no studies have shown where stress physiology fits into that story, nor have lifestyle confounds been ruled out. Finally, a fighting spirit is associated with a better outcome but, again, no one has shown how stress physiology is relevant to that, nor has the critical issue of adherence been ruled out. What does one do with these findings? Beyond doing a lot more research, I'm not entirely sure. However, it is time to discuss what one should *not* do with these findings.

CANCER AND MIRACLES

This leads to a tirade. Once we recognize that psychological factors, stress-reducing interventions, and so on can influence something like cancer, it is often a hopeful, desperate leap to the conclusion that such factors can control cancer. And when that proves to be false, there is a corrosive, poisonous flip side: if you falsely believe you had the power to prevent or cure cancer through positive thinking, you may then come to believe that it is your own fault if you are dying of the disease.

The advocates of a rather damaging overstatement of these psychology-health relationships are not always addled voices from the lunatic fringe. They include influential health practitioners whose medical degrees appear to lend credence to their extravagant claims. I will focus my attention here on the claims of Bernie S. Siegel, a Yale University surgeon who has been wildly effective at disseminating his ideas to the public as the author of a best-seller.

The premise of Seigel's magnum opus, *Love, Medicine and Miracles* (New York: Harper & Row, 1986), is that the most effective way of stimulating the immune system is through love, and that miraculous healing happens to patients who are brave enough to love. Siegel purports to demonstrate this.

As the book unfolds, you note that it is a strange world that Siegel inhabits. When operating on anesthetized patients, "I also do not hesitate to ask the [anesthetized] patient not to bleed if circumstances call for it" (p. 49), he asserts. In his world, deceased patients come back as birds (p. 222), there are unnamed countries in which individuals consistently live for a century (p. 140), and best of all, people who have the right spirituality not only successfully fight cancer, but can drive cars that consistently break down for other people (p. 137).

This is relatively benign gibberish, and history buffs may even feel comforted by those among us who live the belief system of medieval peasants. Where the problems become appallingly serious is when Siegel concentrates on the main point of his book. No matter how often he puts in disclaimers saying that he's not trying to make people feel guilty, the book's premise is that cancer (or any other disease) is curable if the patient has sufficient courage, love, and spirit; if the patient is not cured, it is because of insufficient amounts of those admirable traits. As we have just seen, this is not how cancer works, and a physician simply should not go about telling seriously ill people otherwise.

His book is full of descriptions of people who get cancer because of their uptightness and lack of spirituality. He speaks of one woman who was repressed in her feelings about her breasts: "*Naturally*, [my emphasis] Jan got breast cancer" (p. 85—this seems an indication that Siegel is aware of the literature on cancer-prone personality, but this represents rather a caricature of those mostly careful studies). Of another patient: "She held all her feelings inside and developed leukemia" (p. 164). Or, in an extraordinary statement: "Cancer generally seems to appear in response to loss. . . . I believe that, if a person avoids emotional growth at this time, the impulse behind it becomes misdirected into malignant physical growth" (p. 123).

Naturally, those who do have enough courage, love, and spirit can defeat cancer. Sometimes it takes a little prodding from Siegel. He advises on page 108 that people with serious diseases consider the ways in which they may have wanted their illness because we are trained to associate sickness with reward [Siegel cites our receiving cards and flowers (p. 110)]. Sometimes Siegel has to be a bit more forceful with a recalcitrant cancer patient. One woman was apparently inhibited about drawing something

Siegel requested her to, being embarrassed about her poor drawing skills. "I asked [her] how she expected to get over cancer if she didn't even have the courage to do a picture" (p. 81). You know whose fault it was if she eventually died.

But once the good patients overcome their attitude problems and get with the program, miracles just start popping up everywhere you look. One patient with the proper visualizing techniques cured his cancer, his arthritis, and, as long as he was at it, his 20-year problem with impotency as well (p. 153). Of another, he writes: "She chose the path of life, and as she grew, her cancer shrank" (p. 113). Consider the following exchange (p. 175):

> I came in, and he said, "Her cancer's gone."
>
> "Phyllis," I said, "Tell them what happened."
>
> She said, "Oh, you know what happened."
>
> "I know that I know," I said, "But I'd like the others to know."
>
> Phyllis replied, "I decided to live to be a hundred and leave my troubles to God."
>
> I really could end the book here, because this peace of mind can heal anything.

According to Siegel, cancer is curable with the right combination of attributes, and those people without them may get cancer and die of it. An incurable disease is the fault of the victim. He tries to soften his message now and then: "Cancer's complex causes aren't all in the mind," he says (p. 103), and on page 75 he tells us he's interested in a person gaining understanding of his or her role in a disease rather than in creating guilt. But when he gets past his anecdotes about individual patients and states his premise in its broadest terms, its poisonousness appears unmistakable: "The fundamental problem most patients face is an inability to love themselves" (p. 4); "I feel that all disease is ultimately related to a lack of love" (p. 180).

Siegel has a special place in his book for children with cancer and for the parents of those children trying to understand why it has occurred. After noting that developmental psychologists have learned that infants have considerably greater perceptual capacities than previously believed, Siegel says he "wouldn't be surprised if cancer in early childhood was linked to messages of parental conflict or disapproval perceived even in the womb"

(p. 75). In other words, if your child gets cancer, consider the possibility that you caused it.

And perhaps most directly: "There are no incurable diseases, only incurable people" (p. 99). (Compare the statement by the psychiatrist and stress researcher Herbert Weiner: "Diseases are mere abstractions; they cannot be understood without appreciating the person who is ill." Superficially, Weiner's and Siegel's notions bear some resemblance to each other. The former, however, is a scientifically sound statement of the interactions between diseases and individual makeups of sick people; the latter seems to me an unscientific distortion of those interactions.)

Since at least the Middle Ages, there has been a philosophical view of disease that is "lapsarian" in nature, characterizing illness as the punishment meted out by God for sin (all deriving from humankind's lapse in the Garden of Eden). Its adherents obviously predated any knowledge about germs, infection, the workings of the body. This view has mostly passed (although see the endnotes for an extraordinary example of this thinking that festered in the Reagan administration), but as you read through Siegel's book, you unconsciously wait for its reemergence, knowing that disease has to be more than just not having enough groovy New Age spirituality, that God is going to be yanked into Siegel's world of blame as well. Finally, it bubbles to the surface on page 179: "I suggest that patients think of illness not as God's will but as our deviation from God's will. To me it is the absence of spirituality that leads to difficulties." Cancer, thus, is what you get when you deviate from God's will, a view likely to be enormously offensive to many who are religiously inclined, to say nothing of those who have some understanding of actual disease processes.

Oh, and one other thing about Siegel's views. He runs a cancer program called Exceptional Cancer Patients, which incorporates his many ideas about the nature of life, spirit, and disease. To my knowledge there have been only two published studies of his program and its effects on survival time. Here is the conclusion of the first: "Preliminary findings suggest a strong beneficial effect of the program on survival, which is statistically significant. However, this observed effect is due largely to a selection bias caused by the failure to match on the duration of the lag period between cancer diagnosis and program entry.* Correcting for this bias in the analysis results in a small, nonsignificant program effect." In other words, by the rules of statistical analysis that biomedical science uses to tell the difference between a real effect and one due merely to a random blip in the data, the program has no effect on survivorship. The subsequent study also found no protective effect of his program. One last word from Siegel, on pages 185–186 of his book, washing his hands of the first study (the second had not yet been published): "I prefer to deal with individuals and effective techniques, and let others take care of the statistics."

Siegel is certainly not alone in this style of thinking; I analyze his ideas at length simply because he is such an effective and credentialled promulgator of these potentially damaging and ill-conceived ideas. Stress can influence immune resistance and the likelihood of getting certain diseases. It appears that, at least in the laboratory setting, stress can influence some aspects of tumor growth. However, these influences are simply not all that strong. It is bad enough to have cancer, without being led by some perversion of psychoneuroimmunology into thinking that it is your fault that you have it and that it is within your power to cure it.

This topic is one that I will return to in the final chapter of the book when I discuss stress management theories. Obviously, a theme of this book is just how many things can go wrong in the body because of stress and how important it is for everyone to recognize this. However, it would be utterly negligent to exaggerate the implications of this idea. Every child cannot grow up to be president; it turned out that merely by holding hands and singing folk songs we couldn't end all war, and hunger does not disappear just by visualizing a world without it. Everything bad in human health now is not caused by stress, nor is it in our power to cure ourselves of all our worst medical nightmares merely by reducing stress and thinking healthy thoughts full of courage and spirit and love. Would that it were so. And shame on those who would sell this view.

---

* This is a technical way of saying that the scientists who conducted the study compared patients in Siegel's program with a random collection of cancer patients as the control group and, for some unknown reason, Siegel's patients tend to join his program atypically early after their cancer diagnosis. Thus, even if Siegel's program has no special benefits for prolonging survival, his patients will seemingly have survived their cancer a longer time.

POSTSCRIPT: A GROTESQUE PIECE OF MEDICAL HISTORY

The notion that the mind can influence the immune system, that emotional distress can change resistance to certain diseases, is fascinating; psychoneuroimmunology exerts a powerful pull. Nevertheless, it sometimes amazes me just how many psychoneuroimmunologists are popping up. They are even beginning to speciate into subspecialties. Some study the issue only in humans, others in animals; some analyze epidemiological patterns in large populations, others study single cells. During breaks at scientific conferences, you can even get teams of psychoneuroimmunological pediatricians playing volleyball against the psychoneuroimmunological gerontologists. I am old enough, I admit frankly, to remember a time when there were no such things as psychoneuroimmunologists. Now, like an aging Cretaceous era dinosaur, I watch these new mammals proliferating. There was even a time when it was not common knowledge that stress caused immune tissues to shrink—and as a result, medical researchers carried out some influential studies and misinterpreted their findings, which indirectly led to the deaths of thousands of people.

By the nineteenth century, scientists and doctors were becoming concerned with a new pediatric disorder. On certain occasions parents would place their apparently perfectly healthy infant in bed, tuck the blankets in securely, leave for a peaceful night's sleep—and return in the morning to find the child dead. "Crib death," or sudden infant death syndrome (SIDS), came to be recognized during that time. When it happened, one initially had to explore the unsettling possibility that there was foul play or parental abuse, but that was usually eliminated, and one was left with the mystery of healthy infants dying in their sleep for no discernible reason.

Today, scientists have made some progress in understanding SIDS. It seems to arise in infants who, during the third trimester of fetal life, have some sort of crisis where their brains do not get enough oxygen, causing certain neurons in the brain stem that control respiration to become especially vulnerable. But in the nineteenth century, no one had a clue as to what was going on.

Some pathologists began a logical course of research in the 1800s. They would carefully autopsy SIDS infants and compare them with the normal infant autopsy material. Here is where the subtle, fatal mistake occurred. "Normal infant autopsy material." Who gets autopsied? Who gets practiced on by interns in teaching hospitals? Whose bodies wind up being dissected in gross anatomy by first-year medical students? Usually, it has been poor people.

The nineteenth century was the time when men with strong backs and a nocturnal bent could opt for a career as "resurrectionists"—grave robbers, body snatchers, who would sell corpses to anatomists at the medical schools for use in study and teaching. Overwhelmingly, the bodies of the poor, buried without coffins in shallow mass graves in potter's fields, were taken; the wealthy, by contrast, would be buried in triple coffins. As body-snatching anxiety spread, adaptations evolved for the wealthy. The "patent coffin" of 1818 was explicitly and expensively marketed to be resurrectionist-proof, and cemeteries of the gentry would offer a turn in the dead-house, where the well-guarded body could genteelly putrefy past the point of interest to the dissectors, at which time it could be safely buried. This period, moreover, gave rise to the verb "burking," after the aging resurrectionist who pioneered the practice of luring beggars in for a charitable meal and then strangling them for a quick sale to the anatomists. (Ironic-ending department: William Burke and his sidekick, after their execution, were handed over to the anatomists. Their dissection included particular attention to their skulls, with an attempt to find phrenological causes of their heinous crimes.)

All very helpful for the biomedical community, but with some drawbacks. The poor tended to express a riotous displeasure with the medico–body snatcher complex (to coin a phrase). Frenzied crowds lynched resurrectionists who were caught, attacked the homes of anatomists, burned hospitals. Concerned about the mayhem caused by unregulated preying on the bodies of the poor, governments moved decisively to supervise the preying. In the early nineteenth century, various European governments acted to supply adequate bodies to the anatomists, put the burkers and resurrectionists out of business, and keep the poor in line—all with one handy little law: anyone who died destitute in a poorhouse or a pauper's hospital would now be turned over to the dissectors.

Doctors were thus trained in what the normal human body looked like by studying the bodies and tissues of the poor. Yet the bodies of poor people are changed by the stressful circumstances of their poverty. In the "normal" six-month-old autopsy population, the infants had typically died of chronic diarrheal disorders, malnutrition, tuberculosis. Prolonged, stressful diseases. Their thymus glands had shrunk.

We now return to our pathologists, comparing the bodies of SIDS infants with those of "normal" dead infants. By definition, if children had been labeled as having died of SIDS, there was nothing else wrong with them. No prior stressors. No shrinking of the thymus gland. The researchers begin their studies and discover something striking: SIDS kids had thymuses much larger than those of "normal "dead infants. This is where they got things backward. Not knowing that stress shrinks the thymus gland, they assumed that the thymuses in the "normal" autopsy population were normal. They concluded that some children have an *abnormally* large thymus gland, and that SIDS is caused by that large thymus pressing down on the trachea and one night suffocating the child. And soon this imaginary disorder had a fancy name, "status thymicolymphaticus."*

This supposed biological explanation for SIDS provided a humane substitute for the usual explanation at the time, which was to assume that the parents were either criminal or incompetent, and some of the most progressive physicians of the time endorsed the "big thymus" story (including Rudolph Virchow, a hero of Chapter 15). The trouble was, the physicians decided to make some recommendations for how to *prevent* SIDS, based on this nonsense. It seemed perfectly logical at the time. Get rid of that big thymus. Maybe do it surgically, which turned out to be a bit tricky. Soon, the treatment of choice emerged: shrink the thymus through irradiation. Estimates are that in the ensuing decades it caused tens of thousands of cases of cancers in the thyroid gland, which sits near the thymus.

What recommendations does one offer from the history of status thymicolymphaticus? I could try for some big ones. That

---

* And fevered debate in the medical community raged as to whether SIDS was due merely to the (imaginary) large size of the thymus in status thymicolymphaticus, or whether there were (imaginary) constitutional abnormalities in the functioning of the thymus.

so long as all people are not born equal and certainly don't get to live equally, we should at least be dissected equally. How about something even more grandiose, such as that something should be done about infants getting small thymuses from economic inequality.

Okay, I'll aim for something on a more manageable scientific scale. For example, while we expend a great deal of effort trying to do extraordinary things in medical research—sequencing the human genome, transplanting neurons, building artificial organs—we still need smart people to study some of the moronically simple problems, like "how big is a normal thymus?" Because they are often not so simple. Maybe another lesson is that confounds can come from unexpected quarters—bands of very smart, very subtle public health researchers wrestle with that idea for a living. Perhaps the best moral is that when doing science (or perhaps when doing anything at all in a society as judgmental as our own), be very careful and very certain before pronouncing something to be the norm—because at that instant, you have made it supremely difficult to ever again look objectively at an exception to that supposed norm.

9

STRESS-INDUCED ANALGESIA

In Joseph Heller's classic novel about World War II, *Catch-22*, the antihero, Yossarian, is in bed with a woman with whom he has an unlikely argument about the nature of God. Unlikely, because they are both atheists, which would presumably lead to agreement about the subject. However, it turns out that while he merely does not believe in the existence of a God and is rather angry about the whole concept, the God that she does not believe in is one who is good and warm and loving, and thus she is offended by the vehemence of his attacks.

"How much reverence can you have for a Supreme Being who finds it necessary to include such phenomena as phlegm and tooth decay in His divine system of creation? What in the world was running through that warped, evil, scatological mind of His when He robbed old people of the power to control their bowel movements? Why in the world did He ever create pain?"

"Pain?" Lieutenant Scheisskopf's wife pounced upon the word victoriously. "Pain is a useful symptom. Pain is a warning to us of bodily dangers."

"And who created the dangers?" Yossarian demanded. He laughed caustically. "Oh, He was really being charitable to us when He gave us pain! Why couldn't He have used a doorbell instead to notify us, or one of his celestial choirs? Or a system of blue-and-red neon tubes right in the middle of each person's forehead. Any jukebox manufacturer worth his salt could have done that. Why couldn't He?"

"People would certainly look silly walking around with red neon tubes in the middle of their foreheads."

"They certainly look beautiful now writhing in agony or stupefied with morphine, don't they?"

Unfortunately, we lack neon lights in the middle of our foreheads, and in the absence of such innocuous signs, we probably do need pain perception. Pain can hurt like hell, but it can inform us that we are sitting too close to the fire, or that we should never again eat the novel item that just gave us food poisoning. It effectively discourages us from trying to walk on an injured limb that is better left immobilized until it heals. And in our westernized lives, it is often a good signal that we had better see a doctor before it is too late. People who congenitally lack the ability to feel pain (a condition known as *pain asymbolia*) are a mess; their feet may ulcerate, their knee joints may disintegrate, and their long bones may crack because they don't know how much force to step down with; they burn themselves unawares; in some cases, they've even lost a toe without knowing it.

Pain is useful to the extent that it motivates us to modify our behaviors in order to reduce whatever insult is causing the pain, because invariably that insult is damaging our tissues. Pain is useless and debilitating, however, when it is telling us that there is something dreadfully wrong that we can do nothing about. We must praise evolution or God for providing us with a physiological system that lets us know when our stomachs are empty. Yet at the same time we must deeply rue the same source for providing us with a physiological system that can wrack a terminal cancer patient with unrelenting pain.

Pain, until we get the lights on our foreheads, will remain a necessary but highly problematic part of our natural physiology. What is surprising is how malleable pain signals are—how readily the intensity of a pain signal is changed by the sensations, feelings, and thoughts that coincide with the pain. One example of this modulation, the blunting of pain perception during some circumstances of stress, is the subject of this chapter.

George Cruikshank, 1819: The Headache, *hand-colored etching.*

THE BASICS OF PAIN PERCEPTION

The sensation of pain originates in receptors located throughout our body. Some are deep within the body, telling us about muscle aches; tendon pulls; fluid-filled, swollen joints. Others, in our skin, can tell us that we have been cut, burned, abraded, poked. Often, these skin receptors respond to the signal of local tissue damage. Cut yourself with a paring knife, and you will slice open various cells of microscopic size that then spill out their proverbial guts; and typically, within this cellular soup now flooding out of the area of injury is a variety of chemical messengers that trigger pain receptors into action.

Some pain receptors carry information only about pain (for example, the ones responding to cuts); others carry information about both pain and everyday sensations. For example, by way of various tactile receptors on my back, I am greatly pleased to have my back scratched and rubbed by my wife. However, as evidence that there are limits to all good things, I would not at all enjoy it if she vigorously scratched with coarse sandpaper. In much the same way, we often are pleased to have our thermal receptors stimulated by warm sunlight, but are rarely so when scalded by boiling water. Sometimes pain consists of everyday sensations writ large.

Regardless of the particular type of pain and the particular receptor activated, all these receptors send nervous projections to the spinal cord. There they funnel their information to other specialized pain neurons, which in turn send the information to different areas of the brain.

Activate these pathways, for example, by stomping on a tack with your bare foot. One part of your cortex receives information that something painful has happened; another part figures out what part of the body has been insulted. While they are communicating about this, a far more rapid part of the system effects the reflexive withdrawal of your foot (this part is mostly based in the spinal cord rather than the brain, which allows the withdrawal reflex to be faster). Other regions of your brain are activating your autonomic nervous system and speeding up your heart, and another part signals the hypothalamus to start secreting CRF, which ultimately triggers glucocorticoid secretion from the adrenals.

MODULATION OF PAIN PERCEPTION

A striking aspect of the pain system is how readily it can be modulated by other factors. The strength of a pain signal, for example, can depend on what other sensory information is funneled to the spine at the same time. This, it turns out, is why it feels great to have a massage when you have sore muscles. Chronic, throbbing pain can be inhibited by certain types of sharp, brief sensory stimulation.

The physiology behind this is one of the most elegant bits of wiring that I know of in the nervous system, a circuit sorted out some decades ago by the pain physiologists Patrick Wall and Ronald Melzack. It turns out that the nervous projections—the fibers carrying pain information from your periphery to the spinal cord—are not all of one kind. Instead, they come in different classes. Probably the most relevant dichotomy is between fibers that carry information about acute, sharp, sudden pain and those that carry information about slow, diffuse, constant, throbbing pain. Both project to spinal cord neurons and activate them, but in different ways (see part A of the figure on page 164).

Two types of neurons found in the spinal cord are being affected by painful information (see part B of the illustration). The first ("X") is the same neuron diagrammed before, which relays pain information to the brain. The second neuron ("Y") is a local one called an *interneuron*. When Y is stimulated, it inhibits the activity of X.

As things are wired up, when a sharp, painful stimulus is felt, the information is sent on the fast fiber. This stimulates both neurons X and Y. As a result, X sends a painful signal up the spinal cord, and an instant later, Y kicks in and shuts X off. Thus the brain senses a brief, sharp burst of pain, such as after stepping on a tack.

By contrast, when a dull, throbbing pain is felt, the information is sent on the slow fiber. It communicates with both neurons X and Y, but differently from the way it does on the fast fiber. Once again the X neuron is stimulated and lets the brain know that something painful has occurred. This time, however, the slow fiber *inhibits* the Y neuron from firing. Y remains silent, X keeps firing, and your brain senses a slow, throbbing pain, such as after you've burned yourself.

Suppose that you have some sort of continuous, throbbing pain—sore muscles, an insect bite, a painful blister. How can you stop the throbbing? Briefly stimulate the fast fiber. This adds to the pain for an instant, but by stimulating the Y interneuron, you shut the system down for a while. And that is precisely what we often do in all of those circumstances. Experiencing a good vigorous mauling massage inhibits the dull throbbing pain of sore muscles for a while. An insect bite throbs and itches unbearably, and we often scratch hard right around it to dull the pain. Or we'll pinch ourselves. In all these cases, the slow chronic pain pathway is shut down for up to a few minutes.

This model has had important clinical implications. For one thing, it has allowed scientists to design treatments for people

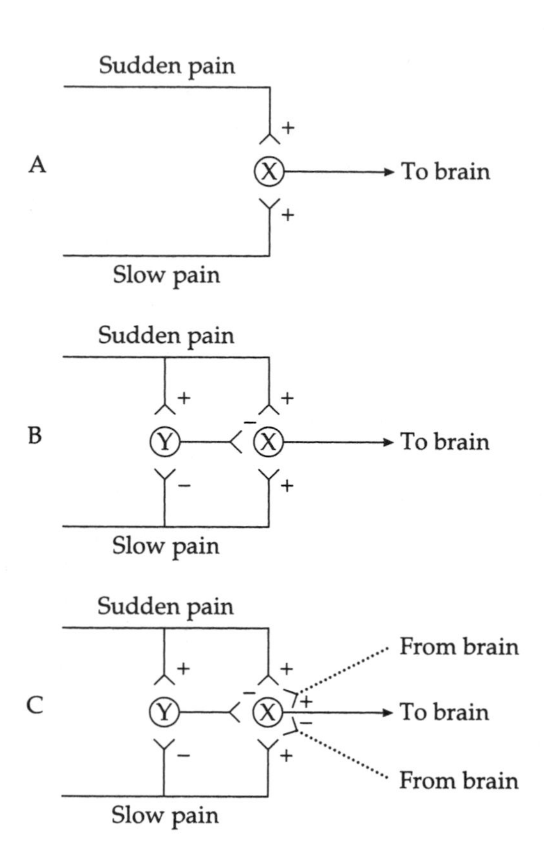

A schematic diagram of how pain information is passed to the brain, and how it can be modulated by the brain. (A) A neuron (X) in the spinal cord sends a signal to the brain that something painful has happened, once it is stimulated by a pain fiber. Such pain fibers can carry information about sudden pain or slow, diffuse pain. (B) A more realistic version of how the system actually works, showing why sudden and slow pain information is differentiated. In the case of sudden pain, the sudden pain fiber stimulates neuron X, causing a pain signal to be relayed to the brain. The sudden pain fiber also stimulates an interneuron (Y) that inhibits neuron X, after a brief delay. Thus, neuron X sends a pain signal to the brain for only a short time. In contrast, the slow pain fiber stimulates neuron X and inhibits interneuron Y. Thus, Y does not inhibit X, and X continues to send a pain signal to the brain, producing a slow, diffuse pain. (C) Both stimulatory and inhibitory fibers come from the brain and send information to neuron X, modulating its sensitivity to incoming pain information. Thus, the brain can sensitize neuron X to a painful signal, or blunt its sensitivity.

with severe chronic pain syndromes (for example, a patient who has had a nerve root crushed in his back). By implanting a little electrode into the fast pain pathway and attaching it to a stimulator on the person's hip, they enable the patient to buzz that pathway now and then to turn off the chronic pain; works wonders in many cases.

This dual pathway also explains a clinical feature of severe type 1 diabetes. In cases when energy delivery is disrupted for lack of insulin, nerves may become damaged. In general it is the fast fibers, which take more energy to operate than the lower-maintenance slow fibers, that are damaged. Thus, the person loses the ability to ever shut down the *Y* interneuron in that pathway, and what would be a mild chronic pain for anyone else becomes a constant throbbing one for a diabetic.

The Wall and Melzack model explains how sensory information can modulate how much pain you feel. The vagaries of your thoughts and feelings can influence pain sensitivity as well. At one extreme, people will have sex in all sorts of unlikely places, happily abrading their rear ends on all sorts of rough surfaces, and hardly notice a thing. Or consider that most paradoxically euphoric of pains, childbirth. At the other extreme, if you are skittish enough about dentists, you're going to decide it hurts the instant your teeth are so much as touched with a pick.

A study conducted in the 1980s provides a particularly striking example of the modulation of pain sensitivity by other sensory information. The scientist examined a decade's worth of records at a suburban hospital, noting how many painkillers were requested by patients who had just had gallbladder surgery; patients who had views of trees from their windows, he found, requested significantly less pain medication than those who looked out on blank walls. Other studies of chronic pain patients show that manipulating psychological variables such as the sense of control over events also dramatically changes the quantity of painkillers that they request. (This important finding will be elaborated upon in the final chapter of the book.)

Your mind's capacity to affect pain sensitivity is obviously going to be important in the effects of stress on pain perception. How can the brain create these effects? We complicate our wiring diagram shown on the facing page even further. Descending projections from the brain, terminating in the spinal cord, influence how sensitive neuron *X* is to a painful signal.

Look at a blank wall out of your hospital window, sit in the dentist's chair in terror of every move, and these descending pathways make neuron *X* very jumpy. Look at a stand of trees out your window, think about sex instead of the bristly carpet you're lying on, and your brain makes neuron *X* less responsive to an incoming pain signal.

STRESS-INDUCED ANALGESIA

Both "ascending" and "descending" influences modulate the workings of the pain pathway in the spine. The descending projections help mediate stress-induced analgesia, the phenomenon that, when you experience stress in just the right way, you don't feel much pain.

Chapter 1 recounted anecdotal cases of people who, highly aroused during battle, did not notice a severe injury. One of the first to note this phenomenon of stress-induced analgesia was an anesthesiologist, Henry Beecher, who examined injured soldiers as a battlefront medic in World War II and compared them with civilian populations. He found that for injuries of similar severity, approximately 80 percent of civilians requested morphine, while only a third of the soldiers did. He cited the French physician Dupuytren, who, more than a century earlier, noted the same pattern and pointed out that for the wounded soldier facing a medic in a field tent, the news of an injury is almost a relief—things could easily have been worse; there might have been no medic; at least I'm out of the battle now; and so on.

Few of us experience stress-induced analgesia in the midst of battle. For us, it is more likely to happen during some sporting event where, if we are sufficiently excited and involved in what we are doing, we can easily ignore an injury. On a more everyday level, stress-induced analgesia is experienced by the droves who exercise. Invariably the first stretch is agony, as you search for every possible excuse to stop before you suffer the coronary that you are now fearing. Then suddenly, about half an hour into this self-flagellation, the pain melts away. You even start feeling oddly euphoric. The whole venture seems like the most wonderful self-improvement conceivable, and you plan to work out like this daily until your hundredth birthday. (All vows, of course, are forgotten the next day when you start the painful process all over again.)

Traditionally many hard-nosed laboratory scientists, when encountering something like stress-induced analgesia, would relegate it to the "psychosomatic" realm, dismissing it as some fuzzy aspect of "mind over matter." The analgesia, however, is an extremely concrete phenomenon, as real as how your brain distinguishes between light and dark.

One bit of evidence for that assertion is that stress-induced analgesia occurs in other animals as well, not just in humans emotionally invested in the success of their nation's army or their office's softball team. This can be shown in animals with the "hot-plate test." Put a rat on a hot plate; then turn it on. Carefully time how long it takes for the rat to feel the first smidgen of discomfort, when it picks up its foot for the first time (at which point the rat is removed from the hot plate). Now do the same thing to a rat that has been stressed—forced to swim in a tank of water, put in a cage with a threatening rat, whatever. It will take longer for this rat to notice the heat of the plate: stress-induced analgesia.

The best evidence that such analgesia is a real phenomenon is the neurochemistry that has been discovered to underlie it. The tale begins in the 1970s, with the subject in which every ambitious, cutting-edge neurochemist of the time was interested—the various opiate drugs that were being used recreationally in vast numbers: heroin, morphine, opium. All those compounds have similar chemical structures and are made in certain plants in similar ways. In the early 1970s, three different groups of neurochemists almost simultaneously demonstrated that these opiate drugs bound to specific opiate receptors in the brain. And these receptors tended to be located in the parts of the brain that process pain perception. This turned out to solve the problem of how opiate drugs block pain—they activate those descending pathways that blunt the sensitivity of the *X* neuron shown in the illustration on page 164.

Terrific; but two beats later, something puzzling hits you. Why should the brain make receptors for a class of compounds synthesized in poppy plants? The realization rushes in; there must be some chemical—neurotransmitter? hormone?—made in the body that is structurally very similar to opiates. Some kind of endogenous morphine must occur naturally in the brain.

Neurochemists went wild at this point looking for endogenous morphine. It was a great chemistry problem in the abstract. It was a fascinating applied problem to find the brain's natural painkillers. And if one could modify these putative compounds to make synthetic versions that blocked pain without being addictive, there was a pot of gold to be earned.

In the ensuing years, competing teams of scientists found exactly what they were looking for: endogenous compounds with chemical structures reminiscent of the opiate drugs. They turned out to come in three different classes—enkephalins, dynorphins, and the most famous of them all, endorphins (a contraction for "endogenous morphines"). The opiate receptors were discovered to bind these endogenous opioid compounds, just as predicted. Furthermore, the opioids were synthesized and released in parts of the brain that regulated pain perception. (*Opiate* refers to analgesics not normally made by the body, such as heroin or morphine. *Opioid* refers to those made by the body itself. Because the field began with the study of the opiates—since no one had discovered the opioids as yet—the receptors found then were called *opiate receptors*. But clearly, their real job is to bind the opioids.)

Chapter 7 introduced the finding that the endorphins and enkephalins also regulate sex hormone release. An additional intriguing finding concerning opioid action emerged: release of these compounds explained how acupuncture worked. Until the 1970s, many western scientists had heard about the phenomenon, but most had written it off, dumping it into a bucket of anthropological oddities—inscrutable Chinese herbalists sticking needles into people, Haitian witch doctors killing with voodoo curses, Jewish mothers curing any and all diseases with their secret-recipe chicken soup. Then, right around the time of the explosion in opiate research, Nixon opened China to the west, and documentation started coming out about the reality of acupuncture. Furthermore, scientists noted that Chinese veterinarians used acupuncture to do surgery on animals, thereby refuting the argument that the painkilling characteristic of acupuncture was one big placebo effect ascribable to cultural conditioning (no cow on earth will go along with unanesthetized surgery just because it has a heavy investment in the cultural mores of the society in which it dwells). Then, as the corker, a prominent western journalist (James Reston of *The New York Times*) got appendicitis in China, underwent surgery using acupuncture as anesthesia, and survived just fine.

Acupuncture stimulates the release of large quantities of endogenous opioids, for reasons no one really understands. The best demonstration of this is what is called a *subtraction experiment*: using a synthetic drug such as naloxone, block the opiate receptors, thus canceling out the activity of any endogenous opioids that are released. When such a receptor is active, acupuncture no longer effectively dulls the perception of pain.

All of this is a prelude to the discovery that stress releases opioids as well. This finding was first reported in 1977 by Roger Guillemin. Fresh from winning the Nobel prize for the discoveries described in Chapter 2, he demonstrated that stress triggers the release of one type of endorphin, beta-endorphin, from the pituitary gland.

The rest is history. We all know about the famed runner's high that kicks in after about half an hour and creates that glowing, irrational euphoria as you edge closer to collapse, just because the pain has gone away. During exercise, beta-endorphin pours out of the pituitary gland, finally building up to levels in the bloodstream around the thirty-minute mark that will cause analgesia. The other opiates, especially the enkephalins, are mobilized as well, mostly within the brain and spine. They activate the descending pathway originating in the brain to shut off the *X* neurons in the spinal cord, and they work directly at the spinal cord to accomplish the same thing. All sorts of other stressors produce similar effects. Surgery, low blood sugar, exposure to cold, examinations, spinal taps, childbirth—all do it (although it turns out that certain stressors cause analgesia through "nonopioid-mediated" pathways. No one is quite sure how those work, nor whether there is a systematic pattern as to which stressors are opioid-mediated).

Wonderful, very useful. Exactly the sort of stress-response needed to keep that zebra functioning when it is injured yet still must get away from the lion. To follow the structure laid out in previous chapters, this represents the good news. So what's the bad news? How does an excess of opioid release make us sick in

Vic Boff, New York Polar Bear Club member known as "Mr. Iceberg," sitting in the snow after a swim during the blizzard of 1978.

the face of the chronic psychological stressors that we specialize in? Does chronic stress make you an endogenous opioid addict? Does it cause so much of the stuff to be released that you can't detect useful pain anymore? What's the downside in the face of chronic stress?

Here the answer is puzzling because it differs from all the other physiological systems examined in this book. When Hans Selye first began to note that chronic stress causes illness, he thought that illness occurs because an organism runs out of the stress-response—that the various hormones and neurotransmitters are depleted, and the organism is left vulnerable to the pummelings of the stressor, undefended. As we've seen in previous chapters, the modern answer is that the stress-response doesn't become depleted; instead, one gets sick because the stress-response itself eventually becomes damaging.

Opioids turn out to be the exception to the rule. Stress-induced analgesia does not go on forever, and the best evidence ascribes this to diminishing secretion of the particular opioids that are effective at blocking pain perception. You are not permanently out of business, but it takes a while for supply to catch up with demand.

To my knowledge, there is no stress-related disease that results from too much opioid release during sustained stressors. From the standpoint of this book and our propensity toward chronic psychological stressors, that is good news—one less stress-related disease to worry about. From the standpoint of pain perception and the world of real physical stressors, the eventual depletion of the opioids means that the soothing effects of stress-induced analgesia are just a short-term fix. And for the elderly woman agonizing through terminal cancer, the soldier badly injured in combat, the zebra ripped to shreds but still alive, the consequence is obvious. The pain will soon return.

STRESS AND MEMORY

I'm old now, very old. I seen a lot of things in my time and by now, I've forgotten a lot of them but, I tell you, that was one day that I'll remember forever like it was yesterday.

I was 24, maybe 25. It was a cold spring morning. Raw, gray. Gray sky, gray slush, gray people. I was looking for a job again and not having much luck, my stomach complaining about the bad rooming house coffee that was last night's dinner and today's breakfast. I was feeling pretty hungry, and I suspect I was starting to look pretty hungry too, like some half-starved animal that picks through a garbage can, and that couldn't make much of an impression in an interview. And neither could the shabby jacket I wearing, that last one I hadn't hocked.

I was plodding along, lost in my thoughts, when some guy comes sprinting around the corner, yelling with excitement, hands up in the air. Before I could even get a good look at him, he was shouting in my face. He was babbling, yelling about something being "classic," something called "classic." I couldn't understand what he was talking about, and then he sprinted off. What the hell, crazy guy, I thought.

But around the next corner, I see more people running around, yelling. Two of them, a man and woman, come running up to me and, by now, I tell you, I knew that something was up. They grabbed me by the arms, shout-

> ing "We won! We won!! We're getting it back!" They were pretty excited but at least making more sense than the first guy, and I finally figured out what they were saying. I couldn't believe it. I tried to speak, but I got all choked up, so I hugged them like they were my brother and sister. The three of us ran into the street, where a big crowd was forming—people coming out of the office buildings, people stopping their cars, jumping out. Everyone screaming and crying and laughing, people shouting, "We Won!, We Won!" Somebody told me a pregnant woman had gone right into labor, another that some old man had fainted right away. I saw a bunch of Navy guys, and one of them stepped right up and kissed this woman, a total stranger, leaning her way back—someone snapped a picture of them kissing, and I heard it became famous afterward.
>
> And the weird thing is how long ago this was—the couple who first told me are probably long gone, but I can still see their faces, remember how they were dressed, the smell of the guys aftershave. I can still remember half the faces in that crowd, the colors of the buildings, the feel of the breeze that was blowing the confetti that people were tossing out the windows above. Still vivid. The mind's a funny thing. Well anyway, like I was saying, that's a day I'll always remember, the day they brought back the original Coke.

We've all had similar experiences. Your first kiss. Your wedding ceremony. The moment when the war ended. And the same for the bad moments as well. The fifteen seconds when those two guys mugged you. The time the car spun out of control and just missed the oncoming truck. Where you were when the earthquake hit, when Kennedy was shot, when *Challenger* exploded. All etched forever in your mind, when it's inconceivable that you can recall the slightest thing about incidents in the twenty-four hours before that life-changing event. Arousing, exciting, momentous occasions, including stressful ones, get filed away very readily. Stress can enhance memory.

At the same time, we've all had the opposite experience. You're in the middle of the final exam, exhausted and frazzled, and you simply can't remember a fact that would come effortlessly at any other time. You're in some intimidating social

A day to remember!

circumstance, and, of course, at the critical moment, you can't remember the name of the person you have to introduce. The first time I was "brought home" to meet my future wife's family, I was nervous as hell; during a word game after dinner, I managed to blow the lead of the team consisting of my future mother-in-law and me by my utter inability at one critical juncture to remember the word *casserole*. And some of these instances of failed memory revolve around infinitely greater traumas—the combat vet who went through some unspeakable battle catastrophe, the survivor of childhood sexual abuse—for whom the details are lost in an amnestic fog. Stress can disrupt memory.

By now, this dichotomy should seem quite familiar. Stress enhances some physiological process; stress disrupts the same—think timecourse, think thirty-second sprints across the savanna versus decades of grinding worry. Short-term stressors of mild to moderate severity enhance cognition, while enormous or very prolonged ones are disruptive to it. And the latter finding has recently been shown to have some very disturbing implications for what chronic stress might do to our own memories.

A PRIMER ON HOW MEMORY WORKS

In order to appreciate how stress affects memory, we need to know something about how memories are formed (consolidated), how they are retrieved, how they can fail.

The first key point is that memory is not monolithic but made up of different types. One particularly important dichotomy distinguishes short-term versus long-term memories. With the former, you look up a phone number, sprint across the room convinced you're about to forget it, punch in the number. and then it's gone forever. Short-term memory is your brain's equivalent of juggling some balls in the air for thirty seconds. In contrast, long-term memory refers to remembering what you had for dinner last night, the name of the U.S. president, how many grandchildren you have, where you went to college. Neuropsychologists are coming to recognize that there is a specialized subset of long-term memory. Remote memories are ones stretching back to your childhood—the name of your village, your native language, the smell of your grandmother's baking. They appear to be stored in some sort of archival way in your brain separate from more recent long-term memories. Often, in patients with a dementia that devastates most long-term memory, the more remote facets can remain intact.

Another important distinction in memory is that between *explicit* (also known as *declarative*) and *implicit* (previously called *procedural*) memory. Explicit memory concerns distinct facts and your conscious awareness of knowing them: I am a mammal, today is Monday, my dentist has particularly thick eyebrows. Things like that. In contrast, implicit memories are about knowing how to *do* things, even without having to think consciously about them: shifting the gears on a car, riding a bicycle, doing the fox trot. Memories can be transferred between explicit and implicit forms of storage. For example, you are learning a new, difficult passage from a piece of piano music. Each time that stretch approaches, you must consciously, explicitly remember what to do—tuck your elbow in, bring your thumb way underneath after that trill. And one day, while playing, you realize you just barreled through that section flawlessly, without having to think about it: you did it with implicit, rather than explicit, memory. And memory can be dramatically disrupted if you intermix implicit and explicit memories. Here's an example that will finally make reading this book worth your while—how to make neurobiology work to your competitive advantage at sports. You're playing tennis against someone who is beating the pants off of you. Wait until your adversary has pulled off some amazing backhand, then offer a warm smile and say, "You are a fabulous tennis player. I mean it; you're terrific. Look at that shot you just made. *How* did you do that? When you do a backhand like that, do you hold your thumb this way or that, and what about your other fingers? And how about your butt, do you scrunch up the left side of it and put your weight on your right toes, or the other way around?" Do it right, and the next time that shot is called for, your opponent (victim) will make the mistake of thinking about it explicitly, and the stroke wont be anywhere near as effective.

Just as there are different types of memory, there are different areas of the brain involved in memory storage and retrieval. One critical site is the cortex, the vast and convoluted surface of the brain. Another is a region tucked just underneath part of the cortex, called the *hippocampus*. (That's Latin for "sea horse," which the hippocampus vaguely resembles if you've been stuck inside studying neuroanatomy for too long instead of going to the seashore. It actually looks more like a jelly roll, but who knows the Latin term for that?) Both of these are regions vital to memory—for example, it is the hippocampus and cortex that are preferentially damaged in Alzheimer's disease. If you want a totally simplistic computer metaphor, think of the cortex as your hard drive, where memories are stored, and your hippocampus as the keyboard, the means by which you place and access memories in the cortex.

There are additional brain regions relevant to a different kind of memory. These structures are diffusely scattered throughout the brain, and are traditionally linked to regulating body movements. What do these sites—such as the cerebellum—have to do with memory? They appear to be relevant to implicit memory, the memory you need to perform reflexive, motoric actions without even consciously thinking about them, where, so to speak, your body remembers how to do something before you do.

The distinction between explicit and implicit memory, and the neuroanatomical bases of that distinction, was first really appreciated because of one of the truly fascinating, tragic figures in neurology, perhaps the most famous neurological patient of all time. This man, known in the literature only by his initials, is missing most of his hippocampus. As an adolescent in the 1950s, "H.M." had a severe form of epilepsy that was centered in his hippocampus, and was resistant to drug treatments available at that time. In a desperate move, a famous neurosurgeon removed a large part of H.M.'s hippocampus, along with much of the surrounding tissue. The seizures mostly abated, and in the aftermath, H.M. was left with a virtually complete inability to turn new short-term memories into long-term ones—mentally utterly frozen in time. Zillions of studies of H.M. have been carried out since, and it has slowly become apparent that despite this profound amnesia, H.M. can still learn how to do some things. Give him some mechanical puzzle to master day after day, and he learns to put it together at the same speed as anyone else, while steadfastly denying each time that he has ever seen it before. Hippocampus and explicit memory are shot; the diffuse motor system and implicit memory remain intact.

This shifts us to the next magnification of examining how the brain handles memories and how stress influences the process—what's going on at the level of clusters of neurons. A longstanding belief among many who studied the cortex was that each individual cortical neuron would, in effect, turn out to have a single task, a single fact that it knew. This was prompted by some staggeringly important work done in the 1960s by a pair named David Hubel and Torstein Wiesel on what was, in retrospect, one of the simpler outposts of the cortex, an area that processed visual information. They found many neurons there that did indeed have specialized tasks—responding to a particular dot of light, or to a particular line of light moving in a certain direction and with a certain orientation. This led people to believe that other parts of the cortex that had even fancier tasks comprised what came to be called "Grandmother" neurons (i.e., somewhere, way up in the most complex and integrative parts of the cortex, there would be one neuron whose job it was to recognize your grandmother's face, and next to it another neuron that recognized your other grandmother, and next to that, a grandfather neuron . . .). With time, it became apparent that there could be very few such neurons in the cortex, in large part because you simply don't have enough cells to go around for such narrow specialization.

Rather than memory and information being stored in single neurons, they are stored in the patterns of excitation of vast arrays of neurons—in trendy jargon, neuronal "networks." How does one of these work? Consider the wildly simplified neural network shown in the accompanying diagram.

The first layer of neurons (neurons 1, 2, and 3) are classical Hubel and Wiesel type neurons, which is to say that each one "knows" one fact for a living. Neuron 1 knows how to recognize Gauguin paintings, 2 recognizes van Gogh, and 3 knows Monet. (Thus, these hypothetical neurons are more "grandmotherly"—specializing in one task—than any real neurons in the brain, but help illustrate well what neural networks are about.) Those three neurons project—send information to—the second layer in this network, comprising neurons A–E. Note the projection pattern: 1 talks to A, B, and C; 2 talks to B, C, and D; 3 talks to C, D, and E.

What "knowledge" does neuron A have? It gets information only from neuron 1 about Gauguin paintings. Similarly, E gets information only from neuron 3 and knows only about Monet. But what about neuron C; what does it know about? It knows about *Impressionism,* the features that these three painters had in common. It has knowledge that does not come from any single informational input, but emerges from the convergence of information feeding into it. Most neurons in your cortex process memory the way C does.

We take advantage of such convergent networks whenever we are trying to pull out a memory that is almost, almost there. Continuing our art history theme, suppose you're trying to remember the name of a painter, that guy, what's his name. He

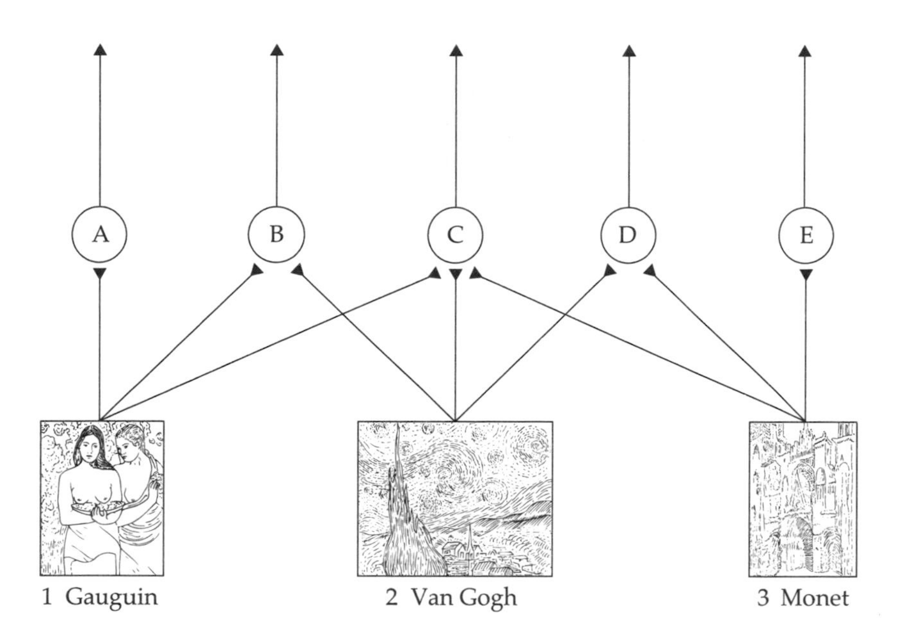

A highly hypothetical neural network involving a neuron that "knows" about Impressionist paintings.

was that short guy with a beard (activating your "short guy" neural network, and your "bearded guy" network). He painted all those Parisian dancers; it wasn't Degas (two more networks pulled in). My high school art appreciation teacher loved that guy; if I can remember her name, I bet I can remember his . . . wow, remember that time I was at the museum and there was that really cute person I tried to talk to in front of one of his paintings . . . oh, what was the stupid pun about that guy's name, about the train tracks being too loose. And with enough of those nets being activated, you finally stumble into the one fact that is at the intersection of all of them: Toulouse-Lautrec, the equivalent of a neuron C.

That's a very rough approximation of how a neural network operates, and neuroscientists have come to think of both learning and storing of memories as involving the "strengthening" of some branches rather than others of a network. How does such strengthening occur? For that, we switch to a final level of magnification, to consider the tiny gaps between the thready branches of two neurons, gaps called *synapses*. When a neuron has heard some fabulous gossip and wants to pass it on, when a wave of electrical excitation sweeps over it, this triggers the release of chemical messengers—neurotransmitters—that float across the synapse and excite the next neuron. There are dozens, probably hundreds, of different kinds of neurotransmitters, and synapses in the hippocampus and cortex disproportionately make use of what is probably the most excitatory neurotransmitter there is, something called *glutamate*.

Besides being superexcitatory, "glutamatergic" synapses have two properties that are critical to memory. The first is that these synapses are nonlinear in their function. What does this mean? In a run-of-the-mill synapse, a little bit of neurotransmitter comes out of the first neuron and causes the second neuron to get a little excited; if a smidgen more neurotransmitter is released, there is a smidgen more excitation, and so on. In glutamatergic synapses, some glutamate is released and nothing happens. A larger amount is released, nothing happens. It isn't until a certain threshold of glutamate concentration is passed that, suddenly, all hell breaks loose in the second neuron and there is a massive wave of excitation. And this is what learning something is about. A professor drones on incomprehensibly in a lecture, a fact goes in one ear and out the other. It is repeated again—and, again, it fails to sink in. Finally, the hundredth time it is repeated, a light bulb goes on, "Aha," and you get it. On a simplistic level, when you finally get it, that nonlinear threshold of glutamate excitation has just been reached.

The second feature is even more important. Under the right conditions, when a synapse has just had a sufficient number of superexcitatory glutamate-driven "aha's," something happens—the synapse becomes persistently more excitable, so that next time, it takes less of an excitatory signal to get the aha. That synapse just learned something; it was "potentiated," or strengthened. And the most amazing thing is that this strengthening of the synapse can persist for a long time. How this process of "long-term potentiation" works is one of the hottest topics in neurobiology, as is the opposite case, that of "long-term depression," and the two are thought to be the cellular underpinnings of learning and forgetting, respectively.

That is all you need to know about how your brain remembers anniversaries and sports statistics and the color of someone's eyes and how to waltz. We can now see how stress makes a mess of it.

IMPROVING YOUR MEMORY DURING STRESS

The first point, of course, is that during short-term stressors that are not too severe, memory is enhanced rather than disrupted. This is relatively easy to show with a laboratory rat, where you can generate an artificial stressor and show that some task associated with it is learned better. It's much tougher to show this with a human, since it's hard to time earthquakes to occur right in the middle of some learning test. However, James McGaugh and colleagues at the University of California at Irvine have shown the phenomenon in people with an elegant study. Read a fairly unexciting story to a group of control subjects—a boy and his mother walk through their town, pass this store and that one, cross the street and enter the hospital where the boy's father works, are shown the x-ray room . . . and so on. Meanwhile, the experimentals are read a story that differs in that the central core of it contains some emotionally laden material—a boy and his mother walk through their town, pass this store and that one, cross the street where . . . the boy is hit by a car! He's rushed to the hospital and taken to the x-ray room. . . . Tested later, the experimental subjects remember their story better than do the controls, but only the middle, exciting part.

That study also indicated how this effect on memory works. Hear the stressful story and a stress-response is initiated. As we by now well know, this includes the sympathetic nervous systems kicking into gear, pouring epinephrine and norepinephrine into the bloodstream. Sympathetic stimulation appears to be critical, because when McGaugh gave subjects a drug to block that sympathetic activation (the beta-blocker propranolol, the same drug used to lower blood pressure), the experimental group did not remember the middle portion of their story any better than the controls remembered theirs. Importantly, it's not that propranolol disrupts memory formation; rather, it disrupts stress-enhanced memory formation (in other words, the experimental subjects did as well as the controls on the boring parts of the story, but simply didn't have the boost in memory for the emotional middle section).

The explanation is probably straight out of Chapters 3 and 4. During sympathetic arousal, you're increasing cardiac tone, shutting down blood flow to unessential organs, dumping glucose into the bloodstream. The net result is that along with increased delivery of oxygen and glucose to your exercising muscle—the idea we've come back to repeatedly—there's increased delivery to your brain as well. And that turns out to be very important, because all that explosive, nonlinear, long-term potentiating, that turning on of light bulbs in your hippocampus with glutamate, doesn't come cheap. It's an amazing thing to see a PET scan (that imaging technique that shows how much energy is being used in different parts of the brain during some circumstance) and see this very tangible demonstration that thought takes energy, huge amounts of it—your brain is about 3 percent of your body's weight and hogs about 20 percent of its energy. We all know a nonsubtle version of this, namely, that you don't do your best thinking when you are about to pass out from hunger. Paul Gold at the University of Virginia has shown that far milder versions of hypoglycemia can disrupt memory formation and retrieval as well. And a stressor is a very good time to be at your best in memory retrieval ("How did I get out of this mess last time?") and memory formation ("If I survive this, I'd better remember just what I did wrong so I don't get into a mess like this again"). So stress acutely causes increased delivery of glucose to the brain, making more energy available to neurons, and thus better memory formation and retrieval.

So the sympathetic arousal during stress indirectly fuels the expensive process of remembering the faces of the crowd shouting ecstatically about Coke Classic. In addition, a mild elevation in glucocorticoid levels (the type you would see during a moderate, short-term stressor) helps memory as well. This appears to be a local effect, right in the hippocampus itself (rather than secondarily to changing business down in your liver or fat cells). The hippocampus turns out to be exquisitely sensitive to glucocorticoids, having more receptors for them than any other

part of the brain. And this makes sense—your hippocampus must be sensitive to either glucocorticoids or to epinephrine and norepinephrine, if it makes its memory-filing decisions on the basis of whether information coming in is emotionally laden. You can show the direct effects of glucocorticoids on hippocampal function by removing the hippocampus of a rat, taking a slice from it and sticking it in a petri dish, where, remarkably, you can artificially induce long-term potentiation in individual synapses (what would be known in the jargon of the business as an "in vitro" equivalent of learning). Throw a little bit of glucocorticoids into the soup that bathes that hippocampal slice, and it is now easier to induce that potentiation. Finally, there are some obscure mechanisms by which moderate, short-term stress makes your sensory receptors more sensitive. Your taste buds, your olfactory receptors, the cochlear cells in your ears all require less stimulation under moderate stress to get excited and pass on the information to your brain. In that special circumstance, you can pick up the sound of a can of soda being opened hundreds of yards away.

AND WHEN STRESS GOES ON FOR TOO LONG

With our "sprinting across the savanna"–"worrying about a mortgage" dichotomy loaded and ready, we can now look at how remembrance goes awry when stressors become too big or too prolonged. Once glucocorticoid levels go from the range seen for mild to moderate stressors to the range typical of big-time stress, the hormone no longer enhances long-term potentiation in that hippocampal slice but begins to disrupt it. For reasons no one really understands, enormous amounts of, or very sustained, sympathetic arousal disrupts that potentiation as well. Furthermore, major stressors appear to enhance the dark side of hippocampal "aha-ing," in a process called *long-term depression*, so that you may start forgetting things more readily, as well.

You also develop problems at the energy end of memory. As noted, within a couple of seconds of the onset of stress, you are enhancing the delivery of glucose to the brain as a whole. By approximately thirty minutes of a sustained stressor, such "glucose utilization" has gone back to baseline. And with the stressor grinding on even longer, you begin *inhibiting* glucose delivery, particularly in the hippocampus. This delayed effect is due to the high glucocorticoid levels that have now accumulated. Hippocampal neurons are taking up about 25 percent less glucose, as are hippocampal "glia," a type of cell that does a lot of behind-the-scenes support for neuron function.

The net result of these various actions is that your memory and concentration are anything but ideal under circumstances of high or prolonged stress. Thus, if you're going to get mildly anxious about a final exam, it is ideal to get anxious starting

Hard-charging businessman Billy Sloan is about to learn that continued stress does inhibit one's memory

about 15 seconds before it begins, rather than just as you wake up at the end of the final lecture of the course.

An obvious question: over and over I've emphasized how important it is during stress to cut down energy delivery to unessential outposts in your body, diverting it instead to exercising muscle. In the previous section, we added your hippocampus to that list of places that are spoon-fed energy with the onset of a stressor. It seems like that would be a clever area to continue to stoke, as the stressor goes on. Why should glucose delivery eventually be inhibited there? Probably because, as time goes by, you are running more on automatic, relying more on the implicit memory outposts in the brain to do things that are reflexive and motoric—the martial arts display you put on to disarm the terrorist or, at least, the coordinated swinging of the softball bat at the company picnic that you've been nervous about. However, to my knowledge, no one has shown that the progressive inhibition of glucose delivery to the hippocampus translates into more glucose delivered to those other sites.

THE PLOT THICKENS

Neuronal Damage Blunting long-term potentiation and starving your hippocampal neurons just when you're trying to come up with a key question on *Jeopardy* is bad enough. With even more sustained stress, something truly insidious begins to happen. Those glucocorticoids actually begin to damage neurons.

If you look back at the diagram about the Impressionism neuron, you'll see that there are arrows indicating how one neuron talks to another, "projects" to it. As mentioned a few paragraphs after that, those projections are quite literal—long multibranched cables coming out of neurons that form synapses with the multibranched cables of other neurons. These cables (known as *axons* and *dendrites*) are obviously at the heart of neuronal communication and neuronal networks. A few years ago, Bruce McEwen and his colleagues at Rockefeller University showed that, in a rat, after as little as a few weeks of stress or of exposure to excessive glucocorticoids, those cables begin to shrivel, to atrophy and retract a bit. They have since shown that the same can occur in the primate brain. And when that happens, synaptic connections get pulled apart and the complexity of your neural networks declines. Fortunately, it appears that at the end of the stressful period, the neurons can dust themselves off and regrow those connections.

This transient atrophy of neuronal connections probably explains a characteristic feature of memory problems during chronic stress. Destroy vast acres of neurons in the hippocampus after a massive stroke or late terminal stage Alzheimer's disease, and memory is profoundly impaired. Memories can be completely lost, and never again will these people remember, for example, something as vital as the names of their spouses. "Weaken" a neural network during a period of chronic stress by retracting some of the complex branches in those neuronal trees, and the memories of Toulouse-Lautrec's name are still there. You simply have to tap into more and more associative cues to pull it out, because any given network is less effective at doing its job. Memories are not lost, just harder to access.

This is certainly bad news for rats, but what about us? Some recent work suggests that this effect applies to the human as

Neurons of the hippocampus of a rat. On the left, healthy neurons, whereas on the right, neurons with their projects atrophied by sustained stress.

well as the primate hippocampus. The former finding concerns a set of diseases called Cushing's syndrome, in which, owing to any of a number of different types of tumors, people secrete tons of glucocorticoids. Understand what goes wrong next in Cushings syndrome and you understand half of this book—the victims develop high blood pressure, diabetes, immune suppression, reproductive problems, the works. And, it's been known for decades that they get memory problems, specifically explicit memory problems. It's known as *Cushingoid dementia.* A few years ago, Monica Starkman and colleagues at the University of Michigan did some very sensitive MRI brain scans on Cushings patients and found that their hippocampi had shrunk. As an important control, the size of the rest of the brain was normal. Only the hippocampus, and the more severe the glucocorticoid excess in the person, the more severe their memory problems and hippocampal atrophy. And, after the person's tumor had been surgically removed and glucocorticoid levels coasted back down to normal, the hippocampal atrophy normalized as well. No one has shown yet that those neuronal cables are actually shriveling in Cushing's patients, as has been shown for rats and primates exposed to high levels of glucocorticoids, but that is the most plausible model available at the moment. And given that the Michigan study is only a single one concerning people with massive levels of hormones, far higher than the body normally ever generates on its own, there certainly is no reason yet to think that simply a lot of traffic jams turns your hippocampus into a desiccated little sea horse. No evidence yet.

Neuroendangerment Which brings up the next thing to worry about. Not only will relatively short periods of stress or glucocorticoid exposure atrophy hippocampal neurons, they will also compromise the ability of those neurons to survive neurological diseases. This is a topic I've studied in my laboratory for the past dozen years, and it revolves around that ability of glucocorticoids to inhibit glucose storage in those neurons. It's not a huge effect—glucocorticoids inhibit about 80 percent of glucose storage in a fat cell but, as noted before, only about 25 percent in a hippocampal neuron. This certainly isn't enough to kill the neuron, but it's enough to make the neuron queasy, a little nauseous and light-headed. And if it happens that along comes some major challenge to the neuron, that neuron is more likely than normal to keel over.

Such glucocorticoid "neuroendangerment"—the hormones are not killing neurons, but are making it harder for them to survive some coincident neurological insult—can be pretty far-reaching. Take a rat, give it a rat's equivalent of a major epileptic seizure, and the higher the glucocorticoid levels at the time of the seizure, the more hippocampal neurons will ultimately die. Same thing for cardiac arrest, where oxygen and glucose delivery to the brain is cut off, or for a stroke, in which a single blood vessel in the brain shuts down. Same for concussive head trauma, drugs that generate oxygen radicals, and a bunch of other neurotoxins as well. Very disturbingly, same thing for the closest there is to a rat neuron's equivalent of being damaged by Alzheimer's disease (exposing the neuron to fragments of an Alzheimer's-related toxin called *beta-amyloid*). Same for a rat hippocampus's equivalent of having AIDS-related dementia (induced by exposing the neuron to a damaging constituent of the

AIDS virus called *gp120*). You can even grow hippocampal neurons in a dish, subject them to similar insults (take the sugar out of their petri-dish soup to mimic hypoglycemia, for example; put the dish in a chamber without oxygen to mimic the oxygen loss that follows cardiac arrest, and so on), and the same thing occurs—the more glucocorticoids, the more dead neurons.*

My lab and others have shown that the relatively mild energy problem caused by that inhibition of glucose storage by glucocorticoids or stress makes it harder for a neuron to contain the eleventy things that go wrong during one of these neurological insults. All these neurological diseases are ultimately energy crises for a neuron: cut off the glucose to a neuron (hypoglycemia), or cut off both the glucose and oxygen (hypoxia-ischemia), or make a neuron work like mad (a seizure), and energy stores drop precipitously. Damaging tidal waves of neurotransmitters and ions flood into the wrong places, oxygen radicals are generated. And if you throw in glucocorticoids on top of that, the neuron is even less able to afford to clean up the mess. Thanks to that stroke or seizure, today's the worst day of that neuron's life, and it goes into the crisis with 25 percent less energy in the bank than usual.

All this suggests that stress may not only exacerbate digestive disorders, cardiovascular disease, and all the rest, but also might very significantly worsen neurological disease as well. Any evidence in humans or primates yet? Nothing but the slightest hint—among humans who have had a stroke, the higher the glucocorticoid levels at the time the person comes into the ER, the more neurological damage and memory loss there ultimately is. Stay tuned for ongoing research in the coming years, because if glucocorticoids do worsen neurological damage in

---

*Chapter 3 described how stress can indirectly give rise to a stroke or cardiac arrest. But for the other neurological problems noted—seizure, head trauma, AIDS-related dementia, and most importantly, Alzheimer's disease—there is no evidence that stress or glucocorticoids *cause* these maladies. Instead, the possibility is that they worsen preexisting cases. This is certainly bad news, and is particularly disturbing for Alzheimer's. Thus, it is important to emphasize that I just noted that glucocorticoids worsen one component of the Alzheimer's problem (having to do with amyloid toxin). In contrast, Alzheimer's appears also to involve some damaging inflammation in the brain, and glucocorticoids are likely to protect against that component. So at this point, the jury is out as to whether glucocorticoids actually worsen the progression of that nightmarish disease.

the human hippocampus, there are some pretty disturbing clinical implications.

Implications for Glucocorticoid Therapy The first concerns the use by neurologists of synthetic versions of glucocorticoids (such as dexamethasone or prednisone) after someone has had a stroke. As we know from Chapter 2, glucocorticoids are classic anti-inflammatory compounds, and are used to reduce the edema, the damaging brain swelling, that often occurs after a stroke. Glucocorticoids do wonders to block the edema that occurs after something like a brain tumor, but it turns out that they don't do much for post-stroke edema, mainly because it's a subtype of brain swelling that happens to be resistant to the action of these steroids. Yet tons of neurologists still use the stuff, despite decades-old warnings of the best people in the field and the finding that the glucocorticoids tend to worsen the neurological outcome. So these recent findings add a voice to that caution—clinical use of glucocorticoids tends to be bad news for neurological diseases that involve a precarious hippocampus. (As a caveat, however, it turns out that huge doses of glucocorticoids appear to be very helpful in reducing damage after a spinal cord injury, for reasons having nothing to do with stress or with much of this book.)

Related to this is the concern that physicians may use synthetic glucocorticoids in other instances where these doses might endanger the hippocampus. A scenario that particularly disturbs me concerns the ability of these hormones to worsen gp120 damage to neurons, and its relevance to AIDS-related dementia. (Remember? The gp120 protein is found in the AIDS virus and appears to play a central role in damaging neurons and causing the dementia). At this point, all that is known is that this damage occurs in neurons in a petri dish—it's not even clear yet if this happens in a rat's brain (although those studies are in progress). But if, many experiments down the line, it turns out that glucocorticoids can worsen the cognitive consequences of HIV infection, this will be very worrisome. That won't be the case just because people with AIDS are under stress. it's also because people with AIDS are often treated with extremely high concentrations of synthetic glucocorticoids to combat other aspects of the disease.

An even more disturbing implication of glucocorticoid-mediated neuroendangerment is that if glucocorticoids turn out to endanger the human hippocampus (making it harder for neurons to survive an insult), you don't need your neurologist to administer synthetic versions of the hormone to add to your troubles after your stroke or seizure. Your body secretes boatloads of the stuff in such circumstances—humans coming into ERs after one of those neurological insults have immensely high levels of glucocorticoids in their bloodstreams. And what we know from rats is that the massive outpouring of glucocorticoids at that time adds to the damage—remove the adrenals of a rat right after a stroke or seizure, or use a drug that will transiently shut down adrenal secretion of glucocorticoids, and less hippocampal damage will result. In other words, what we think of as typical amounts of brain damage after a stroke or seizure is damage being worsened by the craziness of our bodies having stress-responses at the time.

Consider how bizarre and maladaptive this is. Lion chases you; you secrete glucocorticoids in order to divert energy to your thigh muscles—great move. Go on a blind date, secrete glucocorticoids in order to divert energy to your thigh muscles—likely to be irrelevant, but no big deal. Have a grand mal seizure, secrete glucocorticoids in order to divert energy to your thigh muscles—and make the brain damage worse. This is as stark a demonstration as you can ask for that a stress-response is not always what you want your body to be having.

How did such maladaptive responses evolve? The most likely explanation is that the body simply has not evolved the tendency *not* to secrete glucocorticoids during a neurological crisis. Until the last half-century or so, no mammal had much likelihood of surviving something like a stroke, so there was no evolutionary pressure to make the body's response to massive neurological injury more logical.

GLUCOCORTICOID NEUROTOXICITY

Having revealed those disturbing findings, we move to the worst of all—there is now reasonably good evidence that truly prolonged exposure to stress or glucocorticoids can actually kill hippocampal neurons. This is not shriveling them; this is not leaving them dangling on the edge of a cliff and thus more vulnerable to any neurological gust of wind. This is actual killing of neurons.

The first evidence that glucocorticoids could damage the brain came in the late 1960s. Two researchers showed that if guinea pigs are exposed to pharmacological levels of glucocorticoids (that is, higher levels than the body ever normally generates on its own), the brain is damaged. Oddly, damage was mainly limited to the hippocampus. This was right around the time that Bruce McEwen was first reporting that the hippocampus is loaded with receptors for glucocorticoids, and no one really appreciated yet how much the hippocampus was ground zero in the brain for glucocorticoid actions.

Beginning in the early 1980s, various researchers including myself showed that this "glucocorticoid neurotoxicity" was not just a pharmacological effect, but was relevant to normal brain aging in the rat. Collectively, the studies showed that lots of glucocorticoid exposure (in the range seen during stress) or lots of stress itself would accelerate the degeneration of the aging hippocampus. Conversely, to diminish glucocorticoid levels (by removing the adrenals of the rat) was to delay hippocampal aging. And as one might expect by now, these studies showed that the extent of glucocorticoid exposure over the rat's lifetime not only determined how much hippocampal degeneration there would be in old age, but how much memory loss as well.

Where do glucocorticoids and stress get off killing your brain cells? Sure, stress hormones can make you sick in lots of ways, but isn't neurotoxicity going a bit beyond the bounds of good taste? A dozen years into studying the phenomenon, we're not yet sure.

So the final, critical question: can stress or glucocorticoids kill human hippocampal neurons? Up until a few years ago, there was some evidence of glucocorticoid neurotoxicity in the primate hippocampus that looked a lot like the phenomenon in rats, but nothing at the human end. But in the last couple of years, a handful of studies suggest that this could be more relevant to us than originally thought.

The first set of studies concerns people who have been subject to major stressors, namely, people with posttraumatic stress disorder (PTSD)—Vietnam combat vets, childhood sexual abuse survivors. A spate of independent studies using MRIs in the last few years show that sufferers of PTSD have major and selective atrophy of their hippocampus. The more severe the combat exposure, for example, the worse the atrophy.

A similar outcome might occur in people with long-term major depression. As will be seen in Chapter 13, about half of depressives have very significant elevations of glucocorticoids in their bloodstream. And in 1996, Yvette Sheline and colleagues at Washington University School of Medicine used MRIs to show selective atrophy of the hippocampus in people who had suffered major depression—and the longer the duration of the depression, the more severe the atrophy.

Can these be examples of the reversible shriveling of neuronal connections seen with the Cushing's disease patients? Probably not. The Sheline study was not of depressives, but of *ex*-depressives, years to decades after the depression had gone away, and the PTSD patients were decades past their childhood or Vietnam traumas. Whatever is happening is not readily reversible.

This is very tentative stuff, and it's not clear that this is a human example of the glucocorticoid neurotoxicity story. Some problems: (1) as noted, only half of depressives secrete excessive glucocorticoids, and that study did not test whether it was only those subjects who had the hippocampal atrophy (this is hard to do—as mentioned, many of these people were decades past their last depression, and their glucocorticoid levels were not necessarily measured back when). (2) People with PTSD do not secrete excessive glucocorticoid levels (and, in fact, secrete less glucocorticoids than usual, for reasons no one understands). Perhaps the hippocampal atrophy in these folks arises not from the "post" part of their PTSD (when they don't have glucocorticoid excess), but from the trauma itself. However, no one has ever seen what glucocorticoid levels are like in humans who, say, are in the middle of having their battalion slaughtered in front of them. (3) Finally, only about 10 to 20 percent of people who are exposed to combat trauma, for example, actually develop PTSD. Maybe the people who just happen to have small hippocampi are the ones who are likely to succumb to PTSD, rather than trauma itself being what generates hippocampal atrophy. (4) No one has done postmortems on these folks to confirm that there really is loss of neurons—it's simply the most parsimonious explanation so far.

These are tough issues to resolve, and it is far from clear at this point that stress or glucocorticoids can be neurotoxic to the human hippocampus, or even endanger it (that is, make neurological insults more damaging to it). But if they are, it raises some troubling possibilities. The most disturbing one concerns the use of synthetic glucocorticoids. Such "steroids" are not used to treat just brain edema (appropriately or otherwise). About 16 million prescriptions are written annually in the United States for glucocorticoids. Much of the use is benign—a little hydrocortisone cream for some poison ivy, a hydrocortisone injection for a swollen knee, maybe even use of steroid inhalants for asthma (which is probably not a very worrisome route for glucocorticoids to get into the brain). But there are still hundreds of thousands of people taking high-dose glucocorticoids to suppress the inappropriate immune responses in autoimmune diseases (such as lupus, multiple sclerosis, or rheumatoid arthritis). Do these people develop memory problems? Often. Careful work by a number of researchers has shown that even a few days of high-dose glucocorticoids begin to disrupt declarative memory in healthy volunteers. Do these people lose hippocampal neurons? No one yet knows, though a number of groups are trying to find out. Should you avoid taking glucocorticoids now for your autoimmune disease in order to avoid the possibility of accelerated hippocampal aging somewhere down the line? Almost certainly not, in most cases. Potentially, it's a particularly grim and unavoidable side-effect.

We are now fifty, sixty years into thinking about ulcers, blood pressure, and aspects of our sex lives as being sensitive to stress. Most of us recognize the ways in which stress can also disrupt how we learn and remember. This chapter raises the possibility that the effects of stress in the nervous system might extend to even damaging our neurons, and the next chapter continues this theme, in considering how stress might well accelerate the aging of our brains. If these grim findings from animal studies turn out to apply to the human nervous system, this will be quite disturbing. The noted neuroscientist Woody Allen once said, "My brain is my second-favorite organ." My guess is that most of us would rank our brains even higher up on a list.

AGING AND DEATH

Predictably, it comes at the most unpredictable times. I'll be lecturing, bored, telling the same story about neurons I did last year, daydreaming, looking at the ocean of irritatingly young undergraduates, and then it hits, producing almost a sense of wonderment. "How can you just sit there? Am I the only one who realizes that we're all going to die someday?" Or I'll be at a scientific conference, this time barely understanding someone else's lecture, and amid the roomful of savants, the wave of bitterness will sweep over me. "All of you highfalutin medical experts, and not one of you can make me live forever."

It first really dawns on us emotionally sometime around puberty. Woody Allen, once our untarnished high priest of death and love, captures its roundabout assault perfectly in *Annie Hall*. The protagonist is shown, in flashback, as a young adolescent. He is sufficiently depressed for the worried mother to drag him to the family doctor—"Listen to what he keeps saying, what's wrong with him, does he have the flu?" The Allenesque adolescent, glazed with despair and panic, announces in a monotone: "The universe is expanding." It's all there—the universe is expanding; look how big infinity is and how finite we are—and he has been initiated into the great secret of our species: we will die and we know it. With that rite of passage, he has found the mother lode of psychic energy that fuels our most irrational and violent moments, our most selfish and our most altruistic ones, our neurotic dialectic of simultaneously mourning and denying, our diets and exercising, our myths of paradise and resurrection. It's as if we were trapped in a mine,

Morris Zlapo, 1987: Gepetto's Dementia, *collage.*

shouting out for rescuers: save us, we're alive but we're getting old and we're all going to die.

Oh, it doesn't have to be that bad. Perhaps we will grow old with grace and wisdom. Perhaps we will be honored, perhaps surrounded by strong, happy children whose health and fecundity will feel like immortality to us. As I have noted, for many years I have spent part of each year doing stress research on wild baboons in east Africa. The people living there, like many people in the nonwesternized world, clearly think differently about these issues than we do. No one seems to find getting old depressing. How could they?—they wait their whole lives to become powerful elders. My nearest neighbors are Masai tribesmen, nomadic pastoralists. I often patch up their various minor injuries and ills. One day, one of the extremely old men of the village (perhaps sixty years old) tottered into our camp. Ancient, wrinkled beyond measure, tips missing from a few fingers, frayed earlobes, long-forgotten battle scars. He spoke only Masai and not Swahili, the lingua franca of east Africa, so he was accompanied by his more worldly, middle-aged neighbor, who

An elderly hunter-gatherer shaman in the Kalahari desert.

translated for him. He had an infected sore on his leg, which I washed and treated with antibiotic ointment. He also had trouble seeing—"cataracts" was my barely educated guess—and I explained that they were beyond my meager curative powers. He seemed resigned, but not particularly disappointed, and as he sat there cross-legged, naked except for the blanket wrapped around him, basking in the sun, the woman stood behind him and stroked his head. In a voice as if describing last year's weather she said, "Oh, when he was younger, he was beautiful and strong. Soon he will die." That night in my tent, sleepless and jealous of the Masai, I thought "I'll take your malaria and parasites, I'll take your appalling infant mortality rates, I'll take

the chances of being attacked by buffalo and lions. Just let me be as unafraid as you are."

Maybe we will luck out and wind up as respected village elders. Gerontologists studying the aging process find increasing evidence that most of us will age with a fair degree of grace and success; the final chapter of this book tells some of this pleasing story. Nevertheless, there are still so many other outcomes to fear: Wracking pain. Dementia so severe we can't recognize our children. Cat food. Forced retirement. Colostomy bags. Muscles that no longer listen to our commands, organs that betray us, children who ignore us. Mostly that aching sense that just when we finally grow up and learn to like ourselves and to love and play, the shadows lengthen. There is so little time.

Not only are we in the westernized world relatively alone in our fear of aging, but we are relatively alone in undergoing advanced stages of the process. Within the time scale of evolution, widespread aging within a species is a fairly recent invention.

It is not a luxury readily available to most other animals on the planet, nor was it to most of our ancestors or to most of the people in the developing world. Likewise the science of aging, gerontology, is a fairly new discipline. And almost from its beginning, theories about the nature of aging have been intertwined with ideas about stress. In general, these have taken two forms. The first is that advanced age is a time when organisms no longer cope well with stress. The second is that stress, especially prolonged and extreme forms of stress, can accelerate aging. This chapter explores the evidence supporting each of these ideas.

HOW DO AGED ORGANISMS DEAL WITH STRESS?

How do aged organisms deal with stress? Not very well, it turns out. In many ways, aging can be defined as the progressive loss of the ability to deal with stress, and that certainly fits our perception of aged individuals as fragile and vulnerable. This can be stated more rigorously by saying that many aspects of the bodies and minds of old organisms work fine, just as they do in young ones—so long as they aren't pushed. Throw in an exercise challenge, an injury or illness, time pressure, novelty—any of a variety of physical, cognitive, or psychological stressors—and aged organisms are more likely to fall apart. The illustration on the next page shows this in a schematic way. In the absence of stress (of whatever sort), young and old people perform roughly the same way. Under stress, performance declines in both age groups—but faster in the old group.

This difference occurs across all levels of human experience. As one classic example, normal body temperature—98.6 degrees—does not change with age. Nevertheless, it takes the bodies of the elderly longer to restore a normal temperature after being warmed or chilled; or to use the terminology of this book, it takes them longer to reestablish allostasis after temperature stressors. Heart performance follows a similar pattern. Once you eliminate from your study the elderly people who have heart disease and look only at healthy subjects of different ages (so as to study aging, instead of inadvertently studying disease), many aspects of cardiac function remain the same despite age—heart rate, cardiac output, stroke volume, "stiffness" of cardiac muscle,

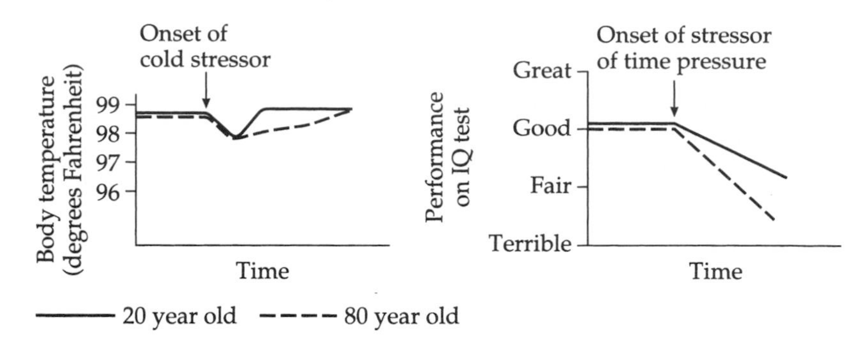

Two ways in which the stress-response may be impaired in an older individual. (Left) Young and old individuals may have the same function under nonstressed conditions (in this case, the same body temperature). However, in the face of a stressor that knocks that measure out of homeostatic balance (chilling the person one degree), it takes the older person longer to reestablish homeostasis. (Right) Young and old individuals may have the same function under nonstressed conditions (performance on an IQ test). However, as the stressor of time pressure is added to the test, performance declines in both individuals—but to a greater extent in the older person.

end diastolic volume, and ejection fraction.* Stress the cardiovascular system with exercise, however, and old hearts do not respond the same as young ones. The maximal work capacity and the maximal heart rate that can be achieved are nowhere near as great as in a young person. Ejection volume decreases more dramatically than in the young, while cardiac muscle stiffness increases more dramatically. All of this is a fancy way of saying that even if some eighty-year-old still has the blood pressure and heart rate of a teenager, it still may not be such a hot idea for her to run a marathon. Or as another example, it takes less of an environmental stressor to suppress the immune system of an aged primate than a young one.

Everything can be fine in the absence of a stressor, but the system's still vulnerable. This idea applies in other arenas of gerontology as well. For example, if the brain is damaged and neurons are killed, remaining neurons can sometimes grow new

---

* *Heart rate* is how often the heart beats per unit time (typically a minute). *Cardiac output* is the amount of blood pumped out in that period. *Stroke volume* is the amount pumped per heartbeat. *Stiffness of cardiac muscle* is an engineering term for expressing how efficiently the heart muscle meets changing blood pressure needs. *End diastolic volume* is the amount of blood in the heart just before it contracts, and *ejection fraction* is how much of that blood is pumped out with the next contraction.

branches to try to compensate for the injury. The phenomenon occurs in old brains but, for the same degree of injury, more slowly than in young brains. Similarly, old and young rat brains contain roughly the same amount of energy (stored in the phosphate bonds of molecules called *ATP*), but when you stress the system by cutting off the flow of oxygen and nutrients, energy levels decline faster in the old brains.

The idea also applies to measures of cognition. What happens to IQ test scores as people get older? (You'll notice that I didn't say "intelligence," just "performance on IQ tests." Whether the latter tells us much about the former is a controversy I don't want to come within a hundred miles of.) The dogma in the field was once that IQ declined with age. Then it was that it did not decline. It depends on how you test it. If you test young and old people and give them lots of time to complete the test, there is little difference. As you stress the system—in this case, by making the subjects race against a time limit—scores fall for all ages, but much farther among older people.

In many ways then, the elderly may function just as the young do, so long as they are not stressed. But throw in a stressor, and some dramatic vulnerabilities become evident. A lot of this relative dysfunction can be explained by a host of problems with the individual hormonal systems involved in the stress-response. In some respects, there is not *enough* of a stress-response. For example, during exercise old people secrete plenty of epinephrine and norepinephrine (a lot more than young people, in fact). This would predict that their cardiovascular systems should kick into high gear during the exercise. Yet, as just mentioned, maximal cardiac output during exercise declines with age. Why? Because the heart and various blood vessels for some reason do not respond as vigorously to the epinephrine and norepinephrine.

An even bigger problem is that aged organisms often have *too much* of a stress-response. For example, older individuals are not so good at turning off epinephrine and norepinephrine secretion after a stressor has finished; it takes longer for their levels of these substances to return to baseline. The same thing is seen with glucocorticoid secretion. Stress young and old rats by holding them still, and they have roughly the same-sized burst of glucocorticoid secretion. But after the immobilization, glucocorticoid levels return to normal in young rats much faster than in old rats.

By now it should be clear that secreting hormones like epinephrine and glucocorticoids when there is no longer a stressor is a bad idea—once the emergency is over, your body should turn off the various alarms as rapidly as possible. Do old rats pay a price for not being able to turn off their stress-responses promptly after the end of stress? A colleague—Thomas Donnelly—and I tested this idea. As discussed in the immunity chapter, if a tumor of a certain type is induced to grow in a rat, it will grow faster if that rat is repeatedly stressed, in part because of the amount of glucocorticoids secreted during stress. We first observed that, for the same pattern of stressors, these tumors grew faster in old rats than in young ones. Was faster growth due to the extra glucocorticoids that old rats secrete because they are sluggish in turning off the stress-response? To test this, we infused some young rats with extra glucocorticoids at the end of each stressful event, in a way that mimicked the "shut-off" problem of old rats—and the young ones grew tumors faster as well.

Aged organisms not only have trouble turning off the stress-response after the end of stress, they also secrete more stress-related hormones even in their normal, nonstressed state. Resting epinephrine and norepinephrine levels, for example, are typically elevated in old rats and aged humans. The same thing is seen with glucocorticoids: resting levels rise with age in the rat and human. The consensus in the field used to be that this did not happen in humans—numerous studies had found the same resting levels of glucocorticoids in old humans as in young. However, these studies were based on a dated concept of what is "old" in a human; they looked at people in their 60s and early 70s. Modern gerontologists do not consider someone aged until the late 70s or 80s, and the more recent studies show a big jump in resting glucocorticoid levels in that age group. A colleague—Jeanne Altmann—and I recently made a similar observation of wild baboons—the really old ones had elevated glucocorticoid levels, showing a jump of about the same magnitude with age as humans. (Studying aging in a wild population of a closely related species like the baboon is particularly useful because it eliminates many of the confounds in studying humans. Baboons of every age eat roughly the same thing and are still physically active; they're all somewhat related, so genetic differences aren't much of a factor; they don't drink alcohol or smoke; and the old ones aren't taking half a dozen different medicines. Furthermore, you don't have to worry much that you are accidentally studying disease instead of aging—an old baboon living in the wild is a supremely healthy beast, because any who weren't got eaten long ago.)

Old individuals of all sorts tend to have the stress-response turned on even when nothing stressful is happening. Do old organisms pay a price for this "normal" hormonal hyperactivity? It seems so. For example, the elevated resting epinephrine and norepinephrine levels probably have a lot to do with the elevated blood pressure often seen during aging. Studies also indicate that the elevated resting glucocorticoid levels in old rats, besides causing all the problems reviewed in Chapter 10, are responsible for the increasing difficulty that neurons have sprouting new branches after an injury.

In previous chapters, we've seen that, ideally, the hormones of the stress-response should be nice and quiet when nothing bad is happening, secreted in tiny amounts. When a stressful emergency hits, your body needs a huge and fast stress-response. At the end of the stressor, everything should shut off immediately. And these traits are precisely what old organisms typically do not have.

CAN STRESS MAKE YOU AGE FASTER?

Intuitively, the idea that stress accelerates the aging process makes sense. We recognize that there is a connection between how we live and how we die. Around the turn of the century, a madly inspired German physiologist, Max Rubner, tried to define this connection scientifically. He looked at all sorts of different domestic species and calculated things like lifetime number of heartbeats and lifetime metabolic rate (not the sort of study that many scientists have tried to replicate). He concluded that there is only so long a body can go on—only so many breaths, so many heartbeats, so much metabolism that each pound of flesh can carry out before the mechanisms of life wear out. A rat, with approximately 400 heartbeats a minute, uses up its heartbeat allotment faster (after approximately two years) than an elephant (with approximately 35 beats per minute and a 60-year life span). Such calculations lay behind ideas about why some species lived far longer than others. Soon the same sort of thinking was applied to how long different individuals *within* a species live—if you squander a lot of your heartbeats and heavy breathing being nervous about blind dates when you're sixteen, there would be that much less metabolic reserve available to you at eighty.

In general, Rubner's ideas about life spans among different species have not held up well in their strictest versions, while the "rate of living" hypotheses about individuals within a species that his ideas inspired have been even less tenable. Nevertheless, they led many people in the field to suggest that a lot of environmental perturbations can wear out the system prematurely. Such "wear and tear" thinking fit in naturally with the era of stressology that Selye introduced, and in his later years Selye theorized that a lifetime of stress depletes an individual of "adaptational" energies, leading to accelerated senescence.

Fascinating, especially the notion of adaptational energies, but no one is really sure what that term means. We can, however, put the issue in more concrete terms: What does being exposed to a lot of glucocorticoids over the life span do to the aging process? We have a list hammered into our heads by now of problems that glucocorticoid excess brings about—fatigue, thinning muscles, adult-onset diabetes, hypertension, osteoporosis, reproductive decline, immune suppression. All of these conditions become more common among the elderly. Can glucocorticoids determine how fast aging occurs? It turns out that, in more than a dozen species, glucocorticoid excess is *the* cause of death during aging.

WHY YOU SELDOM SEE REALLY OLD SALMON

Pictures of heroic wild animals, à la Marlin Perkins: Penguins who stand all winter amid the Antarctic cold, keeping their eggs warm at their feet. Leopards dragging massive kills up trees with their teeth, in order to eat them free of harassment by lions. Desiccated camels marching scores of miles. And most of all,

A male sockeye salmon, after the onset of programmed aging.

salmon leaping over dams and waterfalls to return to the freshwater stream of their birth. Where they spawn a zillion eggs. After which most of them die over the next few weeks.

Why do salmon die so soon after spawning? No one is quite sure, but evolutionary biologists are rife with theories about why this and the rare other cases of "programmed die-offs" in the animal kingdom may make some evolutionary sense. What is known, however, is the proximal mechanism underlying the sudden die-off (not "how come they die, in terms of evolutionary patterns over the millennia," but "how come they die, in the sense of which parts of the body's functioning suddenly go crazy"). It is glucocorticoid secretion.

If you catch salmon right after they spawn, just when they are looking a little green around the gills, you find they have huge adrenal glands, peptic ulcers, and kidney lesions; their immune systems have collapsed, and they are teeming with parasites and infections. Aha, kind of sounds like Selye's rats way back when. Moreover, the salmon have stupendously high glucocorticoid concentrations in their bloodstreams. When salmon spawn, regulation of their glucocorticoid secretion breaks down. Basically, the brain loses its ability to measure accurately the quantities of circulating hormones and keeps sending a signal to the adrenals to secrete more of them. Lots of glucocorticoids can certainly bring about all those diseases with which the salmon are festering. But is the glucocorticoid excess really responsible for their death? Yup. Take a salmon right after spawning, remove its adrenals, and it will live for a year afterward.

The bizarre thing is that this sequence of events not only occurs in five species of salmon, but also among a dozen species of Australian marsupial mice. All the male mice of these species die shortly after seasonal mating; cut out their adrenal glands, however, and they keep living as well. Pacific salmon and marsupial mice are not close relatives. At least twice in evolutionary history, completely independently, two very different sets of species have come up with the identical trick: if you want to degenerate very fast, secrete a ton of glucocorticoids.

CHRONIC STRESS AND THE AGING PROCESS IN THE MAINSTREAM

That is all fine for the salmon looking for the fountain of youth, but we and most other mammals age gradually over time, not in catastrophic die-offs over the course of days. Does stress influence the rate of gradual mammalian aging? This turns out to be the case, in at least one fairly frightening way.

We return to the tendency of very old rats, humans, and primates to have elevated resting levels of glucocorticoids in the bloodstream. Some aspect of the regulation of normal glucocorticoid secretion is disrupted during aging. To get a sense of why this happens, we must become extremely technical and delve into the essential topic of why the water tank on your toilet does not overflow when it's refilling. There is a source of water—a pipe that typically sticks straight up. There is a way for the system to know how much the tank has filled—the flotation device that bobs up higher as the water rises. And there is a connection between how much water there is in the tank and whether more water is released: the flotation device is attached to a valve on the top of the water pipe, and as the device bobs higher, it closes the valve. Engineers who study this sort of thing term that process *negative feedback inhibition* or *end-product inhibition*: increasing amounts of water accumulating in the tank decrease the likelihood of further release of water.

Most hormonal systems, including the glucocorticoids, work by this feedback-inhibition process. The brain secretes corticotropin releasing factor (CRF), which triggers pituitary release of adrenocorticotropic hormone (ACTH), and finally, out come glucocorticoids from the adrenal. The brain needs to know whether to keep secreting more CRF. It does this by sensing the levels of glucocorticoids in the circulation (sampling the hormone from the bloodstream coursing through the brain) to see if levels are at, below, or above a "set point." If levels are low, the brain keeps secreting CRF—just as when water levels in the toilet tank are still low. Once glucocorticoid levels reach or exceed that set point, there is a negative feedback signal and the brain stops secreting CRF. As a fascinating complication, the set point can shift. In the absence of stress, the brain wants different levels of glucocorticoids in the bloodstream from those required when something stressful is happening. (This implies that the quantity of glucocorticoids in the bloodstream necessary to turn off CRF secretion by the brain should vary with different situations, which turns out to be the case.)

This is how the system works normally, as can be shown experimentally by injecting a person with a massive dose of a synthetic glucocorticoid (dexamethasone). The brain senses the sudden increase and says, in effect, "My god, I don't know what is going on with those idiots in the adrenal, but they just secreted *way* too many glucocorticoids." The dexamethasone exerts a negative feedback signal, and soon the person has stopped secreting CRF, ACTH, and her own glucocorticoids. This person would be characterized as "dexamethasone-responsive." If negative feedback regulation is not working very well, however, the person is "dexamethasone-resistant"—she keeps secreting her hormones, despite the whopping glucocorticoid signal in the bloodstream. And that is precisely what happens in old people, old primates, and old rats. Glucocorticoid feedback regulation no longer works very well.

This may explain why very old organisms secrete excessive glucocorticoids (in the absence of stress and during the recovery period after the end of a stressor). Why the failure of feedback regulation? There is a fair amount of evidence that it is due to the degeneration during aging of one part of the brain. The entire brain does not serve as a "glucocorticoid sensor"; instead, that role is served by only a few areas with very high numbers of receptors for glucocorticoids and the means to tell the hypothalamus whether or not to secrete CRF. In the previous chapter, I described how the hippocampus is famed for its role in learning and memory. As it turns out, it is also one of the important negative feedback sites in the brain for controlling glucocorticoid secretion. It also turns out that the hippocampus loses a considerable number of neurons during aging.* And when the hippocampus is damaged, some of the deleterious consequences include a tendency to secrete an excessive amount of glucocorticoids—this could be the reason aged people may have elevated resting levels of the hormone, may have trouble turning off secretion after the end of stress, or may be dexamethasone-resistant. It is as if one of the brakes on the system has been damaged, and hormone secretion rushes forward, a little out of control.

The elevated glucocorticoid levels of old age, therefore, arise because of a problem with feedback regulation that, in turn, is related to neuron loss in the hippocampus. Why does the aging hippocampus lose so many neurons? Laboratory studies with rodents have given a provisional answer—excessive exposure to glucocorticoids over the lifetime damages the hippocampus. This is reviewed in the last chapter.

If you've read carefully, you will begin to note something truly insidious embedded in these findings. When the hippocampus is damaged, the rat secretes more glucocorticoids.

---

* One of the great myths of science folklore is that, from some tender age, twenty-five or so, we all begin to lose massive numbers of neurons. This turns out to be a mistake occasioned by earlier researchers looking at the brains of people with dementia—probably Alzheimer's—and assuming that this represented "normal" aging, instead of a disease associated with aging. But although the extent of neuron loss as we age is not dramatic, the hippocampus tends to be one of the areas most consistently damaged. The other is a region called the *nigrostriatal system*: neuron loss there has something to do with the tremors that many old people get. It is also the region dramatically damaged during Parkinson's disease.

Which should damage the hippocampus further. Which should cause even more glucocorticoid secretion. . . . Each makes the other worse, causing a degenerative cascade that appears to occur in many aging rats, and whose potential pathological consequences have been detailed throughout virtually every page of this book.

Does this degenerative cascade occur in humans? As noted, glucocorticoid levels rise with extreme old age in the human, and Chapter 10 outlines the first evidence that these hormones might have some bad effects on the human hippocampus. The primate and human hippocampus appear to be negative feedback regulators of glucocorticoid release, such that hippocampal damage is associated with glucocorticoid excess, just as in the rodent. So the pieces of the cascade appear to be there in the human, and Chapter 10 raised the possibilities that histories of severe stress, or of heavy use of synthetic glucocorticoids to treat some disease, might accelerate aspects of this cascade.

Does that mean that all is lost, that this sort of dysfunction is an obligatory part of aging? Certainly not. It was not by chance that two paragraphs above, I described this cascade as occurring in "many" aging rats, rather than in "all." Some rats age successfully in a way that spares them this cascade, and even more humans appear to as well these pleasing stories are part of the final chapter of this book.

It is thus not yet clear whether the "glucocorticoid neurotoxicity" story applies to how our brains age. Unfortunately, the answer is not likely to be available for years; the subject is very difficult to study in humans. Nevertheless, from what we know about this process in the rat and monkey, glucocorticoid toxicity stands as a striking example of ways in which stress can accelerate aging. And should it turn out to apply to us as well, it will be an aspect of our aging that will harbor a special threat. If we are crippled by an accident, if we lose our sight or hearing, if we are so weakened by heart disease as to be bed-bound, we cease having so many of the things that make our lives worth living. But when it is our brains that are damaged, when it is our ability to recall old memories or to form new ones that is destroyed, we fear we'll cease to exist as sentient, unique individuals—the version of aging that haunts us most.

George Segal, 1969: Man in a Chair, *wood and plaster.*

Even the most stoic of readers should be pretty frazzled by now, given the detailing in the 11 chapters so far about the sheer number of things that can go wrong with stress. It is time to shift to the second half of the book, which examines stress management, coping, and individual differences in the stress-response. It is time to begin to get some good news.

WHY IS PSYCHOLOGICAL STRESS STRESSFUL?

Some people are born to biology. You can spot them instantly as kids—they're the ones comfortably lugging around the toy microscopes, stowing dead squirrels in the freezer, being ostracized at school for their obsession with geckos.* But all sorts of folks migrate to biology from other fields—chemists, psychologists, physicists, mathematicians.

Several decades after stress physiology began, the discipline was inundated by people who had spent their formative years as engineers. Like physiologists, they thought there was a ferocious logic to how the body worked, but for bioengineers, that tended to mean viewing the body a bit like the circuitry diagram that you get with a new tape recorder: input-output ratios, impedance, feedback loops, servomechanisms. I shudder to even write such words, as I barely understand them; but the bioengineers did wonders for the field, adding a tremendous vigor.

---

* Personally, I used to collect the leftover chicken bones from everyone at the Friday night dinner table, clean them with my knife, and proudly display an articulated skeleton by the end of dessert. In retrospect, I think this was more to irritate my sister than to begin an anatomical quest. A biography of Teddy Roosevelt, however, recently helped me to appreciate that the world lost one of its great potential zoologists when he lapsed into politics. At age eighteen, he had already published professionally in ornithology; when he was half that age, he reacted to the news that his mother had thrown out his collection of field mice, stored in the family icebox, by moping around the house, proclaiming, "The loss to science! The loss to science!"

Suppose you wonder how the brain knows when to stop glucocorticoid secretion—when enough is enough. In a vague sort of way, everyone knew that somehow the brain must be able to measure the amount of glucocorticoids in the circulation, compare that to some desired set point, and then decide whether to continue secreting CRF or turn off the faucet (returning to the toilet tank model of the last chapter). The bioengineers came in and showed that the process was vastly more interesting and complicated than anyone had imagined. There are "multiple feedback domains"; some of the time the brain measures the *quantity* of glucocorticoids in the bloodstream, and sometimes the *rate* at which the level is changing. The bioengineers solved another critical issue: is the stress-response linear or all-or-nothing? Epinephrine, glucocorticoids, prolactin, and other substances are all secreted during stress; but are they secreted to the same extent regardless of the intensity of the stressor (all-or-nothing responsiveness)? The system turns out to be incredibly sensitive to the size of the stressor, demonstrating a linear relationship between the extent of blood pressure drop and the extent of epinephrine secretion, between the degree of hypoglycemia (drop in blood sugar) and glucagon release. The body not only can sense something stressful, but also is amazingly accurate at measuring just how far and how fast that stressor is throwing the body out of allostatic balance.

Beautiful stuff, and important. Selye loved the bioengineers, which makes perfect sense, since at his time the whole stress field must have still seemed a bit soft-headed to some mainstream physiologists. Those physiologists knew that the body does one set of things when it is too cold, and a diametrically opposite set when it is too hot, but here were Selye and his crew insisting that there were physiological mechanisms that respond equally to cold *and* hot? *And* to injury *and* hypoglycemia *and* hypotension? The beleaguered stress experts welcomed the bioengineers with open arms. "You see, it's for real; you can do math about stress, construct flow charts, feedback loops, formulas . . ." Golden days for the business. If the system was turning out to be far more complicated than ever anticipated, it was complicated in a way that was precise, logical, mechanistic. Soon it would be possible to model the body as one big input-output relationship: you tell me exactly to what degree a stressor impinges on an organism (how much it disrupts the allostasis of blood sugar, fluid volume, optimal temperature, and so on), and I'll tell you exactly how much of a stress-response will occur.

This approach, fine for most of the ground that we've covered up until now, will probably allow us to estimate quite accurately what the pancreas of that zebra is doing when the organism is sprinting from a lion. But the approach is not going to tell us which of us will get an ulcer when the factory closes down. Starting in the late 1950s, a new style of experiments in stress physiology began to be conducted that burst that lucid, mechanistic bioengineering bubble. A single example will suffice. An organism is subjected to a painful stimulus, and you are interested in how great a stress-response will be triggered. The bioengineers had been all over that one, mapping the relationship between the intensity and duration of the stimulus and the response. But this time, when the painful stimulus occurs, the organism under study can reach out for its Mommy and cry in her arms. And under these circumstances, this organism shows less of a stress-response.

Nothing in that clean, mechanistic world of the bioengineers could explain this phenomenon. The input was still the same; the same number of pain receptors should have been firing while the child underwent some painful procedure. Yet the output was completely different. A critical realization roared through the research community: the physiological stress-response can be modulated by psychological factors. Two identical stressors with the same extent of allostatic disruption can be *perceived* differently, and the whole show changes from there.

Suddenly the stress-response could be made bigger or smaller, depending on psychological factors. In other words, psychological variables could *modulate* the stress-response. Inevitably, the next step was demonstrated: in the absence of any change in physiological reality—any actual disruption of allostasis—psychological variables alone could *trigger* the stress-response. Flushed with excitement, Yale physiologist John Mason, one of the leaders in this approach, even went so far as to proclaim that all stress-responses were psychological stress-responses.

The old guard was not amused. Just when the conception of stress was becoming systematized, rigorous, credible, along came this rabble of psychologists muddying up the picture. In a series of published exchanges in which they first praised each other's achievements and ancestors, Selye and Mason attempted to shred each other's work. Mason smugly pointed to the growing literature on psychological initiation and modulation of the stress-response. Selye, facing defeat, insisted that *all* stress-responses couldn't be psychological and perceptual: if an organism is anesthetized, it still gets a stress-response when a surgical incision is made.

The psychologists succeeded in getting a place at the table, and as they have acquired some table manners and a few gray hairs, they have been treated less like barbarians. We now have to consider which psychological variables are critical. Why is psychological stress stressful?

PSYCHOLOGICAL STRESSORS

You would expect key psychological variables to be mushy concepts to uncover, but in a series of elegant experiments, a physiologist named Jay Weiss, then at Rockefeller University, demonstrated exactly what is involved. The subject of one experiment is a rat that receives mild electric shocks (roughly equivalent to the static shock you might get from scuffing your foot on a carpet). Over a series of these, the rat develops a prolonged stress-response: its heart rate and glucocorticoid secretion rate go up, for example. For convenience, we can express the long-term consequences by how likely the rat is to get an ulcer, and in this situation, the probability soars. In the next room, a different rat gets the same series of shocks—identical pattern and intensity; its allostasis is challenged to exactly the same extent. But this time, whenever the rat gets a shock, it can run over to a bar of wood and gnaw on it. The rat in this situation is far less likely to get an ulcer. You have given it an *outlet for frustration*. Other types of outlets work as well—let the stressed rat eat something, drink water, or sprint on a running wheel, and it is less likely to develop an ulcer.

We humans also deal better with stressors when we have outlets for frustration—punch a wall, take a run, find solace in a hobby. We are even cerebral enough to *imagine* those outlets and derive some relief: consider the prisoner of war who spends hours imagining a golf game in tremendous detail. I have a friend who passed a prolonged and very stressful illness lying in bed with a mechanical pencil and a notepad, drawing topographic maps of imaginary mountain ranges and taking hikes through them.

A variant of Weiss's experiment uncovers a special feature of the outlet-for-frustration reaction. This time, when the rat gets the identical series of electric shocks and is upset, it can run across the cage, sit next to another rat and . . . bite the hell out of it. Stress-induced displacement of aggression: the practice works wonders at minimizing the stressfulness of a stressor. It's a real primate specialty as well. A male baboon loses a fight. Frustrated, he spins around and attacks a subordinate male who was minding his own business. An extremely high percentage of primate aggression represents frustration displaced onto innocent bystanders. Humans are pretty good at it, too, and we have a technical way of describing the phenomenon in the context of stress-related disease: "He's one of those guys who doesn't get ulcers, he gives them." Taking it out on someone else—how well it works at minimizing the impact of a stressor.

There is an additional way in which we can interact with another organism to minimize the impact of a stressor on us, a way that is considerably more encouraging for the future of our planet than is displacement aggression. Rats rarely use it, but primates are great at it. Put an infant primate through something unpleasant: it gets a stress-response. Put it through the same stressor while in a room full of other primates and . . . it depends. If those primates are strangers, the stress-response gets worse. But if they are friends, the stress-response is decreased. Social support networks—it helps to have a shoulder to cry on, a hand to hold, an ear to listen to you, someone to cradle you and to tell you it will be okay.

The same is seen with primates in the wild. While I mostly do laboratory research on how stress and glucocorticoids affect the brain, I spend my summers in Kenya studying patterns of stress-related physiology and disease among wild baboons living in a national park. The social life of a male baboon can be pretty stressful—you get beaten up as a victim of displaced aggression; you carefully search for some tuber to eat and clean it

Tooker, 1966: Landscape with Figures, *egg tempera on gesso.*

off, only to have it stolen by someone of higher rank; and so on. Glucocorticoid levels are elevated among low-ranking baboons, and among the entire group if the dominance hierarchy is unstable or if a new, aggressive male has just joined the troop. But if you are a male baboon with a lot of friends—you play with kids, or you have frequent nonsexual grooming bouts with females, for example—you have lower glucocorticoid concentrations than males of the same general rank who lack these outlets.

Social support is certainly protective for humans as well. This can be demonstrated even in transient instances of support. In a number of subtle studies, subjects were exposed to a stressor—having to give a public speech or perform a mental arithmetic task, or having two strangers argue with them—with or without a supportive friend present. In each case, social support translated into less of a cardiovascular stress-response. Profound and persistent differences in degrees of social support can influence human physiology as well: within the same family, there are significantly higher glucocorticoid levels among stepchildren than among biological children.

As noted in the chapter on immunity, people with spouses or close friends have longer life expectancies. When the spouse dies, the risk of dying skyrockets. Recall also from that chapter the study of parents of Israeli soldiers killed in the Lebanon war: in the aftermath of this stressor, there was no notable increase in risk of diseases or mortality—except among those who were already divorced or widowed. Some additional examples concern the cardiovascular system. People who are socially isolated have overly active sympathetic nervous systems. Given the likelihood that this will lead to higher blood pressure and more platelet aggregation in their blood vessels (remember that from Chapter 3?), they are more likely to have heart disease—two to five times as likely, as it turns out. And once they have the heart disease, they are more likely to die at a younger age. In a study of patients with severe coronary heart disease, Redford Williams of Duke University and colleagues found that half of those lacking social support were dead within five years—a rate three times higher than was seen in patients who had a spouse or close friend, after controlling for the severity of the heart disease.*

* Recently, I learned about the protective effects of social support in an unexpected way. A local TV station was doing a piece on how stressful rush hour traffic was, and I wound up giving them advice—turning this chapter into a 15-second sound bite. Somewhere along the way we stumbled onto the great idea of getting a certified Type A subject (we eventually found one through a local Type A cardiology clinic) who did the commute each day and measuring his stress hormone levels before and during a commute. The film crew would take some saliva samples from which glucocorticoid levels would be measured. Great. So they get to the guy's house just before his commute, collect some spit in a test tube. Then into traffic—with the film crew increasingly stressed by the worry that there wouldn't be any tie-ups. But soon the snarls began, bumper to bumper. Then the second saliva sample was taken. Laboratory analysis, anxious TV producers awaiting results. Baseline sample at home: highly elevated glucocorticoid levels. Rush hour level: *way down*. Oh no! This seems contrary to what we'd expect. I'm convinced that the explanation for the outcome of that unscientific experiment was the social support. For this guy, who "type A's" his way through rush hour each day, this was fabulous. A chance to be on television, a bunch of people there with him to document what a stressful commute he endures, getting to feel he's the chosen representative of all Type A's, their anointed. He apparently spent the entire ride cheerfully pointing out how horrible it was, how much worse he's seen ("You think this is bad! This isn't bad. You should have been in Troy in '47."). He had a fabulous time. The punch line? Everyone should have a friendly film crew in tow when they're stuck in traffic.

The rat studies of Weiss also uncovered another variable modulating the stress-response. The rat gets the same pattern of electric shocks, but this time, just before each shock, it hears a warning bell. Fewer ulcers. *Unpredictability* makes stressors much more stressful. The rat with the warning gets two pieces of information. It learns when something dreadful is about to happen. And the rest of the time, it learns that something dreadful is *not* about to happen. It can relax. The rat without a warning can always be a half-second away from the next shock. In effect, information that increases predictability tells you that there is bad news, but comforts you that it's not going to be worse—you are going to get shocked soon, but it's never going to be sprung on you without warning. As another variant on the helpfulness of predictability, organisms will eventually habituate to a stressor if it is applied over and over; it may knock physiological allostasis equally out of balance the umpteenth time that it happens, but it is a familiar, predictable stressor by then, and a smaller stress-response is triggered. One classic demonstration involved men in the Norwegian military going through parachute training—as the process went from being hair-raisingly novel to something they could do in their sleep, their anticipatory stress-response went from being gargantuan to nonexistent.

The power of loss of predictability as a psychological stressor is shown in an elegant, subtle study. A rat is going about its business in its cage, and at measured intervals the experimenter delivers a piece of food down a chute into the cage; rat eats happily. This is called an intermittent reinforcement schedule. Now, change the pattern of food delivery so that the rat gets *exactly* the same total amount of food over the course of an hour, but at a random rate. The rat receives just as much reward, but less predictably; and up go glucocorticoid levels. There is not a single physically stressful thing going on in the rat's world. It's not hungry, pained, running for its life—nothing is out of allostatic balance. In the absence of any stressor, loss of predictability triggers a stress-response.

There are even circumstances in which stress-related disease can be *more* likely to occur among people with a lower rate of stressors. You can easily imagine how to design a rat experiment to demonstrate this, but the human version has already been done. During the onset of the Nazi blitzkrieg bombings of England, London was hit every night like clockwork. Lots of stress. In the suburbs the bombings were far more sporadic, occurring perhaps once a week. Fewer stressors, but much less predictability. There was a significant increase in the incidence of ulcers during that time. Who developed more ulcers? The suburban population. (Another measure of the importance of unpredictability: by the third month of the bombing, ulcer rates in all the hospitals had dropped back to normal.) A similar anxious state has often been described by people awaiting execution at an uncertain date. For example Gary Gilmore, the multiple murderer who was executed in 1977 amid a media circus in Utah, expressed relief bordering on euphoria when all the appeals through various courts and the stays of execution were finally exhausted. In such cases, uncertainty can eventually appear even worse than death.

Rat studies also demonstrate a related facet of psychological stress. Give the rat the same series of shocks. This time, however, you study a rat that has been trained to press a lever to avoid electric shocks. Take away the lever, shock it, and the rat develops a massive stress-response. It's as if the rat were thinking, "I can't believe this. I know what to do about electric shocks; give me a blasted lever and I could handle this. This isn't fair." Ulceration city (as well as accelerated growth of implanted tumors). Give the trained rat a lever to press; even if it is disconnected from the shock mechanism, it still helps: down goes the stress-response. So long as the rat has been exposed to a higher rate of shocks previously, it will think that the lower rate now is due to its having control over the situation. This is an extraordinarily powerful variable in modulating the stress-response.

The identical style of experiment with humans yields similar results. Place two people in adjoining rooms, and expose both to intermittent noxious, loud noises; the person who has a button and believes that pressing it decreases the likelihood of more noise is less hypertensive. In one variant on this experiment, subjects with the button who did not bother to press it did just as well as those who actually pressed the button. Thus, the *exercise* of control is not critical; rather, it is the *belief* that you have it. An everyday example: airplanes are safer than cars, yet more of us are phobic about flying. Why? Because, despite the fact that we're at greater risk in a car, most of us in our heart of hearts believe that we are above-average drivers, thus more in control. In an airplane, we have no control at all. My wife and I, neither of us happy fliers, tease each other on flights, exchanging control: "Okay, you rest for awhile, I'll take over concentrating on keeping the pilot from having a stroke."

The variable of control is extremely important; controlling the rewards that you get can be more desirable than getting them for nothing. As an extraordinary example, both pigeons and rats prefer to press a lever in order to obtain food (so long as the task is not too difficult) over having the food delivered freely—a theme found in the activities and statements of many scions of great fortunes, who regret the contingency-free nature of their lives, without purpose or striving.

Some researchers have emphasized that the stressfulness of loss of control and of loss of predictability share a common element. They subject an organism to novelty. You thought you knew how to manage things, you thought you knew what would happen next, and it turns out you are wrong in this novel situation. The potency of this is demonstrated in primate studies in which merely placing the animal into a novel cage suppresses its immune system. Others have emphasized that these types of stressors cause arousal and vigilance, as you search for the new rules of control and prediction. Both views are different aspects of the same issue.

Yet another critical psychological variable in the stress-response has been uncovered. A hypothetical example: two rats get a series of electric shocks. On the first day, one gets ten shocks an hour; the other, fifty. Next day, both get twenty-five shocks an hour. Who becomes hypertensive? Obviously, the one going from ten to twenty-five. The other rat is thinking, "Twenty-five? Piece of cheese, no problem; I can handle that." Given the same degree of disruption of allostasis, a perception that events are improving helps tremendously. A version of this can be observed among the baboons I study in Kenya. In general, when dominance hierarchies are unstable, resting glucocorticoid levels rise. This makes sense, because such instabilities make for stressful times. Looking at individual baboons, however, shows a more subtle pattern: given the same degree of instability, males whose ranks are *dropping* have elevated glucocorticoid levels, while males whose ranks are *rising* amid the tumult don't show this endocrine trait. Similarly, in one classic human study, parents who were told that their children had, for example, a 25 percent chance of dying from cancer showed only a moderate rise in glucocorticoid levels in the bloodstream. How could that be? Because the children were all in remission after a period where the odds of death had been far higher. Twenty-five percent must have seemed like a miracle. Twenty-five shocks an hour, a certain degree of social instability, a one in four chance of your child dying—each can imply either good news or bad, and only the latter seems to stimulate a stress-response.

Thus, there are some powerful psychological factors that can trigger a stress-response on their own or make another stressor seem more stressful: loss of control or predictability, loss of outlets for frustration or sources of support, a perception that things are getting worse. These factors play a major role in explaining how we all go through lives full of stressors, yet differ so dramatically in our vulnerability to them. The final chapter of this book examines the bases of these individual differences in greater detail, serving as a blueprint so that we can analyze how to learn to exploit these psychological variables—how in effect, to manage stress better. As we'll see, many ideas about stress management revolve around issues of control and predictability. However, the answer will not be simply "Maximize control. Maximize predictability. Maximize outlets for frustration." As we will now see, it is considerably more complicated than that.

SOME SUBTLETIES OF PREDICTABILITY

We have already seen how predictability can ameliorate the consequences of stress: one rat gets a series of shocks and develops a higher risk for an ulcer than the rat who gets warnings beforehand. Predictability doesn't always help, however. The experimental literature on this is pretty dense; some human examples of this point make it more accessible.

You're in the dentist's chair, no novocaine; the dentist drills away. Ten seconds of nerve-curling pain, some rinsing, five seconds of drilling, a pause while the dentist fumbles a bit, fifteen seconds of drilling, and so on. In one of the pauses, frazzled and trying not to whimper, you gasp,

"Almost done?" "Hard to say," the dentist mumbles, returning to the intermittent drilling. Think how grateful we are for the dentist who, instead, says, "Two more and we're done." The instant the second burst of drilling ends, down goes blood pressure. By being given news about the stressor to come, you are also implicitly being comforted by now knowing what stressors are not coming.

But here are several scenarios, organized by the salient issues, where foreknowledge doesn't help. (Remember, the stressor is inevitable; the warning cannot change the stressor, just the perception of it.)

How predictable is the stressor, in the absence of a warning? What if, one morning, an omnipotent voice says, "There is no way out of it; a meteor is going to crush your car while you're at work today (but it's the only time it will happen this year)." Not soothing. There's the good news that it's not going to happen again tomorrow, but that's hardly comforting; this is not an event that you anxiously fret over often. At the other extreme, what if one morning an omnipotent voice whispers, "Today it's going to be stressful on the freeway—lots of traffic, stops and go's. Tomorrow, too. In fact, every day this year, except November 9, when there'll hardly be any traffic, people will wave to each other, and a highway patrol cop will stop you in order to share his coffee cake with you." Who needs predictive information about the obvious fact that driving to work is going to be stressful? Thus, warnings are less effective for very rare stressors (you don't usually worry much about meteors) and very frequent ones (they approach being predictable even without the warning).

How far in advance of the stressor does the warning come? Each day, you go for a mysterious appointment: you are led into a room with your eyes closed and are seated in a deep, comfortable chair. Then, with roughly even probabilities but no warning, either a rich, avuncular voice reads you to sleep with your favorite childhood stories, or a bucket of ice water is sloshed over your head. Not a pleasing prospect, I would bet. Would the whole thing be any less unsettling if you were told which treat-

ment you were going to get five seconds before the event? Probably not—there is not enough time to derive any psychological benefits from the information. At the other extreme, how about predictive information long in the future? Would you wish for an omnipotent voice to tell you, "Eleven years and twenty-seven days from now your ice-water bath will last 10 full minutes"? Information either just before or long before the stressor does little good to alleviate the psychological anticipation.

Some types of predictive information can even increase the cumulative anticipatory stressor—vague information, for example. How about, "Attention all shoppers. Someone in this supermarket is going to be mugged soon by a man who looks menacing." Won't every man now look menacing to every person there? Or if the stressor is terrible enough, no degree of warning is likely to be welcome. Would you be comforted by the omnipotent message "Tomorrow an unavoidable accident will mangle your left leg, although your right leg will remain in great shape"?

Collectively, these scenarios tell us that predictability does not always work to protect us from stress. The much more systematic studies with animals suggest that it works only in a midrange of frequencies and intensities of stressors, and with certain lag times and levels of accurate information.

INTERNAL STRATEGIES FOR COPING

These ideas suggest to me that predictability, when it is working as a buffer from stress, helps humans differently from the way it does rats. The warning of impending shocks to the rat has little effect on the size of the stress-response *during* the shocks; instead, allowing the rat to feel more confident about when it *doesn't* have to worry reduces the rat's anticipatory stress-response the rest of the time. Analogously, when the dentist says, "Only two more times and then we're done," it allows us to relax *at the end* of the second burst of drilling. But I suggest, although I cannot prove it, that unlike the case for the rat, proper information will also lower our stress-response *during* the pain. If you were told "only two times more" versus "only ten times more," wouldn't you use different mental strategies to try to cope? With either scenario, you would pull out the comforting thought of "only one more and then it's the last one" at different times; you would save your most distracting fantasy for a different point; you would try counting to zero from different numbers. Predictive information lets us know what internal coping strategy is likely to work best during a stressor.

We often wish for information about the course of some medical problem because it aids our strategizing about how we will cope. A simple example: you have some minor surgery, and you are given predictive information—the first postsurgical day, there is going to be a lot of pain, pretty constant, whereas by the second day, you'll just feel a bit achy. Armed with that information, you are more likely to plan on watching the eight distracting videos on day one and to devote day two to writing delicate haikus than the other way around. And, among other reasons, we wish to optimize our coping strategies when we request the most devastating piece of medical information any of us will ever face: "How much time do I have left?"

SUBTLETIES OF CONTROL

To understand some important subtleties of the effects of control on stress, we need to return to the paradigm of the rat being shocked. It has been previously trained to press a lever to avoid shocks, and now it's pounding away like crazy on a lever. The lever does nothing; the rat is still getting shocked, but with less chance of an ulcer because the rat thinks it has control. To introduce a sense of control into the experimental design decreases the stress-response because, in effect, the rat is thinking, "Ten shocks an hour. Not bad; just imagine how bad it would be if I wasn't on top of it with my lever here." But what if things backfire, and adding a sense of control makes the rat think, "Ten shocks an hour, what's wrong with me? I have a lever here, I should have avoided the shocks, it's my fault." If you believe you have control over stressors that are, in fact, beyond your control, you may consider it somehow to be your fault that the inevitable occurred.

An inappropriate sense of control in the face of awful events can make us feel terrible. Some of our most compassionate words to people experiencing tragedy involve minimizing their perceived sense of control. "It's not your fault, no one could have stopped in time; she just darted out from between the cars." "It's not something you could have done anything about; you tried your best, the economy's just lousy now." "Honey, getting him the best doctor in the world couldn't have cured him." And some of the most brutally callous of society's attempts to shift blame attribute more personal control during a stressor than exists. "She was asking for it" (rape victims have the control to prevent the rape). "It's your fault that your child is schizophrenic" (schizophrenia is generated by poor mothering—a destructive belief that dominated psychiatry for decades before the disease was recognized to be neurochemical). "If they'd only made the effort to assimilate, they wouldn't have these problems" (minorities have the power to prevent their persecution).

The effects of the sense of control on stress are highly dependent on context. In general, if the stressor is of a sort where it is easy to imagine how much worse it could have been, inserting an artificial sense of control helps. "That was awful, but think of how bad it would have been if I hadn't done X." But when the stressor is truly awful, an artificial sense of control is damaging—it is difficult to conceive a yet-worse scenario that you managed to avoid, but easy to be appalled by the disaster you didn't prevent. You don't want to feel as if you could have controlled the uncontrollable when the outcome is awful. People with a strong internal locus of control (in other words, people who think they are the masters of their own ship—that what goes on around them reflects their actions) have far greater stress-responses than do those with external loci when confronted with something uncontrollable. This is a particular risk for the elderly (especially elderly men) as life generates more and more things beyond their control. And as we will see in the final chapter, there is even a personality type whose tendency to internalize control in the face of bad, uncontrollable things greatly increases the risk of a particular disease.

These subtleties about control and predictability help to explain a confusing feature of the stress literature. In general, the less control or predictability, the more at risk you are for a stress-induced disease. Yet an experiment conducted by Joseph Brady in 1958 with monkeys gave rise to the view that more control and more predictability cause ulcers. Half the animals could press a bar to delay shocks ("executive" monkeys); the other half were passively yoked to one of the "executives" such that they received a shock whenever the first one did. In this widely reported study, the executive monkeys were more likely to develop ulcers. Out of these studies came the popular concept of the "executive stress syndrome" and associated images of executive humans weighed down with the stressful burdens of control, leadership, and responsibility. Ben Natelson, of the V.A. Medical Center in East Orange, N.J., along with Jay Weiss, noted some problems with that study. First, it was conducted with parameters where control and predictability are bad news. Second, the "executive" and "nonexecutive" monkeys were not chosen randomly; instead, the monkeys that tended to press the bar first in pilot studies were selected to be executives. Monkeys that press sooner have since been shown to be more emotionally reactive animals, so Brady was inadvertently stacking the executive side with the more reactive, ulcer-prone monkeys. As we will see in Chapter 15, in general, executives of all species are more likely to be giving ulcers than to be getting them.

To summarize, stress-responses can be modulated or even caused by psychological factors, including loss of outlets for frustration and of social support, a perception of things worsening, and under some circumstances, a loss of control and of predictability. These ideas have vastly expanded our ability to answer the question "Why do only some of us get stress-related diseases?" Obviously we differ as to the number of stressors that befall us. And after all the chapters on physiology, you can guess that we differ in how fast our adrenals make glucocorticoids, how many insulin receptors we have in our fat cells, the thickness of our stomach walls, and so on. But in addition to those physiological differences, we can now add another dimension. We differ in the psychological filters through which we perceive the stressors in our world. Two people participating in the same event—a long wait at the supermarket checkout, public speaking, parachuting out of an airplane—may differ dramatically in their psychological perception of the event. "Oh, I'll just read a magazine while I wait" (control for frustration); "I'm nervous as hell, but by giving this after-dinner talk, I'm a shoo-in for that promotion" (things are getting better); "This is great—I always wanted to try sky-diving" (this is something I'm in control of).

In the next chapter we will consider depression as, in part, a failure of some psychological defenses against stress. The chapter after that will consider some personality types that habitually come up with coping responses that are inappropriate to the psychological stressors that occur. Following that is a chapter examining how your place in society, and the type of society that it is, can have profound effects on stress physiology and patterns of disease. And in the final chapter we will examine how stress-management techniques can aid us by teaching how to exploit these psychological defenses.

STRESS AND DEPRESSION

We are morbidly fascinated with the exotica of disease. They fill our made-for-television movies, our tabloids, and the book reports of adolescents thinking about entering medical school someday. Victorians with Elephant Man's disease, murderers with multiple personality disorders, ten-year-olds with progeria, idiot savants with autism, cannibals with kuru. Who could resist? But when it comes to the bread and butter of human misery, try a major depression. It can be life-threatening, it can sabotage careers for years on end, it can demolish the families of sufferers. And it is dizzyingly common—the psychologist Martin Seligman has called it the common cold of psychopathology. Best estimates are that from 5 to 20 percent of us will suffer a major, incapacitating depression at some point in our lives, causing us to be hospitalized or medicated or nonfunctional for a significant length of time.

This chapter differs a bit from those that preceded it, in which the concept of "stress" was at the very forefront. Initially, that may not seem to be the case in our focus on depression. The two appear to be inextricably linked, however, and the concept of stress will run through every page of this chapter. It is impossible to understand either the biology or psychology of major depressions without recognizing the critical role played in the disease by stress.

To begin to understand this connection, it is necessary to get some sense of the disorder's characteristics. We have first to wrestle with a semantic problem. *Depression* is a term that we all use in an everyday sense. Something mildly or fairly upsetting happens to us, and we get "the blues" for a while; then we recover. This is not what psychologists and psychiatrists mean when referring to a major depression. Instead, this vastly crippling disorder leads people to attempt suicide; its victims may lose their jobs, family, and all social contact because they cannot force themselves to get out of bed, or refuse to go to a psychiatrist because they feel they don't deserve to get better. It is a horrific disease, and throughout this chapter I will be referring to this major, devastating form of depression, rather than the transient blues that we may casually signify with the term "feeling depressed."

THE SYMPTOMS

The defining feature of a major depression is loss of pleasure. If I had to define a major depression in a single sentence, I would describe it as a "genetic-neurochemical disorder requiring a strong environmental trigger whose characteristic manifestation is an inability to appreciate sunsets." Depression can be as tragic as cancer or a spinal cord injury. Think about what our lives are about. None of us will live forever, and on occasion we actually believe that; our days are filled with disappointments, failures, unrequited loves. Despite this, almost inconceivably, we not only cope but even feel vast pleasures. I, for example, am resoundingly mediocre at soccer, but nothing keeps me from my game with other faculty members each week. Invariably there comes a moment when I manage to gum up someone more adept than I; I'm panting and heaving and pleased, and there's still plenty more time to play and a breeze and I suddenly feel dizzy with gratitude for my animal existence. What could be more tragic than a disease which, as its defining symptom, robs us of that capacity?

This trait is called *anhedonia*: *hedonism*, "the pursuit of pleasure"; *anhedonia*, "the inability to feel pleasure." Anhedonia is

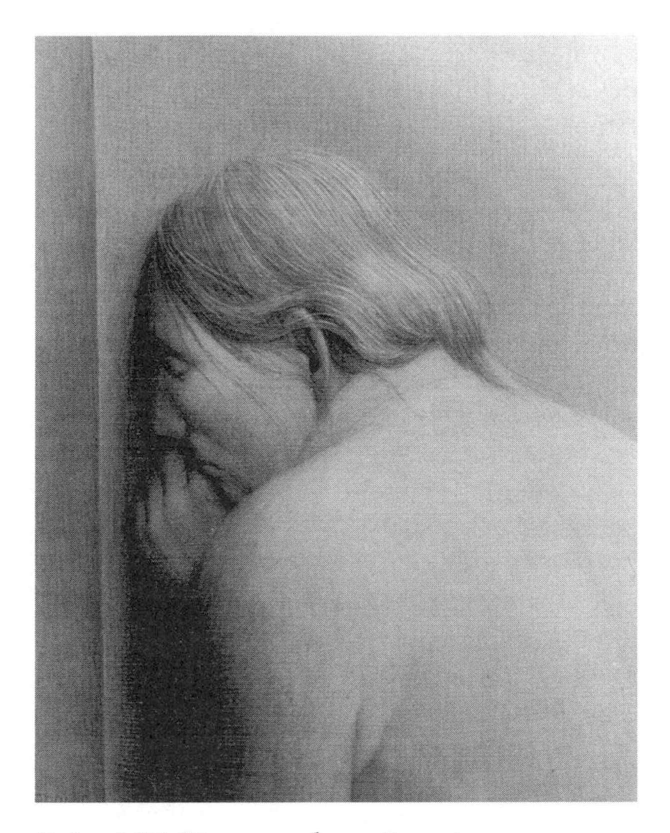

Tooker, 1974: Woman at the wall, *egg tempera on gesso.*

consistent among depressives. A woman has just received the long-sought promotion; a man has just become engaged to the woman of his dreams—and, amid their depression, they will tell you how they feel nothing, how it really doesn't count, how they don't deserve it.

Accompanying major depression are great grief and great guilt. We often feel grief and guilt in the everyday sadnesses that we refer to as "depression." But this occurs to an incapacitating degree in a major depression. In a subset of such patients, such thinking can take on the quality of a delusion. By this, I do not mean the thought-disordered delusions of schizophrenics; instead, delusional thinking in depressives is of the sort where facts are distorted, over- or underinterpreted to the point where one must conclude that things are terrible and getting worse, hopeless.

An example: a middle-aged man, out of the blue, has a major heart attack. Overwhelmed by his implied mortality, the transformation of his life, he slips into a major depression. Despite this, he is recovering from the attack reasonably well, and there is every chance that he will resume a normal life. But each day he's sure he's getting worse.

The hospital in which he is staying is circular in construction, with a corridor that forms a loop. One day, the nurses walk him once around the hospital before he collapses back in bed. The next day, he does two laps; he is getting stronger. That evening, when his family visits, he explains to them that he is sinking. "What are you talking about? The nurses said that you did two loops today; yesterday you only did one." No, no, he shakes his head sadly, you don't understand. He explains that the hospital is being renovated and, um, well, last night they closed off the old corridor and opened a newer, smaller one. And, you see, the distance around the new loop is less than half the distance of the old one, so twice on that is still less than I could do yesterday.

This particular incident occurred with the father of a friend, an engineer who lucidly described radii and circumferences, expecting his family to believe that the hospital had opened up a new corridor through the core of the building in one day. This is delusional thinking; the emotional energies behind the analysis and evaluation are disordered so that the everyday world is interpreted in a way that leads to depressive conclusions—it's awful, getting worse, and this is what I deserve.

Cognitive therapists—like Aaron Beck of the University of Pennsylvania—even consider depression to be primarily a disorder of thought, rather than emotion, in that sufferers tend to see the world in a distorted, negative way. Beck and colleagues have conducted striking studies that provide evidence for this. For example, they might show a subject two pictures. In the first, a group of people are gathered happily around a dinner table, feasting. In the second, the same people are gathered around a coffin. Show the two pictures rapidly or simultaneously; which one is remembered? Depressives see the funeral scene at rates higher than chance. They are not only depressed about something, but see the goings-on around them in a distorted way that always reinforces that feeling. Their glasses are always half empty.

Another frequent feature of a major depression is called *psychomotor retardation*. The person moves and speaks slowly. Everything requires tremendous effort and concentration. She finds the act of merely arranging a doctor's appointment exhausting. Soon it is too much even to get out of bed and get dressed. (It should be noted that not all depressives show psychomotor retardation; some may show the opposite pattern, termed *psychomotor agitation*.) People suffering from major depression also lose interest in sex. If anhedonia is a defining feature of the disorder, what is more likely to be affected than one of the most pleasurable things you can do with your body and mind?

Many of us tend to think of depressives as people who get the same everyday blahs as you and I, but that for them it just spirals out of control. We may also have the sense, whispered out of earshot, that these are people who just can't handle normal ups and downs, who are indulging themselves. (Why can't they just get themselves together?) A major depression, however, is as real a disease as diabetes. Another set of depressive symptoms supports that view. Basically, many things in the bodies of depressives work peculiarly; these are generally called *vegetative symptoms*. You and I get an everyday depression. What do we do? Typically, we sleep more than usual, probably eat more than usual, convinced in some way that such comforts will make us feel better. These traits are just the opposite of the vegetative symptoms seen in most people with major depressions. Eating declines. Sleeping does as well, and in a distinctive manner. While depressives have the trouble falling asleep that one might expect, they also have the problem of "early morning wakening," spending months on end sleepless and exhausted from three-thirty or so each morning. "Do you tend to wake up very early in the morning?" is, in fact, one of the most likely questions you would be asked by an admitting doctor if you arrived at an emergency room severely depressed. Not only is sleep shortened, but the "architecture" of sleep is different as well—the normal pattern of shifting between deep and shallow sleep, the rhythm of the onset of dream states, are disturbed.

An additional vegetative symptom is extremely relevant to this chapter, and was already mentioned in Chapter 10: major depressives often experience elevated levels of glucocorticoids. This is critical for a number of reasons that will be returned to, and helps to clarify what the disease is actually about. When looking at a depressive sitting on the edge of the bed, barely able to move, it is easy to think of the person as energyless, enervated. A more accurate picture is of the depressive as a tightly coiled spool of wire, tense, straining, active—but all inside. As we will see, a psychodynamic view of depression shows the person fighting an enormous, aggressive mental battle. In that regard, depressives bear some resemblance to an animal sprinting across the savanna—no wonder they have elevated levels of stress hormones.

Another feature of depression also confirms that it is a real disease, rather than merely the situation of someone who simply cannot handle everyday ups and downs. There are multiple types of depressions, and they can look quite different. In one variant, unipolar depression, the sufferer fluctuates from feeling extremely depressed to feeling reasonably normal. In another form, the person fluctuates between deep depression and wild, disorganized hyperactivity. This is called *bipolar depression* or, more familiarly, *manic depression* (to complicate things further, there are even subtypes of manic depression). Here we run into another complication because, just as we use *depression* in an everyday sense that is different from the medical sense, *mania* has an everyday connotation as well. We may use the term to refer to madness, as in made-for television homicidal maniacs. Or we could describe someone as being in a manic state when he is buoyed by some unexpected good news—talking quickly, laughing, gesticulating. But the mania found in manic depression is of a completely different magnitude. Let me give an example of the disorder: a woman comes into the emergency room. She's bipolar, completely manic, hasn't been taking her medication. She's on welfare, doesn't have a cent to her name, and in the last week she's bought *three* Cadillacs with money from loan sharks. And, get this, she doesn't even know how to drive. People in manic states will go for days on three hours of sleep a night and feel rested, will talk nonstop for hours at a time, will be vastly distractible, unable to concentrate amid their racing thoughts. In outbursts of irrational grandiosity, they will behave in ways that are foolhardy or dangerous to themselves and others—at the extreme, poisoning themselves in attempting to prove their immortality, burning down their homes, giving away their life savings to strangers. It is a profoundly destructive disease.

The strikingly different subtypes of depression and their variability suggest not just a single disease, but a heterogeneity of diseases that have different underlying biologies. Another feature of the disorder also indicates a biological abnormality. A patient comes to a doctor in the tropics. The patient is running a high fever that abates, only to come back a day or two later, abate again, return again, and so on every 48 to 72 hours. The doctor will recognize this instantly as malaria, because of the rhythmicity of the disorder. It has to do with the life cycle of the malarial parasite as it moves from red blood cells to the liver and spleen. The rhythmicity screams biology. In the same way, certain subtypes of depression have a rhythm. A manic-depressive may be manic for five days, severely depressed for the following week, then mildly depressed for half a week or so, and, finally, symptom-free for few weeks. Then the pattern starts up again—and may have been doing so for a decade. Good things, bad things happen, but the same cyclic rhythm continues—which suggests just as much deterministic biology as in the life cycle of the malarial parasite. In another subset of depression the rhythm is annual—sufferers get depressed during the winter. Called *seasonal affective disorders* (SADs), these are thought to be related to patterns of exposure to light. *Affective* is the psychiatric term for emotional responses. Again, the rhythmicity appears independent of external life events; there is a biological clock ticking away in there that has something to do with mood, and there is something seriously wrong with its ticking.

THE BIOLOGY OF DEPRESSION

The Neurochemistry Considerable evidence exists that something is awry with the chemistry of the brains of depressives. In order to appreciate that, it is necessary to learn a bit about how brain cells communicate with one another. The illustration on page 237 shows a schematic version of two neurons, the principal type of brain cell. If a neuron has become excited with some thought or memory (metaphorically speaking), its excitement is electrical—a wave of electricity sweeps from the dendrites over the cell body, down the axon to the axon terminals. When the wave of electrical excitation reaches the axon terminal, it releases chemical messengers that float across the synapse. These messengers—neurotransmitters—bind to specialized receptors on the adjacent dendrite, causing the second neuron to become electrically excited.

A minor piece of housekeeping, however: what happens to the neurotransmitter molecule after it has done its job and floats off the receptor? In some cases, it is recycled—taken back up by the axon terminal of the first neuron and repackaged for future use. Or it can be degraded in the synapse and the debris flushed out to sea (the cerebrospinal fluid, then to the blood, and then the urine). If these processes of clearing neurotransmitters out of the way fail (reuptake ceases or degradation stops or both), suddenly a lot more neurotransmitter remains in the synapse, giving a stronger signal to the second neuron than usual. Thus, the proper disposal of these powerful messengers is integral to normal neuronal communication.

There are trillions of synapses in the brain. Do we need trillions of chemically unique neurotransmitters? Certainly not. You can generate a seemingly infinite number of messages with a finite number of messengers; consider how many words we can form with the mere 26 letters in our alphabet. All you need are rules that allow for the same messenger to convey different meanings, metaphorically speaking, in different contexts. At one synapse, neurotransmitter A sends a message relevant to pancreatic regulation, while at another synapse the same neurotransmitter substance may pertain to adolescent crushes. There are many neurotransmitters, probably on the order of a few hundred, but certainly not trillions.

The best neurochemical evidence suggests that depression involves abnormal levels of one or both of a pair of neurotransmitters, norepinephrine and serotonin. Before reviewing the evidence, it's important to clear up a point. You are no doubt thinking, "Wasn't there something about norepinephrine and the sympathetic nervous system many chapters ago?" Absolutely, and that proves the point about the varied roles played

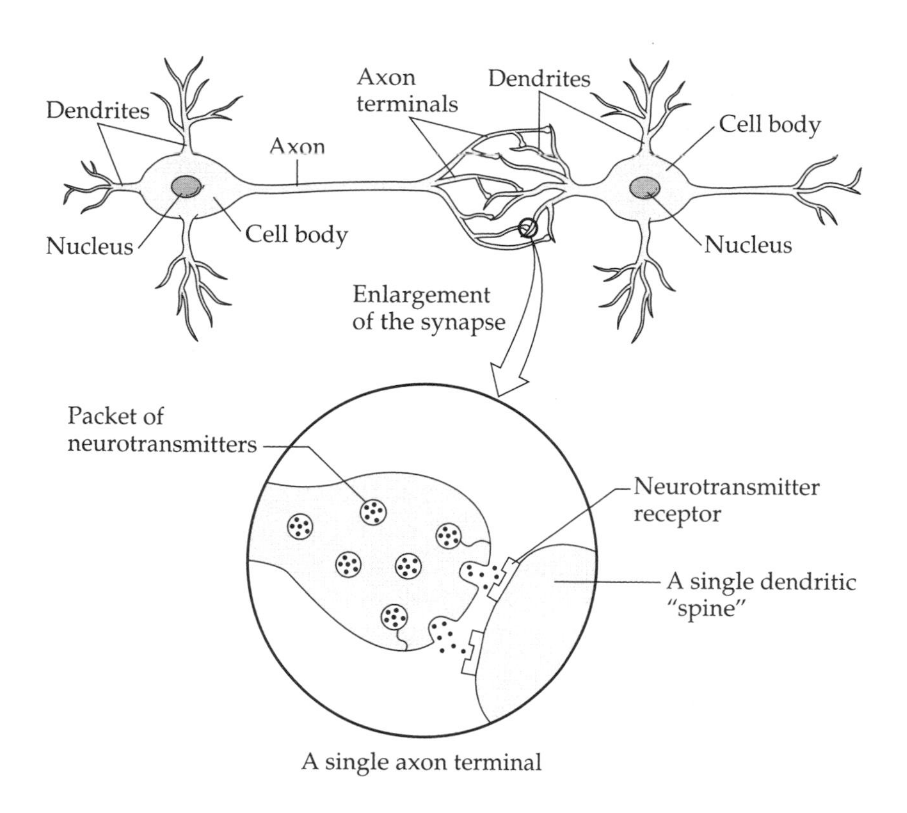

A neuron that has been excited conveys information to other neurons by means of chemical signals at synapses, the contact points between neurons. When the impulse reaches the axon terminal of the signaling neuron, it induces the release of neurotransmitter molecules. Transmitters diffuse across a narrow cleft and bind to receptors in the adjacent neuron's dendritic spine.

by any given neurotransmitter. In one part of the body (the heart, for example), norepinephrine is a messenger concerning arousal and the four F's, while in a different part of the nervous system, norepinephrine seems to have something to do with the symptoms of depression.

Why is it likely that there is something wrong with norepinephrine, serotonin, or both in depression? The best evidence is that most of the drugs that lessen depression increase the amount of signaling by these neurotransmitters. One class of antidepressants, called *tricyclics* (a reference to their biochemical structure), stops the recycling, or reuptake, of norepinephrine and serotonin into the axon terminals. The result is that the neurotransmitter remains in the synapse longer, and is likely to hit their respective receptors a second or third time. Another class of drugs, called *MAO inhibitors*, blocks the degradation of norepinephrine and serotonin in the synapse [by inhibiting the action of a crucial enzyme in that degradation, monoamine oxidase (MAO)]. The result, again, is that more of the messenger remains in the synapse to stimulate the dendrite of the receiving neuron. These findings generate a pretty straightforward conclusion: if you use a drug that increases the amount of norepinephrine and serotonin in synapses throughout the brain, and as a result, someone's depression gets better, there must have been too little of those neurotransmitters in the first place. Case closed.

Naturally, not so fast. As a first issue of confusion, is the problem in depression too little norepinephrine, too little serotonin, or both? The tricyclics and MAO inhibitors work on both neurotransmitter systems, making it impossible to tell which one is critical to the disease. People used to think norepinephrine was the culprit, when it was thought that those classical antidepressant drugs worked only on the norepinephrine synapse. These days, the pendulum has shifted in the direction of serotonin, mainly because of the development of a new generation of reuptake inhibitors that work only on serotonin synapses (*selective serotonin reuptake inhibitors*, or SSRIs, of which Prozac is the most famous). However, there still remains some reason to think that norepinephrine is part of the story as well, as some of the newest antidepressants appear to work exclusively on norepinephrine after all.*

A second piece of confusion is actually quite major. Is the defect in depression with serotonin (or norepinephrine, or both) really one of too *little* neurotransmitter in the synapse? You would think this was settled—the effective antidepressant drugs increase the amounts of these neurotransmitters in the synapse

---

* The current herbal rage, St. John's wort, is reputed to have antidepressant qualities without the frequent side effects seen with the traditional antidepressants. It seems likely that the drug should be at least somewhat efficacious, as it does appear to inhibit the uptake of serotonin and norepinephrine. However, it remains unclear just how effective it is, whether it does indeed have fewer side effects, and, if so, why that should be.

and alleviate depression; thus, the problem had to be too little of the stuff to begin with. However, some clinical data suggest that this might not be so simple.

The stumbling block has to do with timing. Expose the brain to some tricyclic antidepressant, and norepinephrine and serotonin signaling in the synapses changes within hours. However, give that same drug to a depressed person, and it takes weeks for the person to feel better. Something doesn't quite fit. Two theories have arisen in recent years that might reconcile this problem with timing, and they are both extremely complicated.

Revisionist theory 1, the "it's not too little neurotransmitter, it's actually too much norepinephrine" hypothesis. First, some orientation. If somebody constantly yells at you, you stop listening. Analogously, if you inundate a cell with lots of a neurotransmitter, the cell will not "listen" as carefully—it will "down-regulate" (decrease) the number of receptors for that neurotransmitter, in order to decrease its sensitivity to that messenger. If, for example, you double the amount of serotonin reaching the dendrites of a cell and that cell down-regulates its serotonin receptors by 50 percent, the changes roughly cancel out. If the cell down-regulates less than 50 percent, the net result is more serotonin signaling in the synapse; if more than 50 percent, the result is actually less signaling in the synapse. In other words, how strong the signal is in a synapse is a function both of how loudly the first neuron yells (the amount of neurotransmitter released) and of how sensitively the second neuron listens (how many receptors it has for the neurotransmitter).

OK, ready. This revisionist theory states that the original problem is that there is actually too much norepinephrine, serotonin, or both in parts of the brains of depressives. What happens when you prescribe antidepressants that increase signaling of these neurotransmitters even further? At first, that should make the depressive symptoms worse. (Some psychiatrists argue that this actually does occur.) Over the course of a few weeks, however, the dendrites say, "This is intolerable, all this neurotransmitter; let's down-regulate our receptors a whole lot." If this occurs and, critical to the theory, more than compensates for the increased neurotransmitter signal, the depressive problem of excessive norepinephrine signaling goes away: the person feels better.

Revisionist theory 2, "it really is too little norepinephrine or serotonin after all." This theory is even more complicated than the first, and also requires orientation. Not only do dendrites contain receptors for neurotransmitters, but it also turns out that on the axon terminals of the "sending" neuron, as well, there are receptors for the very neurotransmitters being released by that neuron. What possible purpose could these so-called autoreceptors serve? Neurotransmitters are released, float into the synapse, bind to the standard receptors on the second neuron. Some neurotransmitter molecules, however, will float back and wind up binding to the autoreceptors. They serve as some sort of feedback signal; if, say, 5 percent of the released neurotransmitter reaches the autoreceptors, the first neuron can count its metaphorical toes, multiply by 20, and figure out how much neurotransmitter it has released. Then it can make some decisions—should I release more neurotransmitter or stop now? Should I start synthesizing more? and so on. If this process lets the first neuron do its bookkeeping on neurotransmitter expenditures, what happens if the neuron down-regulates a lot of these autoreceptors? Underestimating the amount of neurotransmitter it has released, the neuron will inadvertently start increasing the amount it synthesizes and discharges.

With this as background, here's the reasoning behind the second theory (that there really is too little norepinephrine or serotonin in a part of the brain of depressives). Give the antidepressant drugs that increase signaling of these neurotransmitters. Because of the increased signaling, over the course of weeks there will be down-regulation of norepinephrine and serotonin receptors. Critical to this theory is the idea that the autoreceptors on the first neuron will down-regulate to a greater extent than the receptors on the second neuron. If that happens, the second neuron may not be listening as well, but the first one will be releasing sufficient extra neurotransmitter to more than overcome that. The net result is enhanced norepinephrine or serotonin signaling, and depressive symptoms abate. [This mechanism may explain the efficacy of electroconvulsive therapy (ECT, or "shock therapy"). For decades psychiatrists have used this technique to alleviate major depressions, and no one has quite known why it works. It turns out that, among its many effects, ECT decreases the number of norepinephrine autoreceptors, at least in experimental animal models.]

If you are confused by now, you are in some good company, as the entire field is extremely unsettled. Serotonin or norepinephrine? Too much or too little signaling? If it is, for example, too little serotonin signaling, is it because too little serotonin is being released into synapses, or because there is some defect blunting the sensitivity of serotonin receptors? (To give you a sense of how big a can of worms that one is, there are currently recognized fifteen different types of serotonin receptors, with differing functions, efficacies, and distributions in the brain.) Maybe there are a variety of different neurochemical routes for getting to a depression, and different pathways are associated with different subtypes of depression (unipolar versus manic depression, or one that is triggered by outside events versus one that runs with its own internal clockwork, or one dominated by psychomotor retardation versus one dominated by suicidalism . . .). This is a very reasonable idea, but the evidence for it is still scant.

Amid all those questions, another good one—why does having too much or too little of serotonin or norepinephrine cause a depression? There are a lot of links between these neurotransmitters and function (between serotonin and impulsivity, for example, or norepinephrine and psychomotor retardation). But despite those hints and lots of theorizing, there is nothing resembling a clear answer to that one either.

The function of another neurotransmitter has caused it to be implicated in depression as well. The neurotransmitter is called *dopamine,* and it has something to do with pleasure. Several decades ago, some neuroscientists made a fundamental discovery. They had implanted electrodes into the brains of rats and stimulated areas here and there, seeing what would happen. By doing so, they found an extraordinary area of the brain. Whenever this area was stimulated, the rat became *unbelievably happy*. How can one tell when a rat is unbelievably happy? You ask the rat to tell you, by charting how many times it is willing to press a lever in order to be rewarded with stimulation in that part of the brain. It turns out that rats will work themselves to death on that lever to get stimulation. They would rather be stimulated there than get food when they are starving, or have sex, or receive drugs even when they're addicted and going through withdrawal. The region of the brain targeted in these studies was promptly called the "pleasure pathway" and has been famous since.

That humans have a pleasure pathway was discovered shortly afterward by stimulating a similar part of the human brain during neurosurgery.* The results are pretty amazing. Something along the lines of "Aaaaah, boy, that feels good. It's kind of like getting your back rubbed but also sort of like sex or playing in the backyard in the leaves when you're a kid and Mom calling you in for hot chocolate and then you get into your pajamas with the feet . . ." Where can we sign up?

Although the neurochemistry of this pleasure pathway is not well understood, there is some indication that certain synapses along the pathway use dopamine. The strongest evidence for this is the ability of drugs that mimic dopamine, such as cocaine, to act as euphoriants. Suddenly, it seems plausible to hypothesize that depression, which is characterized above all by anhedonia, might involve too little dopamine and, thus, dysfunction of those pleasure pathways. Nevertheless, the preponderance of antidepressive drugs that work on serotonin and norepinephrine means that dopamine still ranks only third on the list when thinking about the neurochemistry of depression.

The Neuroanatomy This covers some of the major themes about the chemistry of depression. I introduce an illustration here of what the brain looks like, to consider a second way in which brain function might be abnormal in depressives, in addition to the neurochemistry just discussed. One region regulates processes like your breathing and heart rate. It includes the hypothalamus, which is busy releasing hormones and instructing the autonomic nervous system. If your blood pressure drops drastically, causing a compensatory stress-response, it is the hypothalamus, midbrain, and hindbrain that kick into gear. All sorts of vertebrates have roughly the same connections here.

Layered on top of that is a region called the *limbic system*, the functioning of which is related to emotion. As mammals, we

---

* Because the brain is not sensitive to pain, a lot of such surgery is done on patients who are awake (with their scalps anesthetized, of course). This is helpful because prior to modern imaging techniques, surgeons often had to have the patient awake to guide what they were doing. Place an electrode in the brain, stimulate, the patient flops her arm. Go a little deeper with the electrode, stimulate, and the patient flops her leg. Quick, consult your brain road map, figure out where you are, go an inch deeper, hang a left past the third neuron, and there's the tumor. That sort of thing.

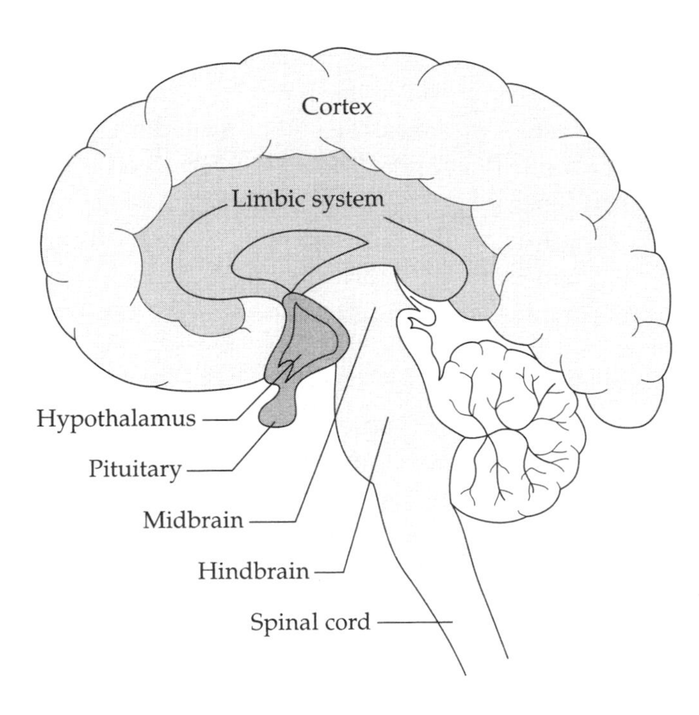

The triune brain.

have large limbic systems; lizards have relatively tiny limbic systems—they are not noted for the complexity of their emotional lives. If you get a stress-response from smelling the odor of a threatening rival, it's your limbic system that is involved.

Above that is the cortex. Everyone in the animal kingdom has some, but it is a real primate specialty. The cortex does abstract cognition, invents philosophy, remembers where your car keys are. The stuff of Chapter 10.

Now think for a second. Suppose you are gored by an elephant. You may feel a certain absence of pleasure afterward, maybe a sense of grief. Throw in a little psychomotor retardation—you're not as eager for your calisthenics as usual. Sleeping and feeding may be disrupted, glucocorticoid levels may be a bit on the high side. Sex may lose its appeal for a while. Hobbies are not as enticing; you don't jump up to go out with friends; you pass up that all-you-can-eat buffet. Sound like some of the symptoms of a depression?

Now what happens during a depression? You think a thought—about your mortality or that of a loved one; you imagine children in refugee camps, the rain forests disappearing and endless species of life evaporating, late Beethoven string quartets—and suddenly you experience some of the same symptoms as after being gored by the elephant. On an incredibly simplistic level, you can think of depression as occurring when your cortex thinks an abstract negative thought and manages to convince the rest of the brain that this is as real as a physical stressor. In this view, people with chronic depressions are those whose cortex habitually whispers sad thoughts to the rest of the brain. Thus, an astonishingly crude prediction: *cut* the connections between the cortex and the rest of a depressive's brain, and the cortex will no longer be able to get the rest of the brain depressed.

Remarkably, it actually works sometimes. The procedure is called a *cingulotomy*, or a *cingulum bundle cut*, and neurosurgeons may perform it on people with vastly crippling depressions that are resistant to drugs, ECT, or other forms of therapy. Afterward, the flow of information from the cortex to the rest of the brain is greatly curtailed (by cutting projection pathways from the former to the latter, you are cutting axons), and depressive symptoms seem to abate. (What else changes after a cingulotomy? If the cortex can no longer send abstract thoughts to the rest of the brain, the person should not only lose the capacity for abstract misery but for abstract pleasure as well, and this is what happens; but cingulotomy is a therapy employed only in patients completely incapacitated by their illness, who spend decades on the back ward of some state hospital, rocking and clutching themselves and feebly attempting suicide with some regularity.)

The Endocrinology So we have a few biological links in depression. There also appear to be hormonal abnormalities in many cases. Some depressives secrete abnormally high levels of glucocorticoids. This adds to the picture that depressed people, sitting on the edge of their beds without the energy to get up, are vigilant and aroused, with a hormonal profile to match—but the battle is inside them. The elevated glucocorticoid levels appear to be due to too much of a stress signal from the brain (back to Chapter 2—remember that the adrenals typically secrete glucocorticoids only when they are commanded to by the brain, via the pituitary), rather than the adrenals just getting some glucocorticoid hiccup all on their own now and then. Moreover, the glucocorticoid hypersecretion is due to what is called *feedback resistance*—in other words, the brain is less effective than it should be at shutting down glucocorticoid secretion. Normally, the levels of this hormone are tightly regulated—the brain senses circulating glucocorticoid levels, and if they get higher than desired (the "desired" level shifts depending on whether events are calm or stressful), the brain stops secreting CRF. Just like the regulation of water in a toilet bowl tank. In depressives, this feedback regulation fails—concentrations of circulating glucocorticoids that should shut down the system fail to do so, as the brain does not sense the feedback signal.*

As another hormonal feature of the story, people who secrete too little thyroid hormone can develop major depressions. This is particularly important because many people, seemingly with depressions of a purely psychiatric nature, turn out to have thyroid disease.

There is another aspect of depression in which hormones may play a role. The incidence rates of major depressions differ greatly, with women suffering far more than men. Why? One theory, from the school of cognitive therapy, concentrates on the ways in which women and men tend to think differently. When something upsetting happens, women are more likely to ruminate over it—think about it or want to talk about it with someone else. And men, terrible communicators that they so often

---

* For those into the minutiae of depression research, such feedback resistance is shown by what is called a *dexamethasone suppression test*. Dexamethasone is a synthetic glucocorticoid. Normally, inject someone with a large pulse of the stuff, and the brain senses it, in effect thinking, "Jeez, I don't know what is going on at the adrenals, but they just secreted a ton of glucocorticoids. Let's not secrete any more CRF." The secretion of CRF, and thus ACTH, and thus adrenal glucocorticoids, goes way down. The system is sensitive to a feedback signal. When there is, instead, feedback resistance, the brain doesn't respond sufficiently to the dexamethasone signal and keeps on secreting CRF. Depressives are often "dexamethasone-resistant"—a technical way of describing why their glucocorticoid levels are elevated.

are, are more likely to want to think about anything but the problem, or even better, go and *do* something—exercise, use power tools, get drunk, start a war. A ruminative tendency, the cognitive psychologists argue, makes you more likely to become depressed.

Another theory about the sex difference is psychosocial in nature. As we will see, much theorizing about the psychology of depression suggests that it is a disorder of lack of power and control, and some scientists have speculated that because women in so many societies traditionally have less control over the circumstances of their lives than do men, they are at greater risk for depression. Both theories are fine, except they fail to explain why women and men have the same rates of bipolar depressions; it is only unipolar depressions that are more common among women.

Both these theories seem particularly weak in their failure to explain a major feature of female depressions—namely, that women are particularly at risk for depressions at certain reproductive points: menstruation, menopause, and most of all, the weeks immediately after giving birth. A number of researchers believe such increased risks are tied to the great fluctuations that occur during menstruation, menopause, and parturition in two main hormones: estrogen and progesterone. As evidence, they cite the fact that women can get depressed when they artificially change their estrogen or progesterone levels (for example, when taking birth-control pills). Critically, both of these hormones can regulate neurochemical events in the brain—including the metabolism of neurotransmitters such as norepinephrine and serotonin. With massive changes in hormone levels (a thousandfold for progesterone at the time of giving birth, for example), current speculation centers on the possibility that the ratio of estrogen to progesterone can change radically enough to trigger a major depression. This is a new area of research with some seemingly contradictory findings, but there is more and more confidence among scientists that there is a hormonal contribution to the preponderance of female depressions.

Strong evidence exists then that there are biological underpinnings to major depression—imbalances in levels of neurotransmitters, and perhaps hormonal abnormalities as well. It's time to see what stress has to do with the disease.

STRESS AND THE ONSET OF DEPRESSION

The first stress-depression link is an obvious one. Statistically, stress and the onset of depression tend to go together. People who are undergoing a lot of significant life stressors are more likely than average to succumb to a major depression, and people sunk in their first major depression are more likely than average to have undergone a recent and significant stressor. However, as noted, some people have the grave misfortune of suffering from repeated depressive episodes, ones that can take on a rhythmic pattern stretching over years. When considering the case histories of those people, stressors emerge as triggers for only the first few depressions. Somewhere around the fourth depression or so, a mad clockwork takes over, and the depressive waves crash, regardless of what is going on in the outside world. What that transition is about will be considered below.

Laboratory studies also link stress and the symptoms of depression. Stress a lab rat, and it becomes anhedonic. Specifically, it takes a stronger electrical current than normal in the rat's pleasure pathways to activate a sense of pleasure. The threshold for perceiving pleasure has been raised, just as in a depressive. Furthermore, sustained stress can bring about one of the main endocrine markers of depression—chronic stress causes dexamethasone resistance in animals and humans, by blunting the sensitivity with which the brain can detect circulating glucocorticoids. (Dexamethasone resistance is by now familiar to those of you who have been assiduously reading the footnotes.) So stress can trigger many of the features of depression.

Critically, glucocorticoids can do the same. A key point in Chapter 10 was how glucocorticoids and stress could disrupt memory. Part of the evidence for that came from people with Cushing's syndrome (as a reminder, that is a condition in which any of a number of different types of tumors wind up causing vast oversecretion of glucocorticoids), as well as from people prescribed high doses of glucocorticoids to treat a number of ailments. It has also been known for decades that a significant subset of Cushingoid patients and people prescribed exogenous glucocorticoids become clinically depressed, independent of memory problems. This has been a bit tricky to demonstrate—

does someone with Cushing's syndrome or taking high pharmacological doses of synthetic glucocorticoids get depressed because glucocorticoids cause that state, or is it because they recognize they have a depressing disease? You show that it is the glucocorticoids which are the culprits by demonstrating higher depression rates in this population than among people with, for example, the same disease and the same severity but not receiving glucocorticoids. At this stage, there's also not much of a predictive science to this phenomenon—for example, no clinician can reliably predict beforehand which patient is going to get depressed when put on high-dose glucocorticoids, let alone at what dose, and whether it is when the dose is raised or lowered to that level. Nonetheless, have lots of glucocorticoids in the bloodstream and the risk of a depression increases.

The similarity in neurochemistry supports the stress-depression link as well. The preceding section detailed, at great length, the considerable confusion about the disease—depression seems to have something or other to do with serotonin or norepinephrine or dopamine. To the extent that this is the case, the glucocorticoid angle fits well, in that the hormones can alter features of all three neurotransmitter systems—the amount of neurotransmitter synthesized, how fast it is broken down, how many receptors there are for each neurotransmitter, how well the receptors work, and so on. Moreover, stress has been shown to cause many of the same changes as well. It is not clear which of those glucocorticoid effects are most important in understanding the links between stress and depression—in large part because it is not clear which neurotransmitter or neurotransmitters are most important. However, it is probably safe to say is that whatever neurochemical abnormalities wind up being shown definitively to underlie depression, there is precedent for stress and glucocorticoids causing those same abnormalities.

The glucocorticoid-depression link has some important implications. When I first introduced that link at the beginning of the chapter, it was meant to give some insight into the flavor of what a depression is like—a person looks like an enervated sea sponge, sitting there motionless on the edge of his bed, but he's actually boiling, in the middle of an internal battle. Tacit in that description was the idea that undergoing a depression is actually immensely stressful, and, therefore, among other things, stimulates glucocorticoid secretion. The data just reviewed suggest the opposite scenario—stress and glucocorticoid excess can be a *cause* of depression, rather than merely a consequence.

If that is really the case, then a novel clinical intervention should work: take one of those depressives with high glucocorticoid levels, find some drug that works on the adrenals to lower glucocorticoid secretion, and the depression should lessen. And, very excitingly, in just the last few years, that has been shown. The approach is filled with problems. You don't want to suppress glucocorticoid levels too much because, umpteen pages into this book, it should be apparent by now that those hormones are pretty important. Moreover, the "adrenal steroidogenesis inhibitors," as those drugs are called, can have some nasty side effects. Nevertheless, the demonstration that antiglucocorticoids can act as antidepressants in at least some cases is immensely exciting—not only teaching us something about the bases of depression, but possibly opening the way for a whole new generation of medications for the disease.

Some investigators have built on these observations with a fairly radical suggestion. For those biological psychiatrists concerned with the hormonal aspects of depression, the traditional glucocorticoid scenario was outlined above: depressions are stressful and raise glucocorticoid levels; when someone is treated with antidepressants, the chemistry of serotonin (or norepinephrine, etc.) is normalized, lessening the depression and, oh by the way, making life feel less stressful, with glucocorticoid levels returning to normal as a by-product. The new scenario is the logical extension of the inverted causality just discussed as well: for any of a number of reasons, glucocorticoid levels rise in someone (because the person is under a lot of stress, because something about the regulatory control of glucocorticoids is awry in that person), causing changes in the chemistry of serotonin (or norepinephrine, etc.) and a depression; antidepressants work by normalizing glucocorticoid levels, thereby normalizing the brain chemistry and alleviating the depression.

For this view to be supported, it has to be shown that the primary mechanism of action of the different classes of antidepressants is to work on the glucocorticoid system, and that changes in glucocorticoid levels precede the changes in brain chemistry or depressive symptoms. A few researchers have presented evidence that antidepressants work to rapidly alter numbers of glucocorticoid receptors in the brain, altering regulatory control of the system and lowering glucocorticoid levels; other researchers have not observed this. As usual, more research is needed. But even if it turns out that in some patients, depression is driven by elevated glucocorticoid levels (and recovery from depression thus mediated by reduction of those levels), that can't be the general mechanism of the disease in all cases: only about half of depressives actually have elevated glucocorticoid levels—in the other half, the glucocorticoid axis seems to work perfectly normally. Perhaps this particular stress-depression link is relevant only during the first few rounds of someone's depression (before the endogenous rhythmicity kicks in), or only in a subset of individuals.

We have now seen ways in which stress and glucocorticoids are intertwined with the biology of depression. That intertwining is made even tighter when considering the psychological picture of the disease.

STRESS AND THE PSYCHODYNAMICS OF MAJOR DEPRESSIONS

I have to begin with Freud. I know it is terribly stylish to dump on Freud these days, and I suppose some of it is deserved, but there is much that he still has to offer. There are few other scientists I can think of who, eighty years after they have completed the bulk of their work, are still considered important and correct enough for anyone to want to bother pointing out their errors instead of just consigning them to the library archives.

Freud was fascinated with depression and focused on the issue that we began with—why is it that most of us can have occasional terrible experiences, feel depressed, and then recover, while a few of us collapse into major depression ("melancholia")? In his classic essay "Mourning and Melancholia" (1917), Freud began with what the two have in common. In both cases, he felt, there is the loss of a love object. (In Freudian terms, such an "object" is usually a person, but can also be a goal or an ideal.) In Freud's formulation, in every loving relationship there is ambivalence, mixed feelings—elements of hatred as well as love. In the case of a small, reactive depression—mourning—you are able to deal with those mixed feelings in a healthy manner: you lose, you grieve, and then you recover. In the case of a major melancholic depression, you have become obsessed with the ambivalence—the simultaneity, the irreconcilable nature of the intense love alongside the intense hatred. Melancholia—a major depression—Freud theorized, is the internal conflict generated by this ambivalence.

This can begin to explain the intensity of grief experienced in a major depression. If you are obsessed with the intensely mixed feelings, you grieve doubly after a loss—for your loss of the loved individual *and* for the loss of any chance now to ever resolve the difficulties. "If only I had said the things I needed to, if only we could have worked things out"—for all of time, you have lost the chance to purge yourself of the ambivalence. For the rest of your life, you will be reaching for the door to let you into a place of pure, unsullied love, and you can never reach that door.

It also explains the intensity of the guilt often experienced in major depression. If you truly harbored intense anger toward the person along with love, in the aftermath of your loss there must be some facet of you that is celebrating, alongside the grieving. "He's gone; that's terrible but . . . thank god, I can finally live, I can finally grow up, no more of this or that." Inevitably, a metaphorical instant later, there must come a paralyzing belief that you have become a horrible monster to feel any sense of relief or pleasure at a time like this. Incapacitating guilt.

This theory also explains the tendency of major depressives in such circumstances to, oddly, begin to take on some of the traits of the lost loved/hated one—and not just any traits, but invariably the ones that the survivor found most irritating. Psychodynamically, this is wonderfully logical. By taking on a trait, you are being loyal to your lost, beloved opponent. By picking an irritating trait, you are still trying to convince the world you were right to be irritated—you see how you hate it when I do it; can you imagine what it was like to have to put up with that for years? And by picking a trait that, most of all, *you* find irritating, you are not only still trying to score points in your argument with the departed, but you are punishing yourself for arguing, as well.

Out of the Freudian school of thought has come one of the more apt descriptions of depression—"aggression turned inward." Suddenly the loss of pleasure, the psychomotor retardation, the impulse to suicide all make sense. As do the elevated glucocorticoid levels. This does not describe someone too lethargic to function; it is more like the actual state of a patient in depression, exhausted from the most draining emotional conflict of his or her life—one going on entirely within. If that doesn't count as psychologically stressful, I don't know what does.

Like other good parts of Freud, these ideas are empathic and fit many clinical traits; they just feel "right." But they are hard to assimilate into modern science, especially biologically oriented psychiatry. There is no way to study the correlation between serotonin receptor density and internalization of aggression, for example, or the effects of estrogen/progesterone ratios on love/hate ratios. The branch of psychological theorizing about depression that seems most useful to me, and is most tightly linked to stress, comes from experimental psychology. Work in this field has generated an extraordinarily informative model of depression.

STRESS, LEARNED HELPLESSNESS, AND DEPRESSION

In order to appreciate the experimental studies underlying this model, recall that in the preceding chapter on psychological stress, we saw that certain features dominated as psychologically stressful: a loss of control and of predictability within certain contexts, a loss of outlets for frustration, a loss of sources of support, a perception of life worsening. In one style of experiments, pioneered by the psychologists Martin Seligman and Steven Maier, animals are exposed to pathological amounts of these psychological stressors. The result is a condition strikingly similar to a human depression.

While the actual stressors may differ, the general approach in these studies always emphasizes repeated stressors with a complete absence of control on the part of the animal. For example, a rat may be subjected to a long series of frequent, uncontrollable, and unpredictable shocks or noises, with no outlets.

After a while, something extraordinary happens to that rat. This can be shown with a test. Take a fresh, unstressed rat, and give it something easy to learn. Put it in a room, for example, with the floor divided into two halves. Occasionally, electricity that will cause a mild shock is delivered to one half, and just beforehand, there is a signal indicating which half of the floor is about to be electrified. Your run-of-the-mill rat can learn this "active avoidance task" easily, and within a short time it readily and calmly shifts the side of the room it sits in according to the signal. Simple. Except for a rat who has recently been exposed to repeated uncontrollable stressors. *That rat cannot learn the task.* It does not learn to cope. On the contrary, it has learned to be helpless.

This phenomenon, called *learned helplessness,* is quite generalized; the animal has trouble coping with all sorts of varied tasks after its exposure to uncontrollable stressors. Such helplessness extends to tasks having to do with its ordinary life, like competing with another animal for food, or avoiding social aggression. One might wonder whether the helplessness is induced by the physical stress of receiving the shocks or, instead, the psychological stressor of having no control over or capacity to predict the shocks. It is the latter. The clearest way to demonstrate this is to "yoke" pairs of rats—one gets shocked under conditions marked by predictability and a certain degree of control; the other rat gets the identical pattern of shocks, but without the control or predictability. Only the latter rat becomes helpless.

Seligman argues persuasively that animals suffering from learned helplessness share many psychological features with depressed humans. Such animals have a *motivational* problem—one of the reasons that they are helpless is that they often do not even attempt a coping response when they are in a new situation. This is quite similar to the depressed person who doesn't even try the simplest task that would improve her life. "I'm too tired, it seems overwhelming to take on something like that, it's not going to work anyway. . . ."

Animals with learned helplessness also have a *cognitive* problem, something awry with how they perceive the world and think about it. When they do make the rare coping response, they can't tell whether it works or not. For example, if you tighten the association between a coping response and a reward, a normal rat's response rate increases (in other words, if the coping response works for the rat, it persists in that response). In contrast, linking rewards more closely to the rare coping responses of a helpless rat has little effect on its response rate. Seligman believes that this is not a consequence of helpless animals somehow missing the rules of the task; instead, he thinks, they have actually *learned not to bother* paying attention. By all logic, that rat should have learned, "When I am getting shocked, there is absolutely nothing I can do, and that feels terrible, but it isn't the whole world; it isn't true for everything." Instead, it has learned, "There is nothing I can do. Ever." Even when control and mastery are potentially made available to it, the rat cannot perceive them.* This is very similar to the depressed human who always sees glasses half empty. As Beck and other cognitive therapists have emphasized, much of what constitutes a depression is centered around responding to one awful thing and overgeneralizing from it—cognitively distorting how the world works.

The learned helplessness paradigm produces animals with other features strikingly similar to those in humans with major depressions. There is a rat's equivalent of anhedonia—the rat stops grooming itself and loses interest in sex and food. The rat's failure even to attempt coping responses suggests that it experiences an animal equivalent of psychomotor retardation.† In some models of learned helplessness, animals mutilate themselves, biting at themselves. Many of the vegetative symptoms

---

* A fascinating variation: another set of parallel experiments has shown that when an organism is *rewarded* consistently with no control or predictability (that is, no matter what the subject does, it is rewarded), it also has difficulty afterward learning coping responses. Seligman calls this a "spoiled brat" design; the investigators who discerned this phenomenon in pigeons used the much more provocative phrase "learned laziness."

† One might wonder if the entire learned helplessness phenomenon is really just about psychomotor retardation. Perhaps the rat is so wiped out after the uncontrollable shocks that it simply doesn't have the energy to perform active avoidance coping tasks. This would shift the emphasis away from learned helplessness as a cognitive state ("there is nothing I can do about this") or an anhedonic emotional state ("nothing feels pleasurable") to one of psychomotor inhibition ("everything seems so exhausting that I'm just going to sit here"). Seligman and Maier strongly object to this interpretation and present data showing that rats with learned helplessness are not only as active as control rats but, more importantly, are also impaired in "passive avoidance tasks"—learning situations where the coping response involves remaining still, rather than actually *doing* something (in other words, situations where a little psychomotor retardation should *help*). Championing the psychomotor retardation view is another major figure in this field, Jay Weiss, who presents an equal amount of data showing that "helpless" rats perform normally on passive avoidance tasks, indicating that the helplessness is a motor phenomenon and not a cognitive or emotional one. This debate has been going on for twenty years, and I sure don't know how to resolve the conflicting views.

appear as well—sleep loss and disorganization of sleep architecture, elevated glucocorticoid levels. Most critically, these animals tend to be depleted of norepinephrine in certain parts of the brain, while antidepressant drugs and ECT speed up their recovery from the learned helplessness state.

Learned helplessness has been induced in rodents, cats, dogs, birds, fish, insects, and primates—including humans. It takes surprisingly little in terms of uncontrollable unpleasantness to make humans give up and become helpless in a generalized way. In one study by Donald Hiroto, student volunteers were exposed to either escapable or inescapable loud noises (as in all of such studies, the two groups were paired so that they were exposed to the same amount of noise). Afterward, they were given a learning task in which a correct response turned off a loud noise; the "inescapable" group was significantly less capable of learning the task. Helplessness can even be generalized to nonaversive learning situations. Hiroto and Seligman did a follow-up study in which, again, there was either controllable or uncontrollable noise. Afterward the latter group was less capable of solving simple word puzzles. Giving up can also be induced by stressors far more subtle than uncontrollable loud noises. In another study, Hiroto and Seligman gave volunteers a learning task in which they had to pick a card of a certain color according to rules that they had to discern along the way. In one group, these rules were learnable; in the other group, the rules were not (the card color was randomized). Afterward, the latter group was less capable of coping with a simple and easily solved task. Seligman and colleagues have also demonstrated that unsolvable tasks induced helplessness afterward in social coping situations.

Thus humans can be provoked into at least transient cases of learned helplessness, and with surprising ease. Naturally, there is tremendous individual variation in how readily this happens—some of us are more vulnerable than others (and you can bet that this is going to be important in considering stress management in the final chapter). In the experiment involving inescapable noise, Hiroto had given the students a personality inventory beforehand. Based on that, he was able to identify the students who came into the experiment with a strongly "internalized locus of control"—a belief that they were the masters of their own destiny and had a great deal of control in their lives—and, in contrast, the markedly "externalized" volunteers, who tended to attribute outcomes to chance and luck. In the aftermath of the uncontrollable stressor, the externalized students were far more vulnerable to learned helplessness.

Collectively, these studies strike me as extremely important in forming links between stress, personality, and depression. Our lives are replete with incidents in which we become irrationally helpless. Some are silly and inconsequential. Once in the African camp that I shared with Laurence Frank, the zoologist whose hyenas figured in the chapter on reproduction, we managed to make a disaster of preparing macaroni and cheese over the campfire. Inspecting the mess, we ruefully admitted that it might have helped if we had bothered to read the instructions on the box. Yet we had both avoided doing that; in fact, we both felt a formless dread about trying to make sense of such instructions. Frank summed it up: "Face it. We suffer from learned cooking helplessness."

But life is full of more significant examples. If a teacher at a critical point of our education, or a loved one at a critical point of our emotional development, frequently exposes us to his or her own specialized uncontrollable stressors, we may grow up with distorted beliefs about what we cannot learn or ways in which we are unlikely to be loved. In one chilling demonstration of this, some psychologists studied inner-city schoolkids with severe reading problems. Were they intellectually incapable of reading? Apparently not. The psychologists circumvented the students' resistance to learning to read by, instead, teaching them Chinese characters. Within hours they were capable of reading more complex symbolic sentences than they could in English. The children had apparently been previously taught all too well that reading English was beyond their ability.

A major depression, these findings suggest, can be the outcome of particularly severe lessons in uncontrollability for those of us who are already vulnerable. This may explain one of the more consistent findings in the literature on depression—loss of a parent early in life puts one at increased risk for depression years later. What could be a more severe lesson that awful things can happen that are beyond our control, a lesson coming at an age when we are first forming our impressions about the nature of the world?

"According to our model," writes Seligman, "depression is not generalized pessimism, but pessimism specific to the effects of one's own skilled actions." Subjected to enough uncontrollable stress, we learn to be helpless—we lack the motivation to try to live because we assume the worst; we lack the cognitive clarity to perceive when things are actually going fine, and we feel an aching lack of pleasure in everything.*

ATTEMPTING AN INTEGRATION

Psychological approaches to depression give us some insight into the nature of the disease. According to one school, it is a state brought about by pathological overexposure to loss of control and of outlets for frustration. In another psychological view, the Freudian one, it is the internalized battle of ambivalences, aggression turned inward. These views contrast with the more biological ones—that depression is a disorder of abnormal neurotransmitter levels, abnormal communication between certain parts of the brain, abnormal hormone ratios. These are quite different ways of looking at the world, and researchers and clinicians from different orientations often don't have a word to say to one another about their mutual interest in depression. Sometimes they seem to be talking radically different languages—psychodynamic ambivalence versus neurotransmitter autoreceptors, cognitive overgeneralization versus steroid hormone ratios in the bloodstream. What I view as the main point of this chapter is that stress is the unifying theme that pulls together these disparate threads of biology and psychology.

---

* Before we leave the issue of learned helplessness, let me acknowledge that these are brutal experiments to subject an animal to. Is there no alternative? Painfully, I think not. You can study cancer in a petri dish—grow a tumor and then see if some drug slows the tumor's growth, and with what other toxicity; you can experiment with atherosclerotic plaque formation in a dish—grow blood-vessel cells and see if your drug removes cholesterol from their sides, and at what dosage. But you can't mimic depression in a petri dish, or with a computer. Millions of us are going to succumb to this nightmarish disorder, the treatments are still not very good, and animal models remain the best methods for seeking improvement. If you are of the school that animal research, while sad, is acceptable, your goal is to do only good science on the smallest number of animals with the least pain.

We have now seen some important links between stress and depression: extremes of psychological stress can cause something in a laboratory animal that looks pretty close to a depression. Moreover, stress is a predisposing factor in human depression as well, and brings about some of the typical endocrine changes of depression. Tightening the link further, glucocorticoids, as a central hormone of the stress-response, can bring about depressionlike states in an animal, and can cause depression in humans. And finally, both stress and glucocorticoids can bring about most of the neurochemical changes that have been implicated in depression.

With these findings in hand, the pieces begin to fit together. Stress, particularly in the form of extremes of lack of control and outlets, causes an array of deleterious changes in a person: cognitively, this involves a distortive belief that there is no control or outlets in *any* circumstance—learned helplessness. On the affective level, there is anhedonia; behaviorally, there is psychomotor retardation. On the neurochemical level, there are likely disruptions of serotonin, norepinephrine, and dopamine signaling; physiologically, there are alterations in, among other things, appetite, sleep patterns, and sensitivity of the glucocorticoid system to feedback regulation. And collectively, we call these array of changes a major depression.

This is terrific. I believe we have a stress-related disease on our hands. But some critical questions remain to be asked.

Why is it that after three or so bouts of major depression the stress-depression link uncouples? This is the business about depressive episodes taking on an internal rhythm of their own, independent of whether the outside world is actually pummeling you with stressors. Why should such a transition occur? At present, there's a lot of theorizing but very little in the way of actual data. Stay tuned.

Why don't we all get depressed? We all get exposed to fairly major stressors at some point or other in our lives. And we typically get pretty down in the aftermath. However, we rarely crash into a major depression. Instead, eventually, we get over it, we put things behind us, we get things in perspective, we move on with our lives . . . we heal and we recover. We've seen some pretty convincing routes by which stressors can lead to a major depression. Why doesn't this happen in all those cases?

To appreciate the answer, we need one additional fact about the neurochemical effects of stress. As we've seen, stress and glucocorticoids can bring about many of the same alterations in serotonin and norepinephrine systems that have been implicated in depression. One of the best documented links is that stress depletes norepinephrine in parts of the brain's limbic system (which, you recall, regulates emotion). No one is sure exactly why the depletion occurs, although it probably has something to do with norepinephrine's being consumed faster than usual (rather than its being made more slowly than usual). Similarly, it is not understood why the depletion is localized to that area of the brain. Nonetheless, stress causes norepinephrine levels in the limbic system to plummet. This got a lot of attention when the dominating theory of the neurochemistry of depression was that there was too little norepinephrine.

Now, for an important fact. Stress not only depletes the brain of norepinephrine but simultaneously initiates the gradual synthesis of *more* norepinephrine. At the same time that norepinephrine content is plummeting, shortly after the onset of stress, the brain is starting to make more of the key enzyme tyrosine hydroxylase, which synthesizes norepinephrine. Glucocorticoids probably have something to do with the induction of new tyrosine hydroxylase, but how is not yet clear. The main point is that, in most of us, stress may cause depletion of norepinephrine, but only transiently. And the guess is that most of the other stress effects of depression-relevant neurochemistry come with a recovery mechanism as well. Thus, while everyday stress brings about some of the neurochemical changes linked to depression along with some of the symptoms—we feel "blue"—at the same time, we are already building in the mechanisms of recovery.

We now have to flip the question around the other way—given this delayed but restorative effect of stress, why should anyone ever be stressed into a major depression? Clearly, on some level that is not yet understood, these restorative mechanisms are not foolproof, and with a big and long enough stressor, the neurochemical defenses are overwhelmed. A depression ensues.

So why are some people more prone toward depressions than others? Put anyone through enough of a stressor, and they are likely to succumb to depression, as has been seen in concentration camps and hellholes of prisoner of war camps. But when faced with the ups and downs of more everyday life, only 5 to 20 percent of us will fall into a depression. Why the enhanced vulnerability for those people?

The answer has not yet been established, but there are enough relevant facts to give some strong hints as to what the answer will likely turn out to be. One of those important facts: depression can have a genetic component, as shown with a variety of pretty complicated studies. But, of vast importance, those studies have shown that just because someone may have one or more genes for depression, that doesn't guarantee they will develop the disease; a rough estimate, derived by cutting across a lot of different styles of genetic approaches, suggests that having a genetic propensity toward depression gives you only about a fifty percent chance of getting the disease. This is one of the critical lessons of behavioral genetics, and it is one that can't be emphasized often enough—genes in this realm are rarely about *inevitability*, but instead about *vulnerability*. And in the specific context of depression, what this means is that an environmental trigger is needed to turn that vulnerability into an overt disease.

This chapter has reviewed the extent to which stress is that trigger. Given that, it is relatively easy to generate a speculative list of what the nuts and bolts might be of that "genetic vulnerability" that makes some individuals more likely to fall into a depression in the face of life's everyday challenges. People who are genetically depression-prone when exposed to stressors:

- May be more likely to perceive the stressor as being stressful, as telling them something about their entire lives instead of about this one particular incident, etc. (stay tuned for the next chapter)
- May secrete more glucocorticoids than most during the stressor
- May have some defect in their glucocorticoid receptor system so that feedback regulation does not work well and glucocorticoids continue to be secreted long after the stressor has ended

- Might have their serotonin and norepinephrine systems atypically vulnerable to the disruptive effects of stress
- May not be able to activate a restorative defense (such as the induction of tyrosine hydroxylase) during a period of stress

And so on. No one knows whether any of these are related to the genetics of depression, but each is quite plausible and would explain critical features of the disease. It would explain why this could be a disorder of vulnerability: in the absence of major stressors, a person with this biochemical abnormality would function normally. When something stressful did come along—the sort of thing that might be extremely difficult for any of us, but not psychiatrically crippling—the person with this biochemical vulnerability would be more likely to sink.

This remains speculative at this point, but whatever the underlying causes of depression, it is a disorder that shows how artificial that hoary old dichotomy between nature and nurture is, between gene and environment. With depression, that dichotomy collapses completely. This is because, as the evidence in this chapter makes abundantly clear, depression is a genetic disorder of being vulnerable to a stressful environment.

PERSONALITY, TEMPERAMENT, AND THEIR STRESS-RELATED CONSEQUENCES

The main point of Chapter 12 was that psychological factors can modulate stress-responses. Perceive yourself in a given situation to have expressive outlets, control and predictive information, for example, and you are less likely to have a stress-response. What this chapter explores is the fact that people *habitually* differ in how they modulate their stress-responses with psychological variables. Your style, your temperament, your personality have much to do with whether you regularly perceive opportunities for control or safety signals when they are there, whether you consistently interpret ambiguous circumstances as implying good news or bad, whether you typically seek out and take advantage of social support. Some folks are good at modulating stress in these ways, and others are terrible. And this turns out to be a very important factor in understanding why some people are more prone toward stress-related diseases than others.

We start with a study in contrasts. Consider Gary. In the prime of his life, he is, by most estimates, a success. He's done okay for himself materially, and he's never come close to going hungry. He's also had more than his share of sexual partners. And he has done extremely well in the hierarchical world that dominates most of his waking hours. He's good at what he does, and what he does is compete—he's already Number 2 and breathing down the neck of Number 1, who's grown complacent and a bit slack. Things are good and likely to get better.

But you wouldn't call Gary satisfied. In fact, he never really has been. Everything is a battle to him. The mere appearance of a rival rockets him into a tensely agitated state, and he views every interaction with a potential competitor as an in-your-face personal provocation. In fact, he views virtually every interaction with a distrustful vigilance. Not surprisingly, Gary has no friends to speak of. His subordinates give him a wide, fearful berth because of his tendency to take any frustration out on them. He behaves the same toward Kathleen, and barely knows their daughter Caitland—this is the sort of guy who is completely indifferent to the cutest of infants. And when he looks at all he's accomplished, all he can think of is that he is still not Number 1.

And Gary's profile comes with some physiological correlates. Elevated basal glucocorticoid levels—a constant low-grade stress-response because life is one big stressor for him. An immune system that you wouldn't wish on your worst enemy. Elevated resting blood pressure, an unhealthy ratio of "good" to "bad" cholesterol, and already the early stages of serious atherosclerosis. And, looking ahead a bit, a premature death in late middle-age.

Contrast that with Kenneth. He's also prime-aged and Number 2 in his world, but he got there through a different route, one reflecting the different approach to life that he's had ever since he was a kid. Someone caustic or jaded might dismiss him as merely being a politician, but he's basically a good guy—works well with others, comes to their aid, and they in turn to his. Consensus builder, team player, and if he's ever frustrated about anything—and it isn't all that certain that he ever is—he certainly doesn't take it out on those around him.

A few years ago, Kenneth was poised for a move to the Number 1 spot, but he did something extraordinary—he walked away from it all. Times were good enough that he wasn't going to starve, and he had reached the realization that there were things in life more important than fighting your way up the hierarchy. So he's spending time with his kids, Sam and Allan, making sure they grow up safe and healthy. He has a best friend in their mother, Barbara, and never gives a thought to what he's turned his back on.

Not surprisingly, Kenneth has a physiological profile quite different from Gary's, basically the opposite on every stress-related measure, and enjoys a robust good health. And he is destined to live to a ripe old age, surrounded by kids, grandkids, and Barbara.

Gary could learn a few pointers from Kenneth about keeping things in perspective. It's not clear if Kenneth could actually verbalize his approach to life—it's not really his style. And even if he could, it's clear that Gary wouldn't understand a word he was saying.

Normally, with these sorts of profiles, you try to protect the privacy of the individuals involved, but I'm going to violate that by including pictures of Gary and Kenneth on the next page. Check them out.

Isn't that something? Some baboons are Type A, avoid ulcers by giving them, see the world as full of water holes that are half empty. And some baboons are the opposite in every way. Talk to any pet owner, and they will give ardent testimonials as to the indelible personality of their parakeet, turtle, or bunny. And they'd usually be at least somewhat right—people have published papers on animal personality, even one about sunfish (some of whom are shy, and some of whom are outgoing social butterflies). Animals are strongly individualistic, and when it comes to primates, there are astonishing differences between individuals in their personalities, temperaments, and coping styles. These differences carry some distinctive physiological consequences and disease risks related to stress. This is not the study of what external stressors have to do with health. This is, instead, the study of the impact on health of how an individual *perceives, responds to,* and *copes with* those external stressors. And the lessons learned from some of these animals can be strikingly relevant to humans.

STRESS AND THE SUCCESSFUL PRIMATE

If you are interested in understanding the stressors in our everyday lives and how some folks cope with them better than others, study a troop of baboons in the Serengeti—big, smart, long-

"Gary."

"Kenneth (with infant)."

lived, highly social animals who live in groups of from 50 to 150. The Serengeti is a great place for them to live, offering minimal problems with predators, low infant mortality rates, easy access to food. Baboons there work perhaps four hours a day, foraging through the fields and trees for fruits, tubers, and edible grasses. This has a critical implication that makes them the perfect study subjects for what I've been interested in over the past two decades when I've snuck away from my laboratory to go there during the summers—the characteristics of the baboons who deal best with stress. If baboons are spending only four hours a day filling their stomachs, that leaves them with eight hours a day to be vile to one another. Social competition, coalitions forming to gang up on other animals, big males in bad moods beating up on someone smaller, snide gestures behind someone's back—just like us.

I am not being facetious. Think about some of the themes of the first chapter—how few of us are getting our ulcers because we have to walk ten miles a day looking for grubs to eat, how few of us become hypertensive because we are about to punch it out with someone over the last gulp from the water hole. We are ecologically buffered and privileged enough to be stressed mainly over social and psychological matters. Because the ecosystem of the Serengeti is so ideal for savanna baboons, they have the same luxury to make each other sick with social and psychological stressors. Of course, like ours, it is a world filled with affiliation, friendships, relatives who support one another; but it is a viciously competitive society as well. If a baboon in the Serengeti is miserable, it is almost always because another baboon has worked hard and long to bring about that state. Individual styles of coping with the social stress appear to be critical. Thus, one of the things I set out to test was whether such styles predicted differences in stress-related physiology and disease. I watched the baboons, collected detailed behavioral data, and then would anesthetize the animals under controlled conditions, using a blowgun. Once they were unconscious, I could measure their glucocorticoid levels, their ability to make antibodies, their cholesterol profiles, and so on, under basal conditions and a range of stressed conditions.*

The cases of Gary and Kenneth already give us a sense of how different male baboons can be. Two males of similar ranks may differ dramatically as to how readily they form coalitional partnerships with other males, how much they like to groom females, whether they play with kids, whether they sulk after losing a fight or go beat up on someone smaller. Two students, Justina Ray and Charles Virgin, and I analyzed years of behav-

---

* The controls are daunting. You have to find an anesthetic that does not distort the levels of hormones that you are measuring. You have to dart every animal at the same time of day to control for daily fluctuations in hormone levels. If you want to get a first blood sample in which hormone levels reflect basal, nonstressed conditions, you can't dart someone who is sick or injured or who has had a fight or intercourse that day. For some of the cholesterol studies, I could not dart anyone who had eaten in the preceding twelve hours. If you are trying to measure resting hormone levels, you can't spend all morning making the same animal nervous as you repeatedly try to dart him; instead you get one shot, and you can't let him see it coming. Finally, once you dart him, you have to obtain the first blood sample rapidly, before hormone levels change in response to the dart. Quite a thing to do with your college education.

You might wonder why these studies are exclusively about males. Because of the difficulties inherent in trying to dart and anesthetize females. At any given time in this baboon population, approximately 80 percent of the adult females are either pregnant or nursing their young. You don't want to dart someone who is pregnant, as there is a good chance that the anesthesia will endanger the pregnancy. And you don't want to dart someone who has a young child holding onto her in a panic as she goes down, or spends a day badly endangered for lack of milk while mom is anesthetized.

ioral data to try to formalize different elements of style and personality among these animals. And we found some fascinating correlations between personality styles and physiology.

Among males who were in the higher-ranking half of the hierarchy, we observed a cluster of behavioral traits associated with low resting glucocorticoid levels independent of their specific ranks. Some of these traits were related to how males competed with one another. The first trait was whether a male could tell the difference between a threatening and a neutral interaction with a rival. How does one spot this in a baboon? Look at a particular male and two different scenarios. First scenario: along comes his worst rival, sits down next to him, and makes a threatening gesture. What does our male subject do next? Alternative scenario: our guy is sitting there, his worst rival comes along and . . . wanders off to the next field to fall asleep. What does our guy do in this situation?

Some males can tell the difference between these situations. Threatened from a foot away, they get agitated, vigilant, prepared; when they instead see their rival is taking a nap, they keep doing whatever they were doing. They can tell that one situation is bad news, the other is meaningless. But some males get agitated even when their rival is taking a nap across the field—the sort of situation that happens five times a day. If a male baboon can't tell the difference between the two situations, on the average his resting glucocorticoid levels are twice as high as those of the guy who can tell the difference—after correcting for rank as a variable. If a rival napping across the field throws a male into turmoil, the latter's going to be in a constant state of stress. No wonder his glucocorticoid levels are elevated. These stressed baboons are similar to the "hot reactor" macaque monkeys that Jay Kaplan has studied. As you will recall from Chapter 3, these are individuals who respond to every social provocation with an overactivation of their stress-response (the sympathetic nervous system) and carry the greater cardiovascular risk.

Next variable. If the situation really is threatening (the rival's a foot away and making menacing moves), does our male sit there passively and wait for the fight, or does he take control of the situation and strike first? Males who sit there passively, abdicating control, have much higher glucocorticoid levels than the take-charge types, after rank is eliminated as a factor in the analysis. We see the same pattern in low-ranking as well as high-ranking males.

A third variable: after a fight, can the baboon tell whether he won or lost? Some guys are great at it; they win a fight, and they groom their best friend. They lose a fight, and they beat up someone smaller. Other baboons react the same way regardless of outcome; they can't tell if life is improving or worsening. The baboon who can't tell the difference between winning and losing has much higher glucocorticoid levels, on the average, than the guys who can, independent of rank.

Final variable: if a male has lost a fight, what does he do next? Does he sulk by himself, groom someone, or beat someone up? Discouragingly, it turns out that the males who are most likely to go beat on someone—and thus displaying *displaced aggression*—have lower glucocorticoid levels, again after rank is eliminated as a variable. This is true for both subordinate baboons and the high-ranking ones.

Thus, after factoring out rank, lower basal glucocorticoid levels are found in males who are best at telling the difference between threatening and neutral interactions; who take the initiative if the situation clearly is threatening; who are best at telling whether they won or lost; and in the latter case, who are most likely to make someone else pay for the defeat. This echoes some of the themes from the chapter on psychological stress. The males who were coping best (at least by this endocrine measure) had high degrees of social control (initiating the fights), predictability (they can accurately assess whether a situation is threatening, whether an outcome is good news), and outlets for frustration (a tendency to give rather than get ulcers). Remarkably, this style is stable over the years of these individuals' lives, and carries a big payoff—males with this cluster of low glucocorticoid traits remain high-ranking significantly longer than average.

Our subsequent studies have shown another set of traits that also predicts low basal glucocorticoid levels. These traits have nothing to do with how males compete with one another. Instead, they are related to patterns of social affiliation. Males who spent the most time grooming females not in heat (not of immediate sexual interest—just good old platonic playmates), who are groomed by them the most frequently, who spend the most time playing with the young—these are the low-glucocorticoid guys. Put most basically (and not at all anthropomorphically), these are male baboons who are most capable of developing friendships. This finding is remarkably similar to those discussed in previous chapters regarding the protective effects of social affiliation against stress-related disease in humans. And as will be discussed in the final chapter of this book, this cluster of personality traits is also stable over time and comes with a very distinctive payoff as well—a male baboon's equivalent of a successful old age.

Thus, among some male baboons, there are at least two routes for winding up with elevated basal glucocorticoid levels—an inability to keep competition in perspective and social isolation. Stephen Suomi at the National Institutes of Health has studied rhesus monkeys and identified another personality style that should seem familiar, and which carries some physiological correlates. About 20 percent of rhesus are what he calls "high-reactors." Just as the baboons who find a rival napping to be an arousing threat, these individual monkeys see challenge everywhere. But in their case, the response to the perceived threat is a shrinking timidity. Put them into a novel environment that other rhesus would find to be a stimulating place to explore, and they react with fear, pouring out glucocorticoids. Place them with new peers, and they freeze with anxiety—shy and withdrawn, and again releasing vast amounts of glucocorticoids. Separate them from a loved one, and they are atypically likely to collapse into a depression, complete with excessive glucocorticoids, overactivation of the sympathetic nervous system, and immunosuppression. And these appear to be life-long styles of dealing with the world, beginning early in infancy.

From where do these various primate personalities arise? When it comes to the baboons, I'll never know. Male baboons change troops at puberty, often moving dozens of miles before finding an adult troop to join. It is virtually impossible to track the same individuals from birth to adulthood, so I have no idea what their childhoods were like, whether their mothers were permissive or stern, whether they were forced to take piano lessons, and so on. But Suomi has done elegant work that indicates both genetic and environmental components to these personality differences. For example, he has shown that an infant monkey has a significant chance of sharing a personality trait with its father, despite the formation of social groups in which the father is not present—a sure hint at a heritable, genetic component. In contrast, the high-reactivity personality in these monkeys can be completely prevented by fostering such animals early in life to atypically nurturant mothers—a powerful vote for environmental factors built around mothering style.

Broadly, these various studies suggest two ways that a primate's personality style might lead down the path to stress-related disease. In the first way, there's a mismatch between the magnitude of the stressors they are confronted with and the magnitude of their stress-response—the most neutral of circumstances is perceived as a threat, demanding either a hostile, confrontational response (as with some of my baboons and Kaplan's macaques) or an anxious withdrawal (as with some of Suomi's monkeys). At the most extreme they even react to a situation that most certainly does not constitute a stressor (for example, winning a fight) the same way as if it were a stressful misery (losing one). In their second style of dysfunction, the animal does not take advantage of the coping responses that might make a stressor more manageable—they don't grab the minimal control available in a tough situation, they don't make use of effective outlets when the going gets tough, and they lack social support.

It would seem relatively straightforward to pull together some sound psychotherapeutic advice for these unhappy beasts. But in reality, it's hopeless. Baboons and macaques are completely distractible during therapy sessions—habitually pulling books off the shelves, for example; they can't keep the days of the week straight and thus constantly miss their appointments; they eat the plants in the waiting room, and so on. Thus, it might be more useful to apply those same insights to making sense of some humans who are prone toward overactivation of the stress-response and increased risk of stress-related disease.

THE HUMAN REALM: A CAUTIONARY NOTE

There are, by now, some fairly impressive and convincing studies linking human personality types with stress-related diseases. Probably the best place to start, however, is with a bit of caution about some reported links that, I suspect, should be taken with a grain of salt.

I've already noted some skepticism about early psychoanalytic theorizing that linked certain personality types with colitis (see Chapter 5). Another example concerns miscarriages and abortions. Chapter 7 reviewed the mechanisms by which stress can cause the loss of a pregnancy, and one hardly needs to have experienced that personally to have an inkling of the trauma involved. Thus, you can imagine the particular agony for women who miscarry repeatedly, and the special state of misery for those who never get a medical explanation for the problem—no expert has a clue what's wrong. Into that breach have charged people who have attempted to uncover personality traits common to women thus labeled as "psychogenic aborters."

Some researchers have identified one subgroup of women with repeated "psychogenic" abortions (accounting for about half the cases) as being "retarded in their psychological development." They are characterized as emotionally immature women, highly dependent on their husbands, who on some unconscious level view the impending arrival of the child as a threat to their own childlike relationship with their spouse. Another personality type identified, at the opposite extreme, are women who are characterized as being assertive and independent, who really don't want to have a child. Thus, a common theme in the two supposed profiles is an unconscious desire not to have the child—either because of competition for the spouse's attention or because of reluctance to cramp their independent lifestyles.

Many experts are skeptical about the studies behind these characterizations, however. The first reason harks back to a caveat I aired early in the book: a diagnosis of "psychogenic" anything (impotency, amenorrhea, abortion, and so on) is usually a diagnosis by exclusion. In other words, the physician can't find any disease or organic cause, and until one is discovered, the disorder gets tossed into the psychogenic bucket. This may mean that it is truly psychologically based, or it may simply mean that the relevant hormone, neurotransmitter, or genetic abnormality has not yet been discovered. Once it *is* discovered, the psychogenic disease is magically transformed into an organic problem—"oh, it wasn't your personality after all." The area of repeated aborting seems to be one that is rife with recent biological insights—in other words, if so many of last decade's psychogenic aborters now have an organic explanation for their malady, that trend is likely to continue.

Another difficulty is that these studies are all retrospective in design: the researchers examine the personalities of women *after* they have had repeated abortions. A study may thus cite the case of a woman who has had three miscarriages in a row, noting that she is emotionally withdrawn and dependent on her husband. But because of the nature of the research design, one can't tell whether these traits are a cause of the miscarriages or a response to them—three successive miscarriages could well exact a heavy emotional price, perhaps making the subject withdrawn and more dependent on her husband. In order to study the phenomenon properly, one would need to look at personality profiles of women *before* they become pregnant, to see if these traits predict who is going to have repeated miscarriages. To my knowledge, this kind of study has not yet been carried out.

As a final problem, none of the studies provides any reasonable speculation as to *how* a particular personality type may lead to a tendency not to carry fetuses to term. What are the mediating physiological mechanisms? What hormones and organ functions are disrupted? The absence of any science in that area makes me pretty suspicious of the claims. Psychological stressors can increase the risk of a miscarriage, but although there is precedent in the medical literature for thinking that having a certain type of personality is associated with an increased risk for miscarriages, scientists are far from being able to agree on *what* personality is associated, let alone whether the personality is a cause or consequence of the miscarriages.

PSYCHIATRIC DISORDERS AND ABNORMAL STRESS-RESPONSES

A number of psychiatric disorders involve personalities, roles, and temperaments that are associated with distinctive stress-responses. We have seen an example of this in the previous chapter on depression—about half of depressives have resting glucocorticoid levels that are dramatically higher than in other people, often sufficiently elevated to cause problems with metabolism or immunity. Or in some cases, depressives are unable to turn off glucocorticoid secretion, their brains being less sensitive to a shut-off signal.

A theme in the previous section on some troubled nonhuman primates is that there is a discrepancy between the sorts of stressors they are exposed to and the coping responses they come up with. Learned helplessness, which we saw to be an underpinning of depression, appears to be another example of such discrepancy. A challenge occurs, and what is the response of a depressive individual? "I can't, it's too much, why bother doing anything, it isn't going to work anyway, nothing I do ever works. . . ." The discrepancy here is that in the face of stressful challenges, depressives don't even attempt to mount a coping response.

A different type of discrepancy is seen with people who are anxiety-prone. What is anxiety? A sense of disquiet, of disease, of the sands constantly shifting menacingly beneath your feet—where constant vigilance is the only hope of effectively protecting yourself. A brutal experimental paradigm, used many years ago, graphically illustrates anxiety in an animal. Teach a dog a learning task concerning a high-pitched tone and a low-pitched one. The high pitch means that it must press a lever in order to avoid getting a mild shock. The low pitch means the opposite, that it must avoid pressing the lever in order to avoid a shock. Easily learned. Then, gradually begin to raise the low pitch and lower the high one, until it becomes less and less clear which one is which—and the dog writhes in anxiety, starting to press the lever, then recoiling in a spasm of uncertainty. The world is full of menace, as the dog scrambles to figure out what the rules are for being safe.

And that is what anxiety is like in a human. A particularly florid example occurs with what are called *panic attacks*, when the person's anxiety boils out into a paralyzing, hyperventilating sense of crisis at some particular moment, often one associated with some prior trauma. Another debilitating version of an anxiety disorder is called *obsessive compulsive disorder* (OCD). In this case, the person tries to keep the sense of unstructured, menacing panic at bay with an endless variety of reassuring rituals.

Unlike depressives, the anxiety-prone person is still attempting coping responses. But the discrepancy is the distorted belief

that stressors are everywhere and perpetual, and that the only hope for safety is constant mobilization of coping responses [whether the bizarre ritual of six hours of washing a day to head off the stressor of germs (something typically seen in OCD), or a mere constant, sickening sense of vigilance for garden-variety anxiety].

Awful. And immensely stressful. Not surprisingly, anxiety disorders are associated with chronically overactive stress-responses. Surprisingly, though, glucocorticoid excess is not the usual response. Instead, it's too much sympathetic activation, an overabundance of circulating catecholamines (epinephrine and norepinephrine).

We have now seen some interesting contrasts between catecholamines and glucocorticoids. Chapter 2 emphasized how the former defend you against stressors by handing out guns from the gun locker within seconds, in contrast to glucocorticoids, which defend you by constructing new weapons over the course of minutes to hours. Or there can be an elaboration of this time-course, in which catecholamines mediate the response to a current stressor while glucocorticoids mediate preparation for the next stressor. When it comes to human psychiatric disorders, it seems that increases in the catecholamines have something to do with still trying to cope and the effort that involves, where overabundance of glucocorticoids seems more of a signal of having given up on coping with that stressor.

TYPE A AND THE ROLE OF UPHOLSTERY IN CARDIOVASCULAR PHYSIOLOGY

In some cardiology circles, personality types that are atypically prone to viewing life's ambiguities as stressful have been given certain labels. One label refers to the mythic king of Corinth who, angering Zeus and being an all-around wise guy, was sentenced to spend the rest of time rolling the same boulder up the side of a mountain; it always rolled back just before reaching the top. People of the "Sisyphus pattern" view life as a joyless struggle—the workaholic executive whose only pleasure in life appears to be checking off items on a "to do" list. One researcher has found that such people appear to be at special risk for sudden cardiac death. Other researchers have noted a connection between certain depressive personality types and cardiac disease. Amid these various links, there is one proposed connection between personality and heart disease which has become so well known that it has suffered the ultimate accolade—namely, being distorted beyond recognition in many people's minds (usually winding up being ascribed to the most irritating behavioral trait that you want to complain about in someone else, or indirectly brag about in yourself). I'm talking, of course, about Type A behavior.

Defining the Type A Personality Two cardiologists, Meyer Friedman and Ray Rosenman, coined the term *Type A* in the early 1960s to describe a collection of traits that they found in some individuals. They didn't describe these traits in terms related to stress (for example, defining Type A people as those who responded to neutral or ambiguous situations as if they were stressful), although I will attempt to do that reframing below. Instead, they characterized Type A people as immensely competitive, over-achieving, time-pressured, impatient, and hostile.

Initially Friedman and Rosenman observed that many of their coronary patients had what they ultimately termed "Type A personalities." This was met with lots of skepticism in the field, as cardiologists spend their time thinking about heart valves and circulating lipids and what someone's liver does about cholesterol, not about how someone deals with a slow line at the supermarket. Thus, there was an initial tendency among many in the field to view the link between the behavior and the disease as the reverse of what Friedman and Rosenman proposed—getting heart disease might make some people act in a more Type A manner. But Friedman and Rosenman did a "prospective" study; they looked at healthy people, examining whether being Type A increased the risk of eventually getting heart disease—which is precisely what they found. The finding made a huge splash, and by the early 1980s, a panel of some of the biggest guns in cardiology convened, checked the evidence, and concluded that being Type A carries at least as much cardiac risk as does smoking or having high cholesterol levels.

Everyone was delighted, and "Type A" entered common parlance. The trouble was, soon thereafter a bunch of very carefully done studies failed to replicate the basic findings of Friedman and Rosenman. Suddenly, Type A wasn't looking predictive of coronary heart disease after all. Then, to add insult to injury, a pair of careful studies came out showing that, once you had coronary heart disease, being Type A was associated with *better* survivorship—being Type A was seemingly good for you (in the notes at the end of the book, I discuss some of the subtle ways this finding might have occurred).

In recent years, there has been some fine-tuning as to which of the Type A traits are most central to heart disease. Work by Redford Williams, a Duke University physician, has paved the way to showing that hostility seems to have a great deal to do with it. For example, when scientists reanalyzed some of the original Type A studies and broke the constellation of traits into individual ones, hostility popped out as the only significant predictor of heart disease. The same result was found in studies of middle-aged doctors who had taken personality inventory tests 25 years earlier as an exercise in medical school. And the same thing was found when looking at American lawyers, Finnish twins, Western Electric employees—a whole range of populations. These various studies have suggested that a high degree of hostility predicts coronary heart disease, atherosclerosis, and higher rates of mortality with these diseases. Many of these studies, moreover, controlled for important variables like age, weight, blood pressure, cholesterol levels, and smoking. Thus, it is unlikely that the hostility–heart disease connection could be due to some other factor (for example, that hostile people are more likely to smoke, and the heart disease arises from the smoking, not the hostility). More recent studies have shown that hostility is associated with a significant overall increase in mortality across all diseases, not just those of the heart.

Friedman and colleagues stick with an alternative view. They suggest that at the core of the hostility is a sense of "time-pressuredness"—"Can you believe that teller, how slowly he's working. I'm going to be here all day. I can't waste my life on some bank line. How did that kid know I was in a rush? I could kill him"—and that the core of the time-pressuredness is rampant insecurity. There's no time to savor anything you've accomplished, let alone enjoy anything that anyone else has done, because you must rush off to prove yourself all over again, and try to hide from the world for another day or at least another minute what a fraud you are. Their work suggests that a persistent sense of insecurity is, in fact, a better predictor of cardiovascular profiles than is hostility, although theirs appears to be a minority view in the field.

Insofar as hostility has something to do with your heart (whether as a primary factor or as a surrogate variable), it remains unclear which aspects of hostility are bad news. For example, the study of lawyers suggested that overt aggressiveness and cynical mistrust were critical—in other words, frequent open expression of the anger that you feel predicts heart disease. In support of that, experimental studies show that the full expression of anger is a powerful stimulant of the cardiovascular system. By contrast, in the reanalysis of the original Type A data a particularly powerful predictor of heart disease was

Type A's in action. The photo on the left shows the early morning parking pattern of a patient support group for Type A individuals with cardiovascular disease—everyone positioned for that quick getaway that doesn't waste a second. On the right, the same scene later in the day.

not only high degrees of hostility, but also the tendency *not* to express it when angry. This latter view is supported by some fascinating work by James Gross of Stanford University. Show a volunteer a film clip that evokes some strong emotion. Disgust, for example (thanks to a gory view of someone's leg being amputated). They writhe in discomfort and distaste and, no surprise, show the physiological markers of having turned on their sympathetic nervous system. Now, show some other volunteers the same film clip but, beforehand, instruct them to try not to express their emotions ("so that if someone were watching, they'd have no idea what you were feeling"). Run them through the blood and guts, and, with the person gripping the arms of the chair and trying to remain stoic and still, the sympathetic activation becomes even greater. Repressing the expression of strong emotions appears to exaggerate the intensity of the physiology that goes along with them.

How Hostility Hurts Your Heart Why would great hostility (of whatever variant) be bad for your heart? Subjectively, we can describe hostile persons as those who get all worked up and angry over incidents that the rest of us would find only mildly provocative, if provocative at all. Similarly, their stress-responses switch into high gear in circumstances that fail to perturb everyone else's.

Give both hostile and nonhostile people a nonsocial stressor (like some math problems) and nothing exciting happens; everyone has roughly the same degree of mild cardiovascular activation. But if you generate a situation with a social provocation, the hostile people dump more epinephrine and norepinephrine into their bloodstreams and wind up with higher blood pressures. All sorts of social provocations have been used in studies: the subjects may be requested to take a test and, during it, be repeatedly interrupted; or they may play a video game in which the opponent not only is rigged to win but acts like a disparaging smart aleck. In these and other cases, the cardiovascular stress-responses of the nonhostile are relatively mild. But blood pressure goes through the roof in the hostile people. (Isn't it remarkable how similar these folks are to Jay Kaplan's hot reactor monkeys, with their exaggerated sympathetic responses to stressors, and their increased risk of cardiovascular disease? Or to my baboons, the ones who can't differentiate between threatening and nonthreatening events in their world? There are card-carrying Type A individuals out there with tails.) Here is that discrepancy again. For anxious people, life is full of menacing stressors that demand vigilant coping responses. For the Type A person, life is full of menacing stressors that demand vigilant coping responses of a particularly hostile nature. This is probably representative of the rest of their lives. If each day is filled with cardiovascular provocations that everyone else responds to as no big deal, life will slowly hammer away at the hearts of the hostile. An increased risk of cardiovascular disease is no surprise.

A pleasing thing is that Type A-ness is not forever. If you reduce the hostility component in Type A people through therapy (using some of the approaches that will be outlined in the final chapter), you reduce the risk for further heart disease. This is great news. I've noticed that many health professionals who treat Type A people are mostly trying to reform these folks. Basically, many Type A people are abusive pains in the keister to lots of folks around them. When you talk to some of the Type A experts, there is an odd tone between their lines that Type A-ness (of which many of them are admittedly perfect examples) is a kind of ethical failing, and the term is a fancied up medical way of describing people who just aren't nice to others. Added to this is a tendency I've noticed for a lot of the Type A experts to be lay preachers, or descendants of clergy. That religious linkage will even sneak in the back door. I once talked with two leaders in the field, one an atheist and the other agnostic, and when they tried to give me a sense of how they try to get through to Type A subjects about their bad ways, they made use of a religious sermon.* I finally asked these two MDs an obvious question—were they in the business of blood vessels or souls? Was the work that they do about heart disease or about ethics? And without a beat they both chose ethics. Heart disease was just a wedge for getting at the bigger issues. I thought this was wonderful. If it takes turning our coronary vessels into our ledgers of sin and reducing circulating lipids as an act of redemption in order to get people to be more decent to each other, more power to them.

Interior Decorating as Scientific Method A final question about this field. How was Type A behavior discovered? A fascinating story. We all know how scientists make their discoveries. There are the discoveries in the bathtub (Archimedes and his discernment of the displacement of water), discoveries in one's sleep (Kekulé and his dream of carbons dancing in a ring to form benzene), discoveries at the symphony [our scientist, strained by overwork, is forced to the concert by a significant other; during a quiet woodwind section, there's the sudden realization, the scribbled equation on the program notes, the rushed "Darling, I must leave this instant for the laboratory (accent on second syl-

---

* I listened to a tape of this sermon, called "Back in the Box," by the Rev. John Ortberg. It concerns an incident from his youth. His grandmother, saintly, kind, nurturant, also happened to be a viciously competitive and skillful Monopoly player, and his summer visits to her were littered with his defeats at the game. He described one year where he practiced like mad, honed his Machiavellian instinct, developed a ruthless jugular-gripping style, and finally mopped up the board with her. After which, his grandmother rose and calmly put the pieces away.

"You know," she said offhandedly, "this is a great game, but when it is all over with, the pieces just go back in the box." Amass your property, your hotels...[the sermon takes off from there]...your wealth, your accomplishments, your awards, your whatevers and eventually it will all be over with and those pieces go back in the box. And all you are left with is how you lived your life.

I listened to this tape while racing to beat red lights on my way to a 5:00 AM commuter train, powerbook ready so as not to miss a moment of work on the train, eating breakfast one-handed while driving, using the time to listen to this sermon on tape. And this sermon, whose trajectory was obvious from the first sentence and which was filled with Jesus and other things that I do not subscribe to, reduced me to tears.

lable á la Anthony Hopkins)," with the rest being history]. But every now and then someone else makes the discovery and comes and tells the scientist about it. And who is that someone? Very often someone whose role in the process could be summed up by an odd imaginary proverb that will probably never end up embroidered into someone's pot holder: "If you want to know if the elephant at the zoo has a stomach ache, don't ask the veterinarian, ask the cage cleaner." People who clean up messes become very attuned to circumstances that change the amount of mess there is. And back in the 1950s that fact caused a guy to just miss changing the course of medical history.

I recently had the privilege of hearing the story from the horse's mouth, Dr. Meyer Friedman. At age 86, he is running the Meyer Friedman Institute at UCSF—in the thick of ongoing research, a self-proclaimed Type A who suffered a heart attack at a relatively early age before making some changes in his life. A remarkably introspective, psychologically minded man, he's one of the two people I described a few paragraphs back (along with his medical director, Bart Sparagon) who voted for being in the business of ethics. And that ethicist's mind-set, when turned on Dr. Friedman himself, is relevant to this story. Dr. Friedman, a man whose demeanor is cordial and gentle to an extreme, illustrates half his examples of Type A-ness with a self-excoriating confessionalism—what an angry, impatient, unappreciative SOB he was in his prime; tales of people he was curt to, whose efforts he never noticed, whose accomplishments he envied. Somewhere in there, it occurs to you that Dr. Friedman might not be the most accurate relater of these tales, the suspicion dawning that this extremely kind and sensitive man has probably been that way his entire life, and that he's way too hard on himself.

And it's in that framework that one finally hears of the discovery of the Type A link. It was the mid-1950s, Friedman and Rosenman with their successful cardiology practice. And they were having an unexpected problem. They were spending a fortune having to reupholster the chairs in their waiting rooms. This is not the sort of issue that would demand a cardiologist's attention. Nonetheless, there seemed to be no end of chairs that had to be fixed. And one day, a new upholsterer came in to see to the problem, took one look at the chairs, and discovered the Type A–cardiovascular disease link. "What the hell is wrong

How it all began . . . almost.

with your patients? People don't wear out chairs this way." It was only the front-most few inches of the seat cushion and of the padded armrests that were torn to shreds, as if some very short beavers spent each night in the office craning their necks to savage the very fronts of the chairs. The patients in the waiting rooms all habitually sat on the edges of their seats, fidgeting, clawing away at the arm rests.

And the rest should have been history: upswelling of music as the upholsterer is seized by the arms and held in a penetrating gaze—"Good heavens, man, do you realize what you've just said?" Hurried conferences between the upholsterer and other cardiologists. Frenzied sleepless nights as teams of idealistic young upholsterers spread across the land, carrying the news of their discovery back to Upholstery / Cardiology Headquarters— "Nope, you don't see that wear pattern in the waiting-room chairs of the urologists, or the neurologists, or the oncologists, or the podiatrists, just the cardiologists. There's something different about people who wind up with heart disease"—and the field of Type A therapy takes off.

Instead, none of that happened. Dr. Friedman sighs. Another confession. "I didn't pay any attention to the man. I was too busy; it went in one ear and out the other." And it wasn't until four, five years later that Dr. Friedman's formal research with his patients began to yield some hints, at which point there was the thunderclap of memory—Oh, my god, the upholsterer, remember that guy going on about the wear pattern? And to this day, no one remembers his name.

MARTHA STEWART'S DISEASE?

This chapter has discussed personality types associated with overactive stress-responses, and argued that a common theme among them is a discrepancy between what sort of stressors life throws at these folks and what sort of coping responses they come up with. This final section is about a newly recognized version of an overactive stress-response. And it's puzzling.

These are not people who are dealing with their stressors too passively, too persistently, too vigilantly, or with too much hostility. They don't appear to have all that many stressors. They claim they're not depressed or anxious, and the psychological tests they are given show they're right. In fact, they describe themselves as pretty happy, successful, and accomplished (and, according to personality tests, they really are). Yet, these people (comprising approximately 5 percent of the population) have chronically activated stress-responses. What's their problem?

Their problem, I think, is one that offers insight into an unexpected vulnerability of our human psyche. The people in question are said to have "repressive" personalities, and we all have

Clifford Goodenough, 1991: Figure Walking in a Landscape, *goldleaf, tempera, oil on masonite.*

met someone like them. In fact, we usually regard these folks with a tinge of envy—"I wish I had their discipline; everything seems to come so easily to them. How do they do it?"

These are the archetypal people who cross all their T's and dot all their i's. They describe themselves as planners who don't like surprises, who live structured, rule-bound lives—walking to work the same way each day, always wearing the same style of clothes—the sort of people who can tell you what they're having for lunch two weeks from Wednesday. Not surprisingly, they don't like ambiguity and strive to set up their world in black and white, filled with good or bad people, behaviors that are permitted or strictly forbidden. And they keep a tight lid on their emotions. Stoic, regimented, hard-working, productive, solid folks who never stand out in a crowd (unless you begin to wonder at the unconventional nature of their extreme conventionality).

Some personality tests, pioneered by the psychologist Richard J. Davidson and colleagues at the University of Wisconsin, identify repressive individuals. For starters, as noted, the personality tests show that these people aren't depressed or anxious. Instead, the tests reveal their need for social conformity, their dread of social disapproval, and their discomfort with ambiguity, as shown by the extremely high rates at which they agree with statements framed as absolutes, statements filled with "never" and "always". No gray tones here.

Intertwined with those characteristics is a peculiar lack of emotional expression. The tests reveal how repressive people "inhibit negative affect"—no expressing of those messy, complicated emotions for them, and little recognition of those complications in others. For example, ask repressors and nonrepressors to recall an experience associated with a specific, strong emotion. Both groups report that particular emotion with equal intensity. However, when asked what else they were feeling, nonrepressors typically report an array of additional, nondominant feelings: "Well, it mostly made me angry, but also a little sad, and a little disgusted too. . . ." Repressors steadfastly report no secondary emotions. Black and white feelings, with little tolerance for subtle blends.

Are these people for real? Maybe not. Maybe beneath their tranquil exteriors, they're actually anxious messes who won't admit to their frailties. Careful study indicates that some repressors are indeed mostly concerned about keeping up appearances. (One clue is that they tend to give less "repressed" answers on personality questionnaires when they can be anonymous.) And so their physiological symptoms of stress are easy to explain. We can cross those folks off the list.

What about the rest of the repressors? Could they be deceiving themselves—roiling with anxiety, but not even aware of it? Even careful questionnaires cannot detect that sort of self-deception; to ferret it out, psychologists traditionally rely on less structured, more open-ended tests (of the "What do you see in this picture?" variety). Those tests show that, yes, some repressors are far more anxious than they realize; *their* physiological stress is also readily explained.

Yet even after you cross the anxious self-deceivers off the list, there remains a group of people with tight, constrained personalities who are truly just fine; mentally healthy, happy, productive, socially interactive. But they have overactive stress-responses. The levels of glucocorticoids in their bloodstream are as elevated as among highly anxious people, and they have elevated sympathetic tone as well. When exposed to a cognitive challenge, repressors show unusually large increases in heart rate, blood pressure, sweating, and muscle tension. And these overaroused stress-responses exact a price. For example, they have relatively poor immune function. Furthermore, coronary disease patients who have repressive personalities are more vulnerable to cardiac complications than are nonrepressors.

Overactive, endangering stress-responses—yet the people harboring them are not stressed, depressed, or anxious. Back to our envious thought—"I wish I had their discipline. How do they do it?" The way they do it, I suspect, is by working like maniacs to generate their structured, repressed world with no ambiguity or surprises. And this comes with a physiological bill.

Davidson and Andrew Tomarken of Vanderbilt University have used electroencephalographic (EEG) techniques to show unusually enhanced activity in a portion of the frontal cortex of repressors. This is a region of the brain involved in inhibiting impulsive emotion and cognition (for example, metabolic activity in this area has been reported to be decreased in violent sociopaths). It's the nearest anatomical equivalent we have to a superego; makes you say you love the appalling dinner, compliment the new hairdo, keeps you toilet trained. It keeps those emotions tightly under control, and as Gross's work showed with emotional repression, it takes a lot of work to keep an especially tight squeeze on those emotional sphincters.

It can be a frightening world out there, and the body may well reflect the effort of threading our way through those dark, menacing forests. How much better it would be to be able to sit, relaxed, on the sun-drenched porch of a villa, far, far from the wild things baying. Yet, what looks like relaxation could well be exhaustion—exhaustion from the labor of having built a wall around that villa, the effort of keeping out that unsettling, challenging, vibrant world. A lesson of repressive personality types and their invisible burdens is that, sometimes, it can be enormously stressful to construct a world without stressors.

15

THE VIEW FROM THE BOTTOM

Toward the end of the first chapter, I voiced a caveat—when I discuss some way in which stress can make you sick, that is merely shorthand for discussing how stress can make you more likely to get diseases that make you sick. That was basically a first pass at a reconciliation between two very different camps that think about poor health. On one hand, you have the mainstream medical crowd that is concerned with reductive biology. For them, poor health revolves around issues of bacteria, viruses, genetic mutations, and so on. At the other extreme are the folks anchored in mind-body issues, for whom poor health is about psychological stress, lack of control and efficacy, and so on. A lot of this book has, as one of its goals, tried to develop further links between those two viewpoints. This has come in the form of showing how sensitive reductive biology can be to some of those psychological factors, and exploring the mechanisms that account for this. And it has come in the form of criticizing the extremes of both camps: on the one hand, trying to make clear how limiting it is to believe that humans can ever be reduced to a DNA sequence, and on the other, trying to indicate the damaging idiocy of denying the realities of human physiology and disease. The ideal resolution harks back to the wisdom of Herbert Weiner, as discussed in Chapter 8, that disease, even the most reductive of diseases, cannot be appreciated without considering the person who is ill.

No comment.

Terrific; we're finally getting somewhere. But this analysis, and most pages of this book up until now have left out a third leg in this stool—the idea that poor health also has something to do with poor jobs in a shrinking economy, or a diet funded by food stamps with too many meals consisting of Coke and Cheetos, or living in a crummy overcrowded apartment close to a toxic waste dump or without enough heat in winter. Let alone living on the streets. Or in a refugee camp. If we can't consider disease outside the context of the person who is ill, we also can't consider it outside the context of the society in which that person has gotten ill, and that person's place in that society.

I recently found support for this view in an unexpected corner. Neuroanatomy is the study of the connections between different areas of the nervous system, and it can sometimes seem like a subject of mind-numbing stamp-collecting—some multisyllabically named part of the brain sends its axons in a projection with another multisyllabic name to eighteen multisyllabic target sites, whereas in the next county over in the brain.... During a period of my errant youth I took particular pleasure in knowing as much neuroanatomy as possible, the more obscure, the better. One of my favorite names was that given to a tiny space that exists between two layers of the meninges, the tough fibrous wrapping found around the brain. It was called the *Virchow-Robin space*, and my ability to toss off that name won me the esteem of my fellow neuroanatomy dorks. I never figured out who Robin was, but Virchow was Rudolph Virchow, a nineteenth-century German pathologist and anatomist. Man, to be honored by having your name attached to some microscopic space between two layers of plastic brain wrap—this guy must have been the king of reductive nuts-and-bolts science to merit that. I'd bet he even wore a monocle, which he'd remove before peering down a microscope.

I found out a bit about Rudolph Virchow. As a young physician, he came of age with two shattering events—a massive typhus outbreak in 1847 that he attempted to combat firsthand and the doomed European revolution of 1848. The first was the perfect case for teaching that appalling living conditions can have as much to do with disease as do microorganisms. And the second taught just how effectively the machinery of power can subjugate. In its aftermath, he emerged not just as someone who was a scientist plus a physician plus a public health pioneer plus a progressive politician. Rather, through a creative synthesis, he saw those roles all as manifestations of a single whole. "Medicine is a social science, and politics nothing but medicine on a large scale," he wrote. And, "Physicians are the natural attorneys of the poor." This is an extraordinarily large vision for a man getting microscopic spaces named for him. And unless one happens to be a very atypical physician these days, this vision must also seem extraordinarily quaint, as sadly quaint as Picasso's thinking he could throw some paint on a canvas, call it *Guernica*, and do something to halt Fascism.

The history of status thymicolymphaticus, the imaginary disease of a supposedly enlarged thymus gland in infants, detailed at the end of Chapter 8, taught us that your place in society can leave its imprint on the corpse you eventually become. The purpose of this chapter is to show how your place in society, and the sort of society it is, can leave their imprint on patterns of disease while you are alive, and to show that part of understanding this imprint incorporates the notion of stress. This will be preparatory for an important notion to be discussed in the final chapter on stress management—that certain techniques for reducing stress work very differently depending on where you dwell in your society's hierarchy.

A strategy that I've employed in a number of chapters is to introduce some phenomenon in the context of animals, often social primates. This has been in order to show some principle in a simplified form before turning to the complexity of humans. I do the same in this chapter, beginning with a discussion of what social rank has to do with health and stress-related diseases among animals. But this time, there is a paradoxical twist that, by the end of this chapter, should seem depressing as hell—this time, it is we humans who provide a brutally simple version and our nonhuman primate cousins the nuance and subtlety.

PECKING ORDERS AMONG BEASTS WITH TAILS

While pecking orders—dominance hierarchies—might have first been discerned among hens, they exist in all sorts of species. Resources, no matter how plentiful, are rarely divvied up evenly. Instead of every contested item's being fought for with bloodied tooth and claw, dominance hierarchies emerge. As formalized systems of inequities, these are great substitutes for continual aggression between animals smart enough to know their place.

Hierarchical competition has been taken to heights of animal complexity by primates. Consider baboons, the kind running around savannas in big social groups of a hundred or so beasts. In some cases, the hierarchy can be fluid, with ranks changing all the time; in other cases, rank is hereditary and lifelong. In some cases, rank can be situationally dependent—A outranks B when it comes to a contested food item, but the order is reversed if it is competition for someone of the opposite sex. There can be circularities in hierarchies—A defeats, B defeats, C defeats A. Ranking can involve coalitional support—B gets trounced by A, unless receiving some well-timed help from C, in which case A is sent packing. And the actual confrontation between two animals can include anything ranging from a near fatal brawl to a

Grooming, a wonderful means of social cohesion and stress reduction, in a society where everyone's back is not scratched equally.

highly dominant individual's doing nothing more than shifting menacingly and giving subordinates the willies.

Regardless of the particulars, if you're going to be a savanna baboon, you probably don't want to be a low-ranking one. You sit there for two minutes digging some tuberous root out of the ground to eat, clean it off and . . . anyone higher-ranking can rip it off from you. You spend hours sweet-talking someone into grooming you, getting rid of those bothersome thorns and nettles and parasites in your hair, and the grooming session can be broken up by someone dominant just for the sheer pleasure of hassling you. Or you could be sitting there, minding your own business, bird-watching, and some high-ranking guy, having a bad day, decides to make you pay for it by slashing you with his canines. (Such third-party "displacement aggression" accounts for a huge percentage of baboon violence. A middle-ranking male gets trounced in a fight, turns and chases a subadult male, who lunges at an adult female, who bites a juvenile, who slaps an infant.) For a subordinate animal, life is filled with a disproportionate share not only of physical stressors but of psychological stressors as well—lack of control, of predictability, of outlets for frustration.

A middle-ranking baboon, who has spent all morning predating an impala, has the kill stolen from him by a high-ranking male.

It's not surprising then, that among subordinate male baboons, resting levels of glucocorticoids are significantly higher than among dominant individuals—for a subordinate, everyday basal circumstances are stressful. And that's just the start of subordinates' problems with glucocorticoids. When a real stressor comes along, their glucocorticoid response is smaller and slower than in dominant individuals. And when it's all passed, their recovery appears to be delayed. All these are features that count as an inefficient stress response.*

More problems: elevated resting blood pressure; sluggish cardiovascular response to real stressors; a sluggish recovery; suppressed levels of the good HDL cholesterol; among male subordinates, testosterone levels that are more easily suppressed by stress than in dominant males; fewer circulating white blood cells; and lower circulating levels of something called *insulin-like growth factor-I,* which helps heal wounds. As should be clear umpteen pages now into this book, all these are indices of bodies that are chronically stressed.

A chronically activated stress-response (elevated glucocorticoid levels, or resting blood pressure that is too high, or an enhanced risk of atherosclerosis) appears to be a marker of being low-ranking in lots of other animal species as well: primates ranging from standard-issue monkeys like rhesus and other macaques to primitive beasts called *promisions* (such as tree shrews and mouse lemurs). Same for rats, mice, hamsters, guinea pigs, wolves, rabbits, pigs. Even fish. Even sugar gliders, whatever they might be.

A critical question: I'm writing as if being low-ranking, and subject to all those physical and psychological stressors, causes chronic overactivation of the stress-response. Could it be the other way around? Could having a second-rate stress-response set you up for being low-ranking?

---

* I spent about a dozen summers with my baboons figuring out the neuroendocrine mechanisms that give rise to the inefficient glucocorticoid system in the subordinate animals. "Neuroendocrine mechanisms" means the steps linking the brain, the pituitary, and the adrenals in the regulation of glucocorticoid release. The question becomes which of the steps—brain, pituitary, adrenals—is the spot where there is a problem. There turn out to be a number of sites where things work differently in subordinate and dominant baboons. And interestingly, the mechanisms that give rise to the pattern in subordinate baboons is virtually identical with those that give rise to the glucocorticoid pattern in humans with major depression.

You can answer this question with studies of captive animals, where you can artificially form a social group. Monitor glucocorticoid levels, blood pressure, and so on, when the group is first formed, and again once rankings have emerged, and the comparison will tell you which direction the causality works—do physiological differences predict who is going to wind up with which rank, or is it the other way around? And the answer, overwhelmingly, is that rank emerges first, and drives the distinctive stress profile.

So we've developed a pretty clear picture. Social subordinance equals being chronically stressed equals an overactive stress-response equals more stress-related disease. Now it's time to see why that's simplistic and wrong.

The first hint is hardly a subtle one. When you stand up at some scientific meeting and tell about the health-related miseries of your subordinate baboons or tree shrews or sugar gliders, invariably, some other endocrinologist who studies the subject in some other species gets up and says, "Well, *my* subordinate animals don't have high blood pressure or elevated glucocorticoid levels." There are lots of species in which social subordinance is not associated with an overactive stress-response.

Why should that be? Why should being subordinate not be so bad in that species? The answer is that in that species, it's not so bad being subordinate, or possibly it's actually a drag being dominant.

An example of the first is seen with a South American monkey called the *marmoset*. Being subordinate among them does not involve the misery of physical and psychological stressors; it isn't a case of subjugation being forcibly imposed on you by big, mean, dominant animals. Instead, it is a very relaxed waiting strategy—marmosets live in small social groups of related "cooperative breeders," where being subordinate typically means you are helping out your more dominant older sibling or cousin and waiting your turn to graduate into that role. And commensurate with this picture, David Abbott at the Wisconsin Regional Primate Research Center has shown that subordinate marmosets don't have overactive stress-responses.

Wild dogs and dwarf mongooses provide examples of the second situation in which subordination isn't so bad. Being dominant in those species doesn't mean a life of luxury, effortlessly getting the best of the pickings and occasionally endowing an art museum. None of that status quo stuff. Instead, being dominant requires the constant reassertion of high rank through overt aggression—one is tested again and again. And as Scott and Nancy Creel at Montana State University have shown, it's not the subordinate animals among those species who have the elevated basal glucocorticoid levels; it's the dominant ones.

David Abbott and I are currently drawing on the collaborative efforts of a large number of colleagues who have studied rank-stress physiology issues in a wide variety of primates. We have tried to formalize what features of a primate society predict whether it is the dominant or the subordinate animals who have the elevated stress-responses. For any given species, we have posed various questions: What are the rewards of being dominant? How much of a role does aggression play in maintaining dominance? How much grief does a subordinate individual have to take? What sources of coping and support (including the presence of relatives) do subordinates of that species have available to them? What covert alternatives to competition are available? If subordinates cheat at the rules, how likely are they to get caught and how bad is the punishment? How often does the hierarchy change, and does that occur through small incremental shifts in ranks or through all-hell-breaking-loose revolutions? Across the dozen different species for which there are decent amounts of data available, the best predictors of elevated glucocorticoid levels appear to be how often subordinate animals are harassed by dominant individuals, and how badly the former are punished if they try to cheat and get caught. Lots of harassment and high costs for cheating, and that's a species where subordinate animals get the stress-related diseases.

So rank means different things in different species. It turns out that rank can also mean different things in different social groups within the same species. Primatologists these days talk about primate "culture," and this is not an anthropomorphic term. For example, chimps in one part of the rain forest can have a very different culture from the folks four valleys over—different frequencies of social behaviors, use of similar vocalizations but with different meanings (in other words, something approaching the concept of a "dialect"), different types of tool use. And intergroup differences influence the rank-stress relationship.

One example is found among female rhesus monkeys, where subordinates normally take a lot of grief and have elevated basal glucocorticoid levels—except in one social group that was studied, which, for some reason, had very high rates of reconciliatory behaviors among animals after fights. Another example concerns male baboons where, as noted, subordinates normally have the elevated glucocorticoid levels—except during a severe drought, when the dominant males were so busy looking for food that they didn't have the time or energy to hassle everyone else (implying, ironically, that for a subordinate animal, an environmental stressor can be a blessing, insofar as it saves you from a more severe social stressor).

A critical intergroup difference in the stress-response concerns the stability of the dominance hierarchy. Consider an animal who is, say, Number 10 in the hierarchy. In a stable system, that individual is getting trounced 95 percent of the time by Number 9 but, in turn, thrashes Number 11 95 percent of the time. In contrast, if Number 10 were winning only 51 percent of interactions with Number 11, that indicates that the two are close to switching positions. In a stable hierarchy, 95 percent of the interactions up and down the ranks reinforce the status quo. Under those conditions, dominant individuals are stably entrenched and have all the psychological perks of their position—control, predictability, and so on. And under those conditions, among the various primate species discussed above, it is the dominant individuals who have the healthiest stress-responses.

In contrast, there are rare periods when the hierarchy becomes unstable—some key individual has died, someone critical has transferred into the group, some pivotal coalitional partnership has formed or come apart—and a revolution results, with animals changing ranks left and right. And under those conditions, it is typically the dominant individuals who are in the very center of the hurricane of instability, subject to the most fighting, the most challenges, most effected by the see-sawing of coalitional politics.* And during such unstable periods among those same primate species, the dominant individuals no longer have the healthiest stress-responses.

* After all, do you think it would have been restful to have been the czar of Russia in 1917?

So while rank is an important predictor of individual differences in the stress-response, the meaning of that rank, the psychological baggage that accompanies it in a particular society, is at least as important. Another critical variable is an animal's personal experience of both its rank and society. For example, consider a period when an immensely aggressive male has joined a troop of baboons, and is raising hell, attacking animals unprovoked left and right. One might predict stress-responses throughout the troop thanks to this destabilizing brute. But instead, the pattern reflects the individual experience of animals—for those lucky enough never to be attacked by this character, there were no changes in immune function. In contrast, among those attacked, the more frequently that particular baboons suffered at this guy's teeth, the more immunosuppressed they were. Thus, you ask the question, "What are the effects of an aggressive, stressful individual on immune function in a social group?" And the answer is, "It depends—it's not the abstract state of living in a stressful society which is immunosuppressive. Instead, it is the concrete state of how often your own nose is being rubbed in that instability."*

As a final variable, it is not just rank that is an important predictor of the stress-response, not just the society in which the rank occurs, or how a member of the society experiences both; it's also personality—the topic of the previous chapter. As we saw, some primates see glasses as half empty and life as full of provocations, and they can't take advantage of outlets or social support—those are the individuals with overactive stress-responses. For them, their rank, their society, their personal experiences might all be wonderfully salutary, but if their personality keeps them from perceiving those advantages, their hormone levels and arteries and immune systems are going to pay a price.

All things considered, this presents a pretty subtle picture of what social rank has to do with stress-related disease among primates. And it's reasonable to expect the picture to be that much more complicated and subtle when considering humans. Time for a surprise.

---

* And all you have to do to appreciate that is consider all the people who have made fortunes black-marketing penicillin or hoarding critical food supplies during wartime.

DO HUMANS HAVE RANKS?

I personally was always picked last for the softball team as a kid, being short, relatively uncoordinated, and typically preoccupied with some book I was lugging around. Thus, having been perpetually ensconced at the bottom of that pecking order, I am skeptical about the notion of ranking systems for humans.

Part of the problem is definitional, in that some supposed studies of human "dominance" are actually examining Type A features—people defined as "dominant" are ones who, in interviews, have hostile, competitive contents to their answers, or who speak quickly and interrupt the interviewer. This is not dominance in a way that any zoologist would endorse.

Other studies have examined the physiological correlates of individual differences in humans who are competing directly against one another in a way that looks like dominance. Some have examined, for example, the hormonal responses in college wrestlers depending on whether they won or lost their match. Others have examined the endocrine correlates of rank competition in the military. One of the most fruitful areas has been to examine ranks in the corporate world. Chapter 12 on psychological stress showed how the "executive stress syndrome" is mostly a myth—people at the top give ulcers, rather than get them. Most studies have instead shown that it is middle management that succumbs to the stress-related diseases. This is thought to reflect the killer combination that these folks are often burdened with, namely, high work demands but little autonomy—responsibility without control.

Collectively, these studies have produced some experimentally reliable correlations. I'm just a bit dubious as to what they mean. For starters, I'm not sure what a couple of minutes of competitive wrestling between two highly conditioned twenty-year-olds teaches us about which sixty-year-old gets clogged arteries. At the other end, I wonder what the larger meaning is of rankings among business executives—while primate hierarchies can ultimately indicate how hard you have to work for your calories, corporate hierarchies are ultimately about how hard you have to work for a big-screen television.

But my skepticism is most strongly anchored in two reasons. First, humans can belong to a number of different ranking systems simultaneously, and ideally are excelling in at least one of them (and thus, may be giving the greatest psychological weight to that one). So, the lowly subordinate in the mailroom of the big corporation may, after hours, be deriving tremendous prestige and self-esteem from being the deacon of his church, or the captain of her weekend softball team, or may be at the top of the class at the adult-extension school. One person's highly empowering dominance hierarchy may be a mere irrelevancy to the person in the next cubicle, and this will greatly skew results.

And most importantly, people put all sorts of spin inside their heads about ranks. Consider a marathon being observed by a Martian scientist studying physiology and rank in humans. The obvious thing to do is keep track of the order in which people finish the race. Runner 1 dominates 5, who clearly dominates 5000. But what if runner 5000 is a couch potato who took up running just a few months ago, who half expected to keel over from a coronary by mile 13 and, instead, finished—sure, hours after the crowds wandered off—but finished, exhausted and glowing. And what if runner 5 had spent the previous week reading in the sports section that someone of their world-class quality should certainly finish in the top 3, maybe even blow away the field. No Martian on earth could predict correctly who is going to feel exultantly dominant afterward.

People are as likely to race against themselves, their own previous best time, as against some external yardstick. This can be seen in the corporate world as well. An artificial example: the kid in the mailroom is doing a fabulous job and is rewarded, implausibly, with a $50,000 a year salary. A senior vice-president screws up big-time and is punished, even more implausibly, with a $50,001 a year salary. By the perspective of that Martian, or even by a hierarchically minded wildebeest, it's obvious that the vice-president is in better shape to acquire the nuts and berries needed to forestall the next famine. But you can guess who is going to be going to work contentedly and who is going to be making angry phone calls to a headhunter from the cell phone in the BMW. And humans can play internal, rationalizing games with rank based on their knowledge of what determined their placement. Consider the following fascinating example: guys who win at some sort of competitive interaction typically show at least a small rise in their circulating testosterone levels—unless they consider the win to have come from sheer luck.

When you put all those qualifiers together, I think the net result is some pretty shaky ground when it comes to considering human rank and its relevance to the stress-response. Except in one realm. If you want to figure out the human equivalent of being a low-ranking social animal, an equivalent that carries with it atypically high rates of physical and psychological stressors, that is ecologically meaningful in that it's not just about how many hours you have to work to buy the new car phone, that is likely to overwhelm most of the rationalizations and alternative hierarchies that one can muster—check out a poor human.

SOCIOECONOMIC STATUS, STRESS, AND DISEASE

If you want to see an example of chronic stress, study human poverty, the place where we began this chapter. Being poor involves lots of physical stressors. Manual labor and a greater risk of work-related accidents. Maybe even two or three exhausting jobs, complete with chronic sleep deprivation. Maybe walking to work, walking to the laundromat, walking back from the market with the heavy bag of groceries—instead of riding in an air-conditioned car. Maybe too little money to afford a new mattress that might help that aching back, or some more hot water in the shower for that daily arthritic throb; and, of course, maybe some hunger thrown in as well. . . . The list goes on and on.

And of course being poor brings disproportionate amounts of psychological stressors as well. Lack of control, lack of predictability: numbing work on an assembly line, an occupational career spent taking orders or going from one temporary stint to the next. The first one laid off when economic times are bad—and studies show that the deleterious effects of unemployment on health begin not at the time the person is laid off, but when the mere threat of it first occurs. Wondering if the money will stretch to the end of the month. Wondering if the rickety car will start in the morning and get you to the job interview on time. How's this for an implication of lack of control: one study of the working poor showed that they were less likely to comply with their doctors' orders to take antihypertensive diuretics (drugs that lower blood pressure by making you urinate) because they weren't allowed to go to the bathroom at work as often as they needed to when taking the drugs.

As another form of psychological stress, poverty brings with it a marked lack of outlets. Feeling a little stressed with life and considering a relaxing vacation, or maybe those classical guitar lessons to get a little peace of mind? Probably not. Or how about quitting that stressful job and taking some time off at home to figure out what you're doing with your life? Not when there's an extended family counting on your paycheck and no money in the bank. Feeling like at least jogging regularly to get some exercise and let off some steam? Statistically, a poor person is far more likely to live in a crime-riddled neighborhood, and jogging may wind up being a hair-raising stressor.

And finally, along with long hours of work and kids to take care of comes a serious lack of social support—if everyone you know is working two or three jobs, you and your loved ones, despite the best of intentions, aren't going to be having much time to sit around being supportive like a bunch of Thirty-Somethings.

All these hardships suggest that low socioeconomic status (SES—typically measured by a combination of income, housing conditions, and education) should be associated with chronic activation of the stress-response. Surprisingly, little has been studied about this issue. In that case, is being poor associated with more stress-related diseases? As a first pass here, let's just ask whether low SES is associated with more diseases, period. And is it ever.

The SES Gradient The health risk of poverty turns out to be a huge effect, the biggest risk factor there is in all of behavioral medicine. For centuries, it's been true that if you want to increase your odds of living a long and healthy life, don't be poor. Poverty is associated with increased risks of cardiovascular disease, respiratory disease, ulcers, rheumatoid disorders, psychiatric diseases, and a number of types of cancer, just to name a few.* In the case of some of those diseases, if you cling to the lowest rungs of the socioeconomic ladder, it can mean five to ten times the prevalence compared with those perched on top. And poverty is a hidden variable in other patterns of disease. For example, SES statistically accounts for a huge percentage of the greater disease risk of African Americans. Moreover, the consequences of poverty are long-lasting. Sociologists Richard Sennett and Jonathan Cobb have shown that having been born to poverty carries tremendous residual psychological and sociological consequences for those who have proved upwardly mobile. The same holds for disease patterns. One remarkable example concerns studies of a group of nuns. They took their vows as young adults, and spent the rest of their lives sharing the same diet, same health care, same housing, and so on. And despite this control of all these variables, in old age, their patterns of disease, of dementia, and of longevity itself were still highly predicted by the SES status they had when they became nuns more than half a century before.

What might be the cause of the SES gradient in health? A century ago, it was about poor people getting more infectious diseases and having an astronomically higher infant mortality rate. But with our shift toward the modern prevalence of slow, degenerative diseases, the possible causes of the gradient have shifted as well. A variety of possibilities come to mind:

- Less access to the health care system.

In the United States, poor people (with or without health insurance) don't have the same access as do the wealthy. This includes fewer preventative check-ups with doctors, a longer lag

* Not all diseases are more prevalent among the poor, and, fascinatingly, some are even more common among the wealthy. Melanoma, for example—suggesting that sun exposure in a lounge chair may have different disease risks than getting your neck red from stooped physical labor (or that a huge percentage of poor people laboring away in the sun have a fair amount of melanin in their skin, if you know what I mean). Or breast cancer. Also, multiple sclerosis and a few other autoimmune diseases. And "hospitalism," a pediatric disease of the 1930s in which infants would waste away in hospitals. It is now understood that it was mostly due to lack of contact and sociality—and kids who would wind up in poorer hospitals were less subject to this, since the hospitals couldn't afford state-of-the-art incubators, necessitating that staff actually hold them.

time for testing when something bothersome has been noted, and less adequate care when something has actually been discovered. As one example of this set of differences in health care delivery: the poorer you are judged to be (based on the neighborhood you live in, your home, your appearance), the less likely paramedics are to try to revive you on the way to the hospital.

- Less access to health-promoting factors, greater exposure to health-decreasing risk factors.

This covers a whole range of conditions. Poorer people are more likely to smoke, to drink in excess, to eat an unhealthy diet, not to exercise. They're more likely to have elevated blood pressure and cholesterol levels, and to be obese. They are less likely to use a seat belt, wear a motorcycle helmet, be able to afford a car that has safety air bags. They are more likely to live near a toxic dump; to have inadequate heat in the winter; to live in crowded conditions, which itself increases exposure to infectious disease; to be beaten by a mugger. The list seems endless.

- Less education.

When people do studies of SES and health, obviously the most accurate way of determining whether someone is poor is to pay attention to the person's income (as opposed to the nature of her education or the fanciness of his clothes or housing). It turns out that paying attention to level of education is virtually as informative as studying income; income and education are very tightly linked. As such, another possible explanation for the SES gradient is that poor people (and thus, typically, poorly educated people) don't understand, don't know about the risk factors they are being exposed to, or the health-promoting factors they are lacking—even if it is within their power to do something, they aren't informed. As one example that boggles me, there are apparently substantial numbers of people who are not aware that cigarettes do bad things to you, and the studies show that these aren't folks too busy working on their doctoral dissertations to note some public health trivia. Other studies indicate that, for example, poor women are the least likely to have heard of Pap tests, and thus are less likely to be screened for cervical cancer. The intertwining of poverty and poor education probably explains the high rates of poor people who, despite their poverty, could still be eating more healthfully, using seat belts or crash helmets, and so on, but don't. And it probably explains why poor people are less likely to comply with some treatment regime that has been prescribed for them that they can actually afford—they are less likely to have understood the instructions or to think that following them is important.

- You guessed it: stress.

The Case for Stress as the Cause of the Gradient Having just suggested several possible causes for the gradient, I'll show why they're probably not major contributors—before I discuss you know what. Most people, including many health professionals, initially assume that the SES gradient with respect to disease can mostly be explained by differences in health care access. Make the health care system equitable, socialize that medicine, and away would go that gradient. But one list shows how little the gradient has to do with health care access, a list of countries in which poverty is robustly associated with increased prevalence of disease: Australia, Belgium, Denmark, Finland, France, Italy, Japan, the Netherlands, New Zealand, the former Soviet Union, Spain, Sweden, the United Kingdom, and, of course, the US of A. Socialize the medical care system, socialize the whole country, turn it into a worker's paradise, and you still get the gradient. In a place like England, the SES gradient has gotten worse over this century, despite the imposition of universal health care.

You could cynically and no doubt correctly point out that systems of wonderfully egalitarian health care access are probably egalitarian in theory only—even the Swedish health care system is likely to be at least a smidgen more attentive to the wealthy industrialist, or the sick doctor, or the famous athlete than to some no-account poor person cluttering up a clinic. Some people are always more equal than others.

But there is a second striking reason why the SES-health relationship couldn't be due just to differential access to health care. It is because the relationship forms the term I've been using, namely, a *gradient*. It's not the case that only poor people have an increased risk of various diseases while everyone else has the same risk. For every step higher up the SES ladder, there is better health—and that's in societies where everyone but the poorest has roughly the same health care access.

More evidence against the access idea: in at least one study of people enrolled in a prepaid health plan, where medical facilities were available to all participants, poorer people had more cardiovascular disease, despite making *more* use of the medical resources.

And a final vote against the health care access argument: the gradient exists for diseases that have nothing to do with access. Take a young person and, each day, scrupulously, give her a good medical examination, check her vitals, peruse her blood, run her on a treadmill, give her a stern lecture about good health habits and then centrifuge her a bit just for good measure, and she is still just as much at risk for some diseases as if she hadn't gotten all that attention. And poor people are still more likely to get those access-proof diseases. Theodore Pincus of Vanderbilt University has carefully documented the existence of an SES gradient for two of those diseases, juvenile diabetes and rheumatoid arthritis.

Thus, the leading figures in this field all seem to rule out health care access as a major part of the story. This is not to rule it out completely (let alone suggest that we not bother trying to establish universal health care access). As evidence, sweaty capitalist America has the worst gradient, while the socialized Scandinavian countries have the weakest ones. But they still have hefty gradients, despite their socialism.

So we turn to a second possible explanation. Is the SES gradient due to too many risk factors and too few protective factors? As reviewed, there's a huge SES disparity in this realm—there's more smoking and less seat belt use among the poor, more toxic dumps and fewer fresh vegetables, and so on. It seems plausible that, collectively, these circumstances are likely to have a pretty hefty impact on disease and longevity. However, that also turns out to have less explanatory power than one might have guessed. One of the most celebrated studies in this field was carried out by Michael Marmot on over ten thousand British civil servants, a vast number of them, running the gamut from unskilled blue-collar workers to high-powered executives. There was an extremely steep SES gradient for cardiovascular disease and mortality, one of the most dramatic examples of the gradient in the literature (in, once again, a country with universal health care); overall, the study showed a fourfold increase in mortality rates when comparing the highest and lowest rungs of the system. A number of risk factors are eliminated immediately, in that even the people at the lowest rungs of the civil service system have adequate pay and steady employment. Marmot and colleagues then carefully controlled for other risk factors—smoking, hypertension, triglyceride levels in the blood—and showed that they accounted for less than a third of the variability in the data. And similar findings have emerged from other studies as well. Socialize the medical system, give everyone regular checkups, control for people doing endangering things about their health, and the poor still get sicker more often.

It's obvious where I'm heading with this one. Okay, let's not be shy, get it out in the open. Maybe stress does have something to do with it as well. The evidence for this does not merely arise from ruling out the alternatives. Study after study shows that the poor do indeed have the highest rate of life stressors, the least social support, just as was speculated a few paragraphs above. And when one examines the SES gradient for individual diseases, the strongest gradients occur for diseases with the greatest sensitivity to stress—heart disease, hypertension, psychiatric disorders—while a class of disease I devoted half a chapter on dissociating from stress—cancer—has the weakest overall gradient.

The late Aaron Antonovsky, a medical sociologist who was one of the first to analyze SES health gradients, argued that the poor lack a strong sense of social "coherence," and that this contributed significantly to their poor health. In his view, healthful (what he called "salutogenic") coherence rested on a number of variables. First, does a person have a sense of being linked to the mainstream of society, of being in the dominant subculture, of being in accord with society's values? Add the poor to the list of the elderly, handicapped, minorities, immigrants, as those disenfranchised in this way. Second, can a person perceive society's messages as information, rather than as noise? In this regard, the poor education that typically accompanies poverty biases toward the latter. Next, has a person been able to develop an ideal set of coping responses (what Antonovsky termed a set of fixed rules and flexible strategies) for dealing with society's challenges? Again, the poor are at a disadvantage in this realm (the final chapter of this book reviews a poverty-related personality style in which there is a distinctive lack of flexible coping strategies, and a major stress-related disease risk as a result). Next, does a person have the resources to carry out plans? Clearly to a lesser extent for the poor. Finally, does a person get meaningful feedback from society—do their messages make a difference? To recognize the extent to which the poor exist without feedback, just consider the varied ways that most of us have developed for not seeing homeless people as we walk past them.

Basically, what Antonovsky was doing was to reframe the main psychological modifiers of the stress-response in the context of society. And most other leaders in the field have reached the conclusion that stress plays a surprisingly large part in the SES gradient.

This is pretty depressing stuff, because of its implications. Nancy Adler and colleagues, of the University of California in San Francisco, have done a great deal of work in this arena, particularly in analyzing the minimal role played by the health care access issue. Writing around the time when universal health insurance (and whether Hillary's hairstyle made her a more or less effective advocate for it) was a daily front-page issue, they concluded that such universal coverage "will have a minor impact on SES-related inequalities in health." This conclusion is anything but reactionary. Instead, it says that if you want to change the SES gradient, it's going to take something a whole lot bigger than rigging up insurance so that everyone can drop in regularly on a friendly small-town doc out of Norman Rockwell. Poverty, and the poor health of the poor, is about much more than simply not having enough money. As Antonovsky showed, it is also about your psychological interactions with society at large and how readily society registers your existence.

This is relevant to an even larger depressing thought, one I've been hinting at since the beginning of this chapter. I initially reviewed what social rank has to do with health in nonhuman primates. Do low-ranking monkeys have a disproportionate share of disease, more stress-related disease? And the answer was, "Well, it's actually not that simple." It depends on the sort of society that animal lives in, its personal experience of that society, its coping skills, its personality, the availability of social support. Change some of those variables and the rank-health gradient can shift in the exact opposite direction. This is the sort of finding that primatologists revel in—look how complicated and subtle my animals are.

And the second half of the chapter looked at humans. Do poor humans have a disproportionate share of disease? And the answer was "Yes, yes, over and over." Regardless of gender or age or race. In societies with universal health care and those without. In societies that are ethnically homogenous and those rife with ethnic tensions. In societies in which illiteracy is widespread and those in which it has been virtually banished. In those in which infant mortality has been plummeting and in some wealthy, industrialized societies in which rates have inexcusably been climbing. And in societies in which the central mythology is a capitalist credo of "Living well is the best revenge" and those in which it is a socialist anthem of "From each according to his ability, to each according to his needs."

What does this dichotomy between our animal cousins and us signify? The primate relationship is nuanced and filled with qualifiers; the human relationship is a sledgehammer that obliterates every societal difference. Are we humans actually less complicated and sophisticated than nonhuman primates? Not even the most chauvinistic primatologists holding out for their marmosets would vote for that conclusion. I think it suggests something else. Agriculture is a fairly recent human invention, and in many ways it was one of the great stupid moves of all time. Hunter-gatherers have thousands of wild sources of food to subsist on. Agriculture changed all that, generating an overwhelming reliance on a few dozen domesticated food sources, making you extremely vulnerable to the next famine, the next locust infestation, the next potato blight. Agriculture allowed for the stratification of society and the invention of classes; it allowed for the stockpiling of surplus resources and thus, inevitably, the unequal stockpiling of them. Thus, it allowed for the invention of poverty. I think that the punch line of the primate-human difference is that when humans invented poverty, they came up with a way of subjugating the low-ranking like nothing ever before seen in the primate world.

MANAGING STRESS

By now, if you are not depressed by all the bad news in the preceding chapters, you probably have only been skimming. Stress can wreak havoc with your metabolism, raise your blood pressure, burst your white blood cells, make you flatulent, ruin your sex life, and if that's not enough, possibly damage your brain.* Why don't we toss in the towel right now?

There is hope. Although it may sneak onto the scene in a quiet, subtle way, once you realize that the hope is there, it may change the way you think about stress. This frequently hits me at gerontology meetings. I'm sitting there, listening to the umpteenth lecture with the same general tone—the kidney expert speaking about how that organ disintegrates with age, the immunology expert on how immunity declines, and so on. There is always a bar graph set to 100 percent for young subjects, with a bar showing that the elderly have only 75 percent of the kidney filtration rate of young subjects, 75 percent of the limb strength, and so on.

Research typically involves the study of populations, rather than single individuals one at a time. The characteristics of various individuals never have exactly the same values—instead,

* An additional pathology, for those who are really trivia fans when it comes to stress-related disease: "alopecia areata." This is the technical term for that extraordinary state of getting so stressed, and terrified by something that your hair turns white or gray overnight. It really does occur.

Henri Matisse, 1910: The Dance, *oil on canvas.*

the bars in a graph represent the average for each age (see the accompanying graph). Suppose one group of three subjects has scores of 19, 20, and 21, for an average of 20. Another group may have scores of 10, 20, and 30. They also have an average score of 20, but the variance of those scores would be much larger. By the convention of science, the bars also contain a measure of how much variation there is within each age group: the size of the "T" above the bar indicates what percentage of the subjects in the group had scores within X number of steps of the average.

One thing that is utterly reliable is that the extent of variance increases a lot with age—the conditions of the elderly are always much more variable than those of the young subjects. What a drag, you say as a researcher, because with that variance your statistics are not as neat and you have to include more subjects in your aged population to get a reliable average. But really *think* about that fact for a minute. Look at the size of the bars for the young and old subjects, look at the size of the T-shaped variance symbols, do some quick calculations, and suddenly the extraordinary realization hits you—amid the population of, say, fifty subjects, for six of them things are *improving* with age. Their

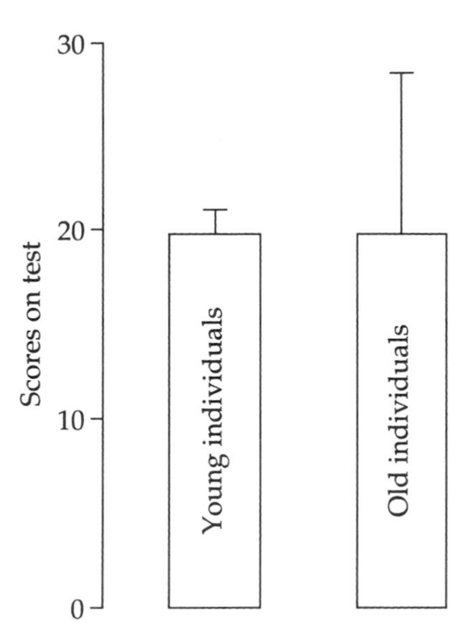

Schematic presentation of the fact that a group of young and old individuals may receive the same average score on a given test, yet the variability in the scores is typically greater among the older population.

kidney filtration rates have gotten better, their blood pressures have decreased, they do better on memory tests. You're on the edge of your seat. Who *are* those six? What are they doing right? And with all scientific detachment abandoned, how can *I* do that, too?

This pattern used to be a statistical irritant to gerontologists. Now it's the trendiest subject in the field. "Successful aging." Not everyone falls apart miserably with age, not every organ system poops out, not everything is bad news.

The same pattern occurs in many other realms in which life tests us. Ten men are released from years spent as political hostages. Nine come out troubled, estranged from friends and family, with nightmares, difficulties readapting to everyday life. Of those nine, one or two succumb to a crippling posttraumatic stress disorder and never function well again. Yet invariably there is one guy who comes out saying, "Yeah, the beatings were awful, the times they put a gun to my head and cocked the trigger were the worst in my life, of course I would never want to do it again, but it wasn't until I was in captivity that I realized what is really important, that I decided to devote the rest of my life to X. I'm almost grateful." How did he do it? Physiological studies of people carrying out dangerous, stressful tasks—parachuting, learning to land on an aircraft carrier in choppy seas, carrying out underwater demolition—show the same pattern: some people have massive stress-responses and others are physiologically unflustered.

A similar phenomenon appears in the most mundane aspects of our everyday stressors. The supermarket line that you pick, naturally, turns out to move the slowest. You're simmering, your sense of irritation made worse by the fact that the person just behind you looks perfectly happy standing there, daydreaming.

Despite the endless ways in which stress can disrupt our well-being, we do not all collapse into puddles of stress-related disease and psychiatric dysfunction. Of course, we are not all exposed to identical external stressors; but given the same stressors, even the same major stressors, we vary tremendously in how our bodies and psyches cope. This final chapter asks the questions born of hope. Who makes up that subset who can cope? How do they do it? And how can we? The previous chapter suggested that some personalities and temperaments aren't well-suited to dealing with stress, and it is easy to imagine the opposite case—some are; this chapter, however, shows that having the "right" personality doesn't explain all successful coping—there's even hope for the rest of us who don't just happen to have that winning personality.

INDIVIDUAL DIFFERENCES IN THE STRESS-RESPONSE: SOME PLEASING EXAMPLES

Successful Aging Probably the best place to start is with successful aging, given how the aging chapter culminated much of the bad news of the first eleven chapters. One particularly grim set of findings had to do with glucocorticoids. Old rats, recall, secrete too much of these hormones—they have elevated levels during basal, nonstressful situations and difficulty shutting off secretion at the end of stress. I discussed the evidence that this could arise from damage to the hippocampus, the part of the brain that (in addition to playing a role in learning and memory) helps inhibit glucocorticoid secretion. Then, to complete the distressing story, it was revealed that glucocorticoids could hasten the death of hippocampal neurons. Furthermore, the tendency of glucocorticoids to damage the hippocampus increases the oversecretion of glucocorticoids, which in turn leads to more hippocampal damage, more glucocorticoids, spiraling downward.

I proposed that "feedforward cascade" model a dozen years ago. It seemed to describe a basic and inevitable feature of aging in the rat, one that seemed important (at least from my provincial perspective, having just spent 80 hours a week studying it in graduate school). I was pretty proud of myself. Then an old friend, Michael Meaney of McGill University, did an experiment that deflated my grandiosity.

Meaney and colleagues studied that feedforward cascade in old rats. But they did something clever first. Before starting the studies, they tested the memory capacity of the rats. As is usual, on the average these old rats had memory problems, compared with young controls. But as usual there was a subset of old rats who were doing just fine, with no memory impairment whatsoever. Meaney and crew split the group of old rats into the impaired and the unimpaired. The latter turned out to show no evidence at all of that degenerative feedforward cascade. They had normal glucocorticoid levels basally and after stress. Their hippocampi had not lost neurons or lost receptors for glucocorticoids. All those awful degenerative features turned out not to be an inevitable part of the aging process. All those rats had to do was age successfully.

What was this subset of rats doing right? Oddly, it might have had something to do with their childhoods. If a rat is handled during the first few weeks of its life, it secretes less glucocorticoids as an adult. This generated a syllogism: If neonatal handling decreases the amount of glucocorticoids secreted as an adult, and such secretion in an adult influences the rate of hippocampal degeneration in old age, then handling a rat in the first few weeks of its life should alter the way it ages years later. Meaney's lab and I teamed up to test this and found exactly that. Do nothing more dramatic than pick a rat up and handle it fifteen minutes a day for the first few weeks of its life, put it back in its cage with the unhandled controls, come back two years later . . . and the handled rat is spared the entire feedforward cascade of hippocampal damage, memory loss, and elevated glucocorticoid levels.

Real rats in the real world don't get handled by graduate students. Is there a natural world equivalent of "neonatal handling" in the laboratory? Meaney has recently shown that rat mothers who spend more time licking and grooming their pups in those critical first few weeks induce the same handling phenomenon. It seems particularly pleasing that this grim cascade of stress-related degeneration in old age can be derailed by something this subtle years earlier. No doubt there are other genetic and experiential factors that bias a rat toward successful or unsuccessful aging, a subject that Meaney still pursues. Of greatest importance for our purposes now, however, is simply that this degeneration is not inevitable.

If the fates of inbred laboratory rats are this variable, how humans fare is likely to be even more diverse. Research on "successful aging" in humans is in its infancy. Some mammoth prospective studies are beginning in which the characteristics of large numbers of middle-aged people will be monitored closely for the rest of their lives—diet, activity, cardiac function, personality profiles, cognitive skills, hormonal secretion, and so on. Ultimately, some of these people will prove to be decrepit at age 70; others will be running marathons at 90. And then the hardy and patient researchers will be able to return to the decades of data and determine what, at the outset, distinguished the people who were destined to be successful agers. With any luck, they will have behavioral and physiological lessons to teach the rest of us.

Coping with Catastrophic Illness In the early 1960s, when scientists were just beginning to investigate whether psychological stress triggers the same hormonal changes that physical stressors do, a group of psychiatrists conducted what has become a classic study. It concerned the parents of children dying of cancer and the high glucocorticoid levels that those parents secreted. Predictably, there was great variance in this measure—some of the parents were secreting immense quantities of glucocorticoids; others were in the normal range. The investigators, in psychi-

Joseph Greenstein, "The Mighty Atom," in old age. An idol of my youth, Greenstein was still performing his feats of strength in Madison Square Garden as an octogenarian. He attributed it to clean, vegetarian living.

atric interviews to explore in depth which parents were holding up best to this horrible stressor, identified a number of coping styles associated with a reduced glucocorticoid stress-response.

One important variable was the ability of parents to displace a major worry onto something less threatening. A father has been standing vigil by his sick child for weeks. It's clear to everyone that he needs to get away for a few days, to gain some distance, as he is near a breaking point. Plans are made for him to leave for a few days, but at the last moment he cancels the vacation, too anxious to go away. Why? At one extreme is the parent who says, "I've seen how rapidly medical crises can develop at this stage. What if my daughter suddenly gets very sick and dies while I am away?" At the other extreme is the parent who is able to attribute the anxiety to something more manageable—"It would be good to get away now, but I worry that she will be lonely while I'm gone." The researchers found that the latter style was associated with lower glucocorticoid levels.

A second variable had to do with denial. When a child went into remission, which frequently happened, did the parents look at her and say to the doctor, "It's over with, there's nothing to worry about, we don't even want to hear the word 'remission,' she is going to be fine"? Or did they peer anxiously at the child, wondering if every cough, every pain, every instant of fatigue was a sign that the disease had returned? During periods of remission, parents who denied that relapse and death were likely and instead focused on the seemingly healthy moment had lower glucocorticoid levels. As we will see shortly, this facet of the study had a surprising postscript.

A final variable was whether the parent had a structure of religious rationalization to explain the illness. At one extreme was the parent who, while obviously profoundly distressed by her child's cancer, was deeply religious and perceived the cancer to be God's test of her family. She even reported an increase in her self-esteem—in effect, "God does not choose just anyone for a task like this; He chose us because He knew we were special and could handle this." At the other extreme was the parent who said, in effect, "Don't tell me that God works in mysterious ways. In fact, I don't want to hear about God at all." The researchers found that if you can look at your child having cancer and decide that God is choosing you for an extra-special assignment, you are likely to have less of a stress-response.

Some of the ramifications of this last facet have generated controversy recently. A number of studies have indicated better health among people with religious faith than among atheists. This has been typically attributed to enhancement of health-improving behaviors and reduction of risky behaviors as a part of religious observance, as well as to the psychological benefits of belonging to a supportive religious community. Herbert Benson, a Harvard cardiologist who has done some research and much proselytizing about the stress reducing effects of relaxation, has added an additional interpretation to those findings. In a recent book, he suggests that belief in a god is itself beneficial to health (independent of the benefits of the social aspects of belonging to a religious community that typically comes with belief). He is certainly not the first scientist to suggest that notion, and previous researchers have even done experimental work in that area. However, Benson has been somewhat of a lightning rod for other scientists criticizing how unscientific this idea supposedly is. Some of the criticism may reflect some additional baggage of Benson's these days—he is currently doing studies to test whether being prayed for by someone improves the health of a hospitalized patient (even if the person is not aware of being prayed for). As a logical extension of this pseudoscientific approach, he and his colleagues keep secret when and where they are conducting these studies, so that some skeptic can't pray for the *opposite* outcome and mess up their data (not to mention the health of the prayed-for guinea pig, if one believes this stuff). An additional cause of the criticism is Benson's suggestion that religious belief is instinctive and that humans are "wired for God." That is viewed by most scientists as drawing on some very unsophisticated ideas regarding behavioral determinism.

Despite those two bits of nonsense, the primary assertion about the protective effects of belief strikes this atheist as perfectly logical, if you believe in the right kind of god or gods. If the god is one who intervenes in human affairs with rules that are mysterious, you at least get the stress-reducing advantages of attribution—it may not be clear what the deity is up to, but you at least know who is responsible for the locust swarm or the winning lottery ticket. If it is an intervening deity with discernible rules, it provides the comfort of both attribution and predictive information—carry out ritual X, or Y is going to happen. If it is a deity who does all the above, *and* will respond to your personal and specific entreaties (most religions do not include that belief), there is an added and wonderfully stress-reducing element of control introduced. And if on top of all that, the deity is viewed as benign, the stress-reducing advantages must be extraordinary. If you can view cancer and Alzheimer's disease, the Holocaust and ethnic cleansing, if you can view the inevitable threshold where your own heart will cease to beat, all in the context of a loving plan, that must constitute the greatest source of support imaginable.

Differences in Vulnerability to Learned Helplessness In Chapter 13, I described the learned helplessness model and its relevance to depression. I emphasized there how generalized the model appears to be: animals of many different species show some version of giving up on life in the face of something aversive and out of their control.

Yet when you look at research papers based on the learned helplessness model, they are like those of any other stress-related field—dozens of bar graphs with T-shaped variance bars indicating large differences in response. For example, of the laboratory dogs put through one learned helplessness paradigm, about one third wind up being resistant to the phenomenon. This is the same idea as the one out of ten hostages who comes out of captivity a mentally healthier person than when he went in. Some folks and some animals are much more resistant to learned helplessness than average. Who are the lucky ones?

In that same chapter, we saw that an important factor in a person's resiliency was whether she had developed an "internalized" locus of control—the perception that she is the master of her destiny—or an "externalized" locus—in which she tends to perceive herself as having little control over the events of everyday life. Those who were internalizing in nature were far more resistant to learned helplessness. In effect, if you generally feel as if you have your hands on the wheel, you view an individual case of uncontrollable bad news to be only that—a single case, and not the whole world.

Why are some dogs naturally resistant to learned helplessness? An important clue: dogs born and raised in laboratories, bred only for research purposes, are more likely to succumb to learned helplessness than those who have come to the lab by way of the pound. Martin Seligman, one of the originators of helplessness research, offers this explanation: if a dog has been out in the real world, experiencing life and fending for itself (as the dogs who wind up in a pound are likely to have done), it has learned about how many controllable things there are in life. When the experience with an uncontrollable stressor occurs, the dog, in effect, is more likely to conclude that "this is awful, but it isn't the entire world." It resists globalizing the stressor into learned helplessness.

More Stress Management Lessons from the Baboons Chapters 14 and 15 introduced social primates, and some critical variables that shaped social success for them: dominance rank, the society in which rank occurs, the personal experience of both, and perhaps most importantly, the role played by personality. In their Machiavellian world, we saw there was more to social success and health for a male than just a lot of muscle or some big sharp canines.

Just as important were social and political skills, the ability to build coalitions, and the ability to walk away from provocations. The personality traits associated with low glucocorticoid levels certainly made sense in the context of effective handling of pschological stressors—the abilities to differentiate threatening from neutral interactions with rivals, to exert some control over social conflicts, to differentiate good news from bad, to displace frustration. And, above all else, the ability to make social connections—grooming, being groomed, playing with infants. So how do these variables play out over time, as these animals age?

Baboons are long-lived animals, sticking around the savanna for anywhere from 15 to 25 years. Which means you don't get to follow an animal from its first awkward bloom of puberty into old age very readily. Nearly twenty years into this project, I'm just beginning to get a sense of the life histories of some of these animals, and the development of their individual differences.

The first hints came a few years into the first of the personality studies. Males with the "low glucocorticoid" personalities were likely to remain in the high-ranking cohort significantly longer than rank-matched males with high glucocorticoid profiles. About three times longer. Among other things, that probably means that the low glucocorticoid guys are out-reproducing the other team. From the standpoint of evolution—passing on copies of your genes, all that jazz—this is a big difference. It suggests that if you were to go away for a couple of zillion millennia, allow that differential selection to play out, and then return to finish your doctoral dissertation, your average baboon will be a descendent of these low-glucocorticoid guys, and the baboon social world will involve a lot of impulse control and gratification postponement. Maybe even toilet training.

And what about the old ages of these individual baboons that are alive today? The most dramatic difference I've uncovered concerns the variable of social affiliation. Your average male baboon has a pretty lousy old age, once he's gotten a paunch and some worn canines and dropped down to the cellar of the hierarchy. Look at the typical pattern of dominance interactions among the males. Usually, Number 3 in the hierarchy is having most of his interactions with Numbers 2 and 4, while Number 15 is mostly concerned with 14 and 16 (except, of course, when 3 is having a bad day and needs to displace aggression onto someone way down). Most interactions then usually occur between animals of adjacent ranks. However, amid that pattern, you'll note that the top-ranking half-dozen or so animals, nevertheless, are spending a lot of time subjecting poor Number 17 to a lot of humiliating dominance displays, displacing him from whatever he is eating, making him get up whenever he settles into a nice shady spot, just generally giving him a hard time. What's that about? Number 17 turns out to have been very high-ranking back when the current dominant animals were terrified adolescents. They remember, and can't believe they can make this decrepit ex-king grovel anytime they feel like it.

So as he ages, your average male baboon gets a lot of grief from the current generation of thugs, and this often leads to a particularly painful way of passing your golden years—the treatment gets so bad that the male picks up and transfers to a different troop. That's a stressful, hazardous journey, with an extremely high mortality rate for even a prime-aged animal—moving across novel terrain, chancing predators on your own. All that to wind up in a new troop, subject to an extreme version of that too-frequently-true truism about primate old age; namely, aging is a time of life spent among strangers.

But what about males who, in their prime, had the low glucocorticoid personality of spending lots of time affiliated with females, grooming, sitting in contact, playing with kids? They just keep doing the same thing. They get hassled by the current rulers, but it doesn't seem to count as much as the social connectedness to these baboons. They don't transfer troops, and continue the same pattern of grooming and socialization for the rest of their lives. And that seems like a pretty good definition of successful aging for any primate.

COPING WITH STRESS: SOME SUCCESS STORIES

Parents somehow shouldering the burden of their child's fatal illness, a low-ranking baboon who has a network of friends, a dog resisting learned helplessness—these are striking examples of individuals who, faced with a less than ideal situation, nevertheless excel at coping. That's great, but what if you don't already happen to be that sort of individual? When it comes to rats who wish to age successfully, a previous section generates the useful bit of advice to make sure you pick the right sort of infancy. And when it comes to humans who wish to cope with stress and achieve successful aging, you should be sure to pick the right parents' genes, and the right parents' socioeconomic status as well. The other cases of successfully coping with stress may not be any more encouraging to the rest of us. What if we happen not to be the sort of baboon who looks at the bright side, the person who holds on to hope when others become hopeless, the parent of the child with cancer who somehow psychologically manages the unmanageable? There are many stories of individuals who have supreme gifts of coping. For us ungifted ones, are there ways to change the world around us and to alter our perceptions of it so that psychological stress becomes at least a bit less stressful?

A first thing to emphasize is that we can change the way we cope, both physiologically and psychologically. As the most obvious example, physical conditioning brought about by regular exercise will lower blood pressure and resting heart rate and increase lung capacity, just to mention a few of its effects. Among Type A people, psychotherapy has been shown to change not only behaviors but also cholesterol profiles, risk of heart attack, and risk of dying, independent of changes in diet or other physiological regulators of cholesterol. Some preliminary studies have also shown that various relaxation techniques or techniques that "alter consciousness" may have beneficial physiological effects. For example, trained practitioners of transcendental meditation are reported to be able to reduce glucocorticoid levels and various indices of their bodies' metabolism, at least while they are actually meditating. As another example, the pain and stressfulness of delivery can be modulated to some extent by relaxation techniques such as Lamaze.* Sheer repetition of certain activities can change the connection between your behavior and activation of your stress-response. In one classic study discussed earlier, Norwegian soldiers learning to parachute were studied

* In the previous edition, this sentence simply read "the pain and stressfulness of delivery can be modulated by relaxation techniques such as Lamaze..." Having seen my wife go through delivery a year ago, I felt the need to throw in a few qualifiers. Maybe I should even have changed it to "can be modulated for about five minutes by Lamaze while the husband frantically and ineffectually checks his class notes." Back to a theme from the end of Chapter 9—stress-induced analgesia appears to get you only so far.

over the course of months of training. At the time of their first jump, they were all terrified; they felt like vats of Jell-O, and their bodies reflected it. Glucocorticoids and epinephrine levels were elevated, testosterone levels were suppressed—all for hours before and after the jump. As they repeated the experience, mastered it, and, most of all, learned not to be terrified, their hormone secretion patterns changed. By the end of training they were no longer turning on their stress-response hours before and after the jump, only at the actual time. They were able to confine their stress-response to an appropriate moment, when there was a physical stressor; the entire psychological component of the stress-response had been habituated away.

All of these examples show that the workings of the stress-response can change over time. We grow, learn, adapt, get bored, develop an interest, drift apart, mature, harden, forget. We are malleable beasts. What are the buttons we can use to manipulate the system in a way that will benefit us?

The issues raised in the chapter on the psychology of stress are obviously critical: control, predictability, social support, outlets for frustration. Seligman and colleagues, for example, have reported some laboratory success in buffering people from learned helplessness when confronted with an unsolvable task—if subjects are first given "empowering" exercises (various tasks that they readily can master and control). But this is a fairly artificial setting. Some classic studies have manipulated similar psychological variables in the real world—some of the grimmest parts of the real world. As we will see, the results have been startling.

SELF-MEDICATION AND CHRONIC PAIN SYNDROMES

Whenever something painful happens to me, amid all the distress I am surprised at being reminded of how painful pain is. That thought is always followed by another, "What if I hurt like this all the time?" Chronic pain syndromes are extraordinarily debilitating. Diabetic neuropathies, crushed spinal nerve roots, severe burns, recovery after surgery can all be immensely painful. This poses a medical problem, insofar as it is often difficult to give enough drugs to control the pain without causing addiction or putting the person in danger of an overdose. As any nurse will attest, this also poses a management problem, as the chronic pain patient spends half the day hitting the call button, wanting to know when his next painkiller is coming, and the nurse has to spend half the day explaining that it is not yet time. A memory that will always make me shudder: at one point, my father was hospitalized for something. In the room next door was an elderly man who, seemingly around the clock, every thirty seconds, would plaintively shout in a heavy Yiddish accent, "Nurse. Nurse! It hurts. It hurts! Nurse!" The first day it was horrifying. The second day it was irritating. By the third day, it had all the impact of the rhythmic chirping of crickets.

Awhile back some researchers got an utterly mad idea, the thought of frothing lunatics. Why not give the painkillers to the patients and let them decide when they need medication? You can just imagine the apoplexy that mainstream medicine would have over that one—patients will overdose, become addicts, you can't let patients do that. It was tried with cancer patients and postsurgical patients, and it turned out that the patients did just fine when they self-medicated. In fact, the total amount of painkillers consumed decreased.

Why should consumption go down? Because when you are lying there in bed, in pain, uncertain of the time, uncertain if the nurse has heard your call or will have time to respond, uncertain of everything, you are asking for painkillers not only to stop the pain but also to stop the uncertainty. Reinstitute control, give the patient the knowledge that the medication is there for the instant that the pain becomes too severe, and the pain often becomes far more manageable.

INCREASING CONTROL IN NURSING HOMES

I can imagine few settings that better reveal the nature of psychological stress than a nursing home. Under the best of circumstances, the elderly tend to have a less active, less assertive coping style than young people. When confronted by stressors, the latter are more likely to try to confront and solve the problem, while the former are more likely to distance themselves from the stressor or adjust their attitude toward it. The nursing home setting worsens these tendencies toward withdrawal and passivity: It's a world in which you are often isolated from the social support network of a lifetime and in which you have little control over your daily activities, your finances, often your own body. A world of few outlets for frustration, in which you are often treated like a child—"infantilized." Your easiest prediction is "life will get worse."

A number of psychologists have ventured into this world to try to apply some of the ideas about control and self-efficacy outlined in Chapter 12. In one study, for example, residents of a nursing home were given more responsibility for everyday decision making. They were made responsible for choosing their meals for the next day, signing up in advance for social activities, picking out and caring for a plant for their room, instead of having one placed there and cared for by the nurses ("Oh, here, I'll water that, dear; why don't you just get back into bed?"). People became more active—initiating more social interactions—and described themselves in questionnaires as happier. Their health improved, as rated by doctors unaware of whether they were in the increased-responsibility group or the control group. Most remarkable of all, the death rate in the former group was half that of the latter.

In other studies, different variables of control were manipulated. Almost unanimously, these studies show that a moderate increase in control produces all the salutary effects just described; in a few studies, physiological measures were even taken, showing changes like reductions in glucocorticoid levels or improved immune function. The forms that increased control could take were many. In one study, the baseline group was left alone, while the experimental group was organized into a residents' council that made decisions about life in the nursing home. In the latter group, health improved and individuals showed more voluntary participation in social activities. In another study, residents in a nursing home were being involuntarily moved to a different residence because of the financial collapse of the first institution. The baseline group was moved in the normal manner, while the experimental group was given extensive lectures on the new home and given control of a wide variety of issues connected with the move (the day of the move, the decor of the room they would live in, and so on). When the move occurred, there were far fewer medical complications for the latter group. The infantilizing effects of loss of control were shown explicitly in another study in which residents were given a variety of tasks to do. When the staff present *encouraged* them, performance improved; when the staff present *helped* them, performance declined.

Another example of these principles: this study concerned visits by college students to people in nursing homes. One nursing-home group, the baseline group, received no student visitors. In a second group, students would arrive at unpredictable times to chat. There were various improvements in functioning and health in this group, testifying to the positive effects of increased social contact. In the third and fourth groups, control and predictability were introduced—in the third group, the residents could decide when the visit occurred, whereas in the fourth they could not control it, but at least were told when the visit would take place. Functioning and health improved even more in both of those groups, compared with the second. Control and predictability help, even in settings where you think it won't make a dent in someone's unhappiness.

STRESS MANAGEMENT: READING THE LABEL CAREFULLY

These studies generate some simple answers to coping with stress that are far from simple to implement in everyday life. They emphasize the importance of manipulating feelings of control, predictability, outlets for frustration, social connectedness, the perception of whether things are worsening or improving. In effect, the nursing home and pain studies are encouraging dispatches from the front lines in this war of coping. Their simple, empowering, liberating message: if manipulating such psychological variables can work in these trying circumstances, it certainly should for the more trivial psychological stressors that fill our daily lives.

This is the message that fills stress management seminars, therapy sessions, and the many books on the topic. Uniformly, they emphasize finding means to gain at least some degree of control in difficult situations, viewing bad situations as discrete events rather than permanent or pervasive ones, finding appropriate outlets for frustration and means of social support and solace in difficult times.

That's great. But it is vital to realize that the story is not that simple. It is critical that one not walk away with the conclusion that in order to manage and minimize psychological stressors, the solution is always to have more of a sense of control, more predictability, more outlets, more social affiliation. These principles of stress management work only in certain circumstances. And only for certain types of people with certain types of problems.

I was reminded of this awhile back. Thanks to this book's having transformed me from being a supposed expert about rats' neurons to being a supposed one about human stress, I was talking to a magazine writer about the subject. She wrote for a women's magazine, the type with articles about how to maintain that full satisfying sex life while being the CEO of a Fortune 500 company. We were talking about stress and stress management, and I was giving an outline of some of the ideas in Chapter 12. All was going well, and toward the end, the writer asked me a personal question to include in the article—what are my outlets for dealing with stress. And I made the mistake of answering honestly—I love my work, I try to exercise daily, and I have a fabulous marriage. Suddenly, this hard-nosed New York writer blew up at me—"I can't write about your wonderful marriage! Don't tell me about your wonderful marriage! Do you know who my readers are? They're forty-five-year-old professionals who are unlikely to ever get married and want to be told how great that is!" It struck me that she was, perhaps, in this category as well. And it also struck me, as I slunk back to my rats and test tubes afterward, what an idiot I had been. You don't counsel Rwandan refugees to watch out about too much cholesterol or saturated fats in their diet. You don't tell an overwhelmed single mother living in some inner-city hellhole about the stress-reducing effects of a daily hobby. And you sure don't tell the readership of a magazine like this how swell it is to have a soul mate for life. "More control, more predictability, more outlets, more social support" is not some sort of mantra to be handed out indiscriminately, along with a smile button.

This lesson is taught with enormous power by two studies that we have already heard about, that seem superficially to be success stories in stress management but turned out not to be. Back to the parents of children with cancer who were in remission. Eventually, all the children came out of remission and died. When that occurred, how did the parents fare? There were those who all along had accepted the possibility, even probability, of a relapse, and there were those who staunchly denied the possibility. As noted, during the period of remission the latter parents tended to be the low glucocorticoid secretors. But when their illusions were shattered and the disease returned, they had the largest increases in glucocorticoid concentrations.

An even more poignant version of this unfortunate ending comes from a nursing home study. Recall the one in which residents were visited once a week by students—either unannounced, at an appointed time predetermined by the student, or at a time of the resident's choice. As noted, the sociality did everyone some good, but the people in the last two groups, with the increased predictability and control, did even better. Wonderful, end of study, celebration, everyone delighted with the clear-cut and positive results, papers to be published, lectures to be given. Students visit the people in the nursing home for a last time, offer an awkward, "It's been great getting to know you, I'll be dropping by, er, sometime soon, best of luck. . . ." What happens then? Do the people whose functioning, happiness, and health improved now decline back to preexperiment levels? No. They drop even farther, winding up worse than before the study.

This makes perfect sense. Think of how it is to get 25 shocks an hour when yesterday you got 10, compared with 25 shocks an hour when yesterday you got 50. Think of what it feels like to have your child come out of remission after you had spent the last year denying the possibility that it could ever happen. In both cases, a perception of things worsening. It is one thing to be in a nursing home, lonely, isolated, visited once a month by your bored children. It is even worse to be in that situation and, having had a chance to spend time with bright, eager young people who seemed interested in you, to find now they aren't coming anymore. All but the most heroically strong among us would slip another step in the face of this loss. It is true that hope, no matter how irrational, can sustain us in the darkest of times. But nothing can break us more effectively than hope given and then taken away capriciously. Manipulating these psychological variables is a powerful but double-edged sword.

When do these principles of injecting a sense of control, of predictability, of outlets, of sociality work and when are they disastrous to apply? There are some rules, first outlined in the latter half of Chapter 12.

To begin, social affiliation is not always the solution to stressful psychological turmoil. We can easily think of people who would be the last ones on earth we would want to be stuck with when we are troubled. We can easily think of troubled circumstances where being with *anyone* would make us feel worse. Physiological studies have demonstrated this as well. Take a rodent or a primate that has been housed alone and put it into a social group. The typical result is a massive stress-response. In the case of monkeys, this can go on for weeks or months while they tensely go about figuring out who dominates whom in the group's social hierarchy.*

In another demonstration of this principle, infant monkeys were separated from their mothers. Predictably, they had pretty sizable stress-responses, with elevations in glucocorticoid levels. The elevation could be prevented if the infant was placed in a group of monkeys—but only if the infant already knew those animals. There is little to be derived in the way of comfort from strangers.

Even once animals are no longer strangers, on the average half of those in any group will be socially dominant to any given individual, and having more dominant animals around is not necessarily a comfort during trouble. Even intimate social affiliation is not always helpful. We saw in the psychoimmunity chapter that being married is associated with all sorts of better

* A few years ago, the U.S. government proposed new guidelines to improve the psychological well-being of primates used for research; one well-intentioned but uninformed feature was that monkeys housed individually during a study should, at least once a week, spend time in a group of other monkeys. That precise social situation had been studied for years as a model of chronic social stress, and it was clear that the regulations would do anything but increase the psychological well-being of these animals. Fortunately, the proposed rules were changed after some expert testimony.

health outcomes. That chapter also noted an obvious but important exception to this general rule, however: being in a *bad* marriage is associated with immune suppression.

Increased degrees of predictability and information about the future are not always good news, either. As noted in Chapter 12, it does little good to get predictive information about very common events (because they are basically inevitable) or very rare ones (because you weren't anxious about them in the first place). It does little good to get predictive information a few seconds before something bad happens (because there isn't time to derive the psychological advantages of being able to relax a bit) or way in advance of the event (because who's worrying anyway?).

In some situations, predictive information can even make things worse—for example, when the information tells you little. Suppose you are incredibly nervous about an important examination you are about to take, and a friend furtively gives you a piece of news: "I just heard that they are really going to be hard on you, absolutely rake you over the coals." How? "I don't know, I couldn't find out that part." I would wager that most of us would be more anxious with that extra information, not less.

An overabundance of information can be stressful as well. One of the places I dreaded most in graduate school was the "new journal desk" in the library, where all the science journals received the previous week were displayed, thousands of pages of them. Everyone would circle around it, teetering on the edge of anxiety attacks. All that available information seemed to taunt us with how out of control we felt—stupid, left behind, out of touch, and overwhelmed.

And manipulating a sense of control is playing with the variable in psychological stress that is most likely to be double-edged. Too much of a sense of control can be crippling, whether the sense is accurate or not. I offer two disparate examples.

When he was a medical student, a friend embarked on his surgery rotation. That first day, nervous, with no idea what to expect, he went to his assigned operating room and stood at the back of a crowd of doctors and nurses doing a kidney transplant. Hours into it, the chief surgeon suddenly turned to him: "Ah, you're the new medical student; good, come here, grab this retractor, hold it right here, steady, good boy." Surgery continued; my friend was ignored as he precariously maintained the uncomfortable position the surgeon had put him in, leaning forward at an angle, one arm thrust amid the crowd, holding the instrument, unable to see what was going on. Hours passed. He grew woozy, faint from the tension of holding still. He found himself teetering, eyes beginning to close—when the surgeon loomed before him. "Don't move a *muscle* because you're going to screw up everything!" Galvanized, panicked, half-ill, he barely held on . . . only to discover that the "you're going to screw up everything" scenario was a stupid hazing trick done to every new med student. He had been holding an instrument over some irrelevant part of the body the entire time, tricked into feeling utterly responsible for the survival of the patient, when in fact, his actions had no consequence at all. (He chose another medical specialty.)

The second example concerns one of the heaviest burdens of psychological stress imaginable. It involved a lone man given an extraordinary sense of control over a situation that simply could not be ameliorated by any action. In 1947, Lord Mountbatten surprised the world by announcing that Britain was pulling out of the British Raj within months, that the Indian subcontinent would get its independence far sooner than had been expected. What should be done about the Hindu and Muslim populations—different cultures, seething with tensions for centuries? The awful decision was made to divide the subcontinent into two countries, Hindu India and Muslim Pakistan.

Where would the boundary be drawn? Endless unworkable solutions were raised and scuttled until everyone agreed to choose an outside arbitrator. The call went to Sir Cyril Radcliffe, an English barrister known for his brilliance, fairness, and most of all, his utter disinterest in and ignorance about the Indian subcontinent. With all the force of imperial noblesse oblige used to twist his arm, he was whisked off to India and given seven weeks to decide the fate of 88 million people and 15,000 square miles of land, working with unreliable maps and unreliable advisers, every Indian he encountered frantic to cajole, bully, threaten, and beg Radcliffe into drawing the border at a particular point.

For weeks Radcliffe worked, trying to determine when to make a decision based on ethnicity, when on topography, when on economics. Should he separate an agricultural region from the irrigation system that watered it, or leave a Muslim village stranded in Hindu India? Separate a factory and the port used to ship its products from the roads by which raw materials were delivered, or sentence a Hindu holy city to control by Muslim Pakistan? Under stupefying time pressure he demarcated thirty miles of border a day, locked away in a cottage under increasing threats of violence, certain that no matter what he did. It would be a disaster in a land too heterogeneous ever to be divided peacefully and too hostile and fragmented ever to be left peacefully as one. His task finished, he left under military protection, the most hated man in the subcontinent.

In the ensuing months, the line Radcliffe drew circumscribed one of the greatest tragedies of history, as 10 million people were forced to flee their ancestral lands for their new national homes. Somewhere between 200,000 and 500,000 people were slaughtered amid ethnic and religious fighting. In our more grandiose moments, some of us might relish the opportunity to affect the lives of 88 million people—but never, absolutely never, when the sense of control is illusory and any action you take leads inevitably to disaster. It is anything but psychologically comforting to believe that you can control the uncontrollable.*

Finally, recall the discussion in Chapter 8 on the links between psychological factors and cancer. There is indeed a link between the two—certain types of emotional stress appear to be associated with an increased risk of cancer in humans and accelerated tumor growth in experimental animals—but the association is a fairly weak one. It is clearly a travesty to lead cancer patients or their families to believe, misinterpreting the power

---

* Remarkably, the story of Cyril Radcliffe not only teaches us about the dangers of having an illusory sense of control but also about the enormous range of individual responses to stress. Apparently Mountbatten had found the one human capable of performing the impossible task who could cope with its stressfulness, utterly unaffected by the human drama. (Perhaps, British gentleman that he was, Radcliffe thought that any emotions he might feel were none of the world's damn business.) A book of Radcliffe's legal and political essays includes the transcript of a broadcast he gave on the BBC in October 1947, a few months after his return and the granting of independence. It was a month in which the British news was dominated daily by sickening reports of massacre after massacre, Muslims and Hindus and Sikhs butchering one another, mass rapes, villages of people burned to death in their homes, ritual mutilation of victims, all amid columns of dying refugees crawling toward their new countries. And what did Radcliffe speak about? The excellent values and standards of the British civil servants who had served in India over the centuries; he lamented how little art and literature there was to commemorate them.

of these studies, that there is more possibility for control over the causes and courses of cancers than actually exists. Doing so is simply teaching the victims of cancer and their families that the disease is their own fault, which is neither true nor conducive to reducing stress in an already stressful situation.

Control is not always a good thing psychologically, and a principle of good stress management cannot be simply to increase the perceived amount of control in one's life. It depends on what that perception implies, as we saw in Chapter 12. Is it stress-reducing to feel a sense of control when something bad happens? If you think, "Whew, that was bad, but imagine how much worse it would have been if I hadn't been in charge," a sense of control is clearly working to buffer you from feeling more stressed. However if you think, "What a disaster and it's all my fault. I should have prevented it," a sense of control is working to your detriment. This dichotomy can be roughly translated into the following rule for when something stressful occurs: the more disastrous a stressor is, the worse it is to believe you had some control over the outcome, because you are inevitably led to think about how much better things would have turned out, if only you had done something more. A sense of control works best for milder stressors. (Remember, this advice concerns the sense of control you *perceive* yourself as having, as opposed to how much control you actually have.)

Having an illusory sense of control in a bad setting can be so pathogenic that one version of it gets a special name in the health psychology literature. It could have been included in the chapter on personality types and stress-related disease, but I saved it until now. As described by Sherman James, an epidemiologist at the University of Michigan, it is called John Henryism. The name refers to the American folk hero who, hammering a six-foot-long steel drill, tried to outrace a steam drill tunneling through a mountain. John Henry beat the machine, only to fall dead from the superhuman effort. As James defines it, John Henryism involves the belief that any and all demands can be vanquished, so long as you work hard enough. On questionnaires, John Henry individuals strongly agree with statements such as "When things don't go the way I want them, it just makes me work even harder," or "Once I make up my mind to do something, I stay with it until the job is completely done." This is the epitome of individuals with an internal locus of control—they believe that, with enough effort and determination, they can regulate all outcomes.

What's so wrong with that? Nothing, if you have the good fortune to live in the privileged, meritocratic world in which one's efforts truly do have something to do with the rewards one gets, and in a comfortable, middle-class world, an internal locus of control does wonders. For example, always attributing events in life to your own efforts (an internal locus of control), is highly predictive of lifelong health among that population of individuals who are the epitome of the privileged stratum of society—a cohort of Harvard graduates. However, in a world of people born into poverty, of limited educational or occupational opportunities, of prejudice and racism, it can be a disaster to be a John Henry, to decide that those insurmountable odds could have been surmounted, if only, if only you worked *even* harder—John Henryism is associated with a marked risk of hypertension and cardiovascular disease. Strikingly, James's pioneering work has shown that the dangers of John Henryism occur predominately among the very people who most resemble the mythic John Henry himself, working class African Americans—a personality type that leads you to believe you can control the aversively uncontrollable.

There's an old parable about the difference between heaven and hell. Heaven, we are told, consists of spending all of eternity in the study of the holy books. In contrast, hell consists of spending all of eternity in the study of the holy books. To a certain extent, our perceptions and interpretations of events can determine whether the same external circumstances constitute heaven or hell, and the second half of this book has explored the means to convert the latter to the former. But the key is, "to a certain extent." The realm of stress management is mostly about techniques to help deal with challenges that are less than disastrous. And it is pretty effective in that sphere. But it just won't work to generate a cult of subjectivity in which these techniques are blithely offered as a solution to the hell of a homeless street person, a Bosnian refugee, someone prejudged to be one of society's Untouchables, or a terminal cancer patient. Occasionally, there is the person in a situation like that with coping powers to make one gasp in wonder, who does indeed benefit from these techniques. Celebrate them, but that's never grounds for turning to the person next to them in the same boat and offering that as a feel-good incentive just to get with the program. Bad science, bad clinical practice, and, ultimately, bad ethics. If any hell can be converted into a heaven, you only have to rouse yourself from eating grapes in your lounge chair in order to inform a victim of some horror whose fault it is if they are unhappy.

SOME CONCLUSIONS AND TENTATIVE PRESCRIPTIONS

When stress management techniques are applied prudently, to problems that are psychologically surmountable, they remain extremely powerful. And the caveats of their cautious application run through the work of the scientists and clinicians giving the best advice in stress management. Judith Rodin, for example, one of the leaders in the field of encouraging greater control among people in nursing homes, emphasizes this in her writing—do not inadvertently overload the elderly with a sense of responsibility for things that they cannot plausibly control; pick the battles carefully. A similar theme is found in the writings of Ellen Langer, a Harvard psychologist who has been a frequent collaborator of Rodin's.

Redford Williams, the Duke University physician perhaps most responsible for focusing attention on the hostility component of Type A personalities, writes extensively about the particular dangers of "unrealistic anger." He warns against trying to assert control over something that does not need correcting or that cannot be corrected—an approach at which Type A's tend to excel.

Similar ideas are emphasized by David Spiegel, the Stanford psychiatrist who surprised both the medical community and himself with his observations that a supportive therapy setting caused a significant extension of survival time in breast cancer patients. "We encourage our patients to hope for the best, but prepare for the worst," he writes, citing research to show that increasing a patient's sense of control over the *future* course of the disease is associated with an increase in spirit, whereas increasing a patient's belief in her control over what *caused* the disease produces the opposite. Once again, if the outcome of something is awful, it does not do great things for morale to be led to believe that you had the power to have prevented it.

Truth and mental health are often viewed as critically intertwined in many branches of psychotherapy. But Richard Lazarus, the University of California (Berkeley) psychologist who is a leading theorist on stress and coping, has written extensively about the occasional advantages of denial and lack of information. In the immediate aftermath of some medical disaster, for example, when a person is too weak to act constructively in a problem-solving way, the last things he needs is detailed information about how bad reality is or advice to start working on a sense of control. It is only later that these strategies are likely to be helpful, rather than hurtful.

This idea is also encompassed in a personality test in a recent book by Martin Seligman, the University of Pennsylvania psychologist who pioneered thinking about learned helplessness. He has since concentrated on the subset of individuals who are resistant to the phenomenon, asking what they are doing right and how we can make use of these attitudes ourselves. To use the title of a book of his, he is now studying "learned optimism." The personality test measures people's attributional style: when something happens, to what do they attribute its cause? How *externalized* are they: do they attribute outcomes to their own actions, to those of others, or to chance? How *pervasive* do they view events as being: is this occurrence an isolated event, or is this how everything works? How *permanent* do they view events as being: is this a one-time occurrence, or are things always going to be this way? Critically, the extent to which people are pervasive, permanent, or externalizing in their beliefs does not by itself predict how they fare in the face of life's challenges. Seligman argues that it depends on whether the news is good or bad. If the news is good, the healthiest people think, "This is always how it is in all sorts of situations, and it came about as a result of my efforts." When the news is bad, those same healthy people use an opposite attributional strategy: "That was awful, but it was due to this one fluky event, out of my control and unlikely to happen again or to apply to other parts of my life." Good news viewed as pervasive, long-lasting, and generated by your own actions; bad news viewed as just the opposite.

What general strategies can help us the most in the face of psychological stressors? Obviously, the first step is to accurately recognize signs of the stress-response and to identify the situations most responsible for it. Once you have done that, there are some ways to proceed:

- One successful strategy is to find an outlet for life's frustrations. Set aside time to do it and do it regularly—you can't save your stress management for the weekends, or for when you are on hold on the phone. Make your outlet a benign one for those around you—one should not give ulcers in order to avoid getting them—and choose one that you find personally compatible. Prayer, meditation, ballroom dancing, psychoanalysis, Bach, competitive sports—each may help some people but not others. Read the fine print and the ingredient list on each new form of supposed antistress salvation, be skeptical of hype, don't believe anyone who says it's been scientifically proved that their brand works best, listen to your own responses with each new exploration and trust them.
- In the face of terrible news beyond control, beyond prevention, beyond healing, those who are able to find the means to deny tend to cope best. Such denial is not only permissible, it may be the only means of sanity. But in the face of lesser problems, one should hope, but protectively and rationally. Find ways to view even the most stressful of situations as holding the promise of improvement but do not deny the possibility that things will not improve. Balance these two opposing trends carefully. Hope for the best and let that dominate most of your emotions, but at the same time let one small piece of you prepare for the worst.
- Those who cope with stress successfully tend to seek control in the face of stressors but do not try to control, in the present, things that have already come to pass. They do not try to control future events that are uncontrollable and do not try to fix things that are not broken or that are broken beyond repair. When faced with the large wall of a stressor, one should not assume there will be a breakthrough, one single, controlling solution that will make the wall disappear. Assume instead that the wall can be scaled by a series of footholds of control, each one small but still capable of giving support.
- It is generally helpful to seek predictable, accurate information. However, such information is not useful if it comes too soon or too late, if it is unnecessary, if there is so much infor-

"Is there anyone here who specializes in stress management?"

mation that it is stressful in and of itself, or if the information is about news far worse than one wants to know.

- It is important to find sources of social affiliation and support. Even in this most obsessively individualistic of societies, most of us yearn to feel part of something larger than ourselves. But one should not mistake true affiliation and support for mere socializing. A person can feel vastly lonely in a crowd or when faced with a supposed intimate who has proved to be a stranger. Be patient; most of us spend a lifetime learning how to be truly good friends and spouses.

Two far more beautiful ways of expressing these ideas about flexibility, resiliency, picking your battles and your weapons carefully include something I once heard at a Quaker meeting:

> In the face of strong winds, let me be a blade of grass.
>
> In the face of strong walls, let me be a gale of wind.

And the prayer of the theologian Reinhold Niebuhr, adopted by Alcoholics Anonymous:

> God grant me the serenity to accept the things I cannot change, courage to change the things I can, and wisdom to know the difference.

Constantin Brancusi, 1912: The Kiss, *limestone.*

Stress is not everywhere. Every twinge of dysfunction in our bodies is not a manifestation of stress-related disease. It is true that the real world is full of bad things that we can finesse away by altering our outlook and psychological makeup, but it is also full of awful things that cannot be eliminated by a change in attitude, no matter how heroically, fervently, complexly, ritualistically we may wish. Once we are actually sick with the illnesses, the fantasy of which keeps us anxiously awake at two in the morning, the things that will save us have little to do with the content of this book. Once we have that cardiac arrest, once a tumor has metastasized, once our brain has been badly deprived of oxygen, little about our psychological outlook is likely to help. We have entered the realm where someone else—a highly trained physician—must use the most high-tech of appropriate medical interventions.

These caveats must be emphasized repeatedly in teaching what cures to seek and what attributions to make when confronted with many diseases. But amid this caution, there remains a whole realm of health and disease that is sensitive to the quality of our minds—our thoughts and emotions and behaviors. And sometimes whether we become sick with the diseases that frighten us at two in the morning will reflect this realm of the mind. It is here that we must turn from the physicians and their ability to clean up the mess afterward, and recognize our own capacity to *prevent* some of these problems beforehand in the small steps with which we live our everyday lives.

Perhaps I'm beginning to sound like your grandmother, advising you to be happy and not to worry so much. This advice may sound platitudinous, trivial, or both. But change the way even a rat perceives its world, and you dramatically alter the likelihood of its getting a disease. These ideas are no mere truisms. They are powerful, potentially liberating forces to be harnessed. As a physiologist who has studied stress for many years, I clearly see that the physiology of the system is often no more decisive than the psychology. We return to the catalogue at the beginning of the first chapter, the things we all find stressful—traffic jams, money worries, overwork, the anxieties of relationships. Few of them are "real" in the sense that that zebra or that lion would understand. In our privileged lives, we are uniquely smart enough to have invented these stressors and uniquely foolish enough to have let them, too often, dominate our lives. Surely we have the potential wisdom to banish their stressful hold.

NOTES

CHAPTER 1: Why Don't Zebras Get Ulcers?

PAGE 1 For years, in lectures, I've rhetorically compared disease patterns in humans with those of zebras, and when sitting down to write this book, it suddenly scared the willies out of me that I wasn't sure about the business with zebras and ulcers. And then where would we be? What good is a book entitled something like *Why Do Zebras Get Ulcers Less Frequently Than We Do and for Some Fairly Different Reasons, Although It's Complicated?* However, according to *Zoo and Wild Animal Medicine,* 2d ed. (M. Fowler; Philadelphia: Saunders, 1986) and phone calls to the zebra vets at the Brookfield, Bronx, National, Philadelphia, and San Diego zoos, ulcers are extremely uncommon in zebras. They occur in animals undergoing severe and unnatural stress (e.g., when they are first transported into a zoo), but that is about the only circumstance. Stated in the framework of this book, when left to their own devices (either in the wild or in reasonably large enclosures in a zoo), zebras don't develop ulcers.

Many of the ideas in this chapter have a long history in stress physiology. The main point was stated well by Walter Cannon over half a century ago: "A highly important change has occurred in the incidence of disease in our country . . . serious infections, formerly extensive and disastrous, have markedly decreased or almost disappeared, . . . meanwhile, conditions involving strain in the nervous system have been greatly augmented" ("The role of emotion in disease," *Annals of Internal Medicine,* vol. 9, no. II, May 1936).

PAGE 2 Viewed through the eclipse of World War II, we seem to remember World War I with odd fondness—Irving Berlin tunes, colorful uniforms, rickety motorcars, and heads of states with silly titles and big mustaches. Eight and a half million people were killed in the pointless bloodbath we know as World War I (D. Fromkin. 1989. *A Peace to End All Peace.* New York: Avon Books, 379). The flu that swept the planet at the same time, by contrast, killed 20 million (W. McNeill. *Plagues and Peoples.* New York: Doubleday Books, 255). "The sum of American sailors and soldiers who died of flu and pneumonia in 1918 is over 43,000, about 80 percent of American battle deaths in the war" (A. Crosby. 1918, 1976. *Epidemic and Peace.* London: Greenwood Press, 36).

PAGE 6 The definitive study on chess players was carried out by the physiologist Leroy DuBeck and his graduate student Charlotte Leedy. They wired up chess players in order to measure their breathing rates, blood pressure, muscle contractions, and so on, and monitored the players before, during, and after major tournaments. They found tripling of breathing rates, muscle contractions, systolic blood pressures that soared to over 200—exactly the sort of thing seen in athletes during physical competition. See the original report, Leedy's thesis, "The effects of tournament chess playing on selected physiological responses in players of varying aspirations and abilities" (Temple University, 1975) or their brief report (Leedy, C., and DuBeck, L. 1971. "Physiological changes during tournament chess." *Chess Life and Review,* 708). In a telephone conversation, DuBeck also tells the story of the international match in the early 1970s between grand masters Bent Larson and Bobby Fischer, in which the former had to be given antihypertensive medication in the middle of his losing match; his blood pressure remained elevated for days afterward. The Kasparov-Karpov report is from *The New York Times,* 20 December 1990. And for that special chess fan out there who just can't get enough of this subject, may I suggest as the perfect gift a copy of Glezerov, V., and Sobol, E. 1987. "Hygienic evaluation of the changes in work capacity of young chess players during training." *Gigiena i Sanitariia* 24, in the original Russian.

PAGE 7 For an entrée to the wonderful world of allostasis, see Sterling, P., and Eyer, J. 1988. "Allostasis: a new paradigm to explain arousal pathology." In *Handbook of Life Stress, Cognition, and Health,* Fisher, S., Reason, J. (eds.). Wiley. Also see McEwen, B., and Stellar, E. 1993. "Stress and the individual. Mechanisms leading to disease." *Archives of Internal Medicine* 153, 2093.

PAGE 8 Selye published numerous autobiographical articles and books, many of which contain the story of the ovarian extract and his discovery of the nonspecific stress-response; a good example is *The Stress of My Life* (New York: Van Nostrand, 1979). The book also contains Selye's claim that he was the first to use the word *stress* in a biomedical, rather than an engineering, sense. Actually, Walter Cannon beat him to it by decades (1914. "The interrelations of emotions as suggested by recent physiological researches." *American Journal of Psychology* 25, 256). This point was brought up in a colorful debate between Selye and John Mason, a psychiatrist whose pioneering work on the psychological stress-response is discussed later (Mason, J., 1975. "A historical view of the stress field." *Journal of Human Stress* 1, 6; part II: 1, 22. Selye, H., 1975. "Confusion and controversy in the stress field." *Journal of Human Stress* 1, 37).

PAGE 15 Descriptions of Addison's disease can be found in all endocrinology textbooks, as it is one of the best-studied endocrine disorders. Shy-Drager is rarer and more recent, first described in 1960. For a description right from the horses' mouths, see Shy, G., and Drager, G., 1960. "A neurological syndrome associated with orthostatic hypotension." *A.M.A. Archives of Neurology* 2, 41–511. Also see Low, P., *Seminars in Neurology,* vol. 7, no. 1 (March 1987), 53; and Bannister, R., and Mathios, C. 1992. *Autonomic Failure* (New York: Oxford University Press).

For a good history of stress physiology, see Weiner, H. 1992. *Perturbing the Organism: The Biology of Stressful Experience* (Chicago: University of Chicago Press).

CHAPTER 2: Glands, Gooseflesh, and Hormones

PAGE 19 The D. H. Lawrence quotation is from *Lady Chatterley's Lover* (Cutchogue, N.Y.: Buccaneer Books, 1983). The idea for this example comes from a colleague, the British immunologist Nick Hall. He regularly lectures to halls of distracted scientists clicking away with their three-color pens; he starts off with some really steamy passage of Lawrence recited in his impressive English accent, and rivets their attention.

PAGE 25 The testicular injection mania began in 1889, with a paper published by the formidable Charles-Edouard Brown-Sequard, entitled "On the physiological and therapeutic role of a juice extracted from the testicles of animals according to a number of facts observed in man" [*Archives de physiologie normale et pathologique* (5e series), 1889, 1, 739].

A lot of the facts Brown-Sequard collected had been observed in one man, himself. Brown-Sequard was arguably the most august physiologist in the world at the time, age seventy-two and with somewhat declining energies. He had theorized that some features of senescence of humans were due to declining gonadal function (the more global statements about such decline as *the* cause of aging came from later followers). He felt that the testes contained some sort of active secreted substance, and he started injecting himself subcutaneously with extracts of testes from dogs and guinea pigs. He was absolutely right that the testes secreted a substance—testosterone (which had not yet been discovered; the term *hormone* did not even exist then)—but his experiment couldn't possibly work, since he made his extracts in water; testosterone, because of its chemical nature, does not dissolve in water.

Despite that, he reported wondrous results (increased physical vitality, increased length of his jet of urine—the latter no doubt being the sort of thing we all hope to retain into our golden years). All placebo. The reproductive physiologist Roger Gosden of Leeds University in the United Kingdom suspects that Brown-Sequard was probably depressed at the time of his experiments and thus was particularly vulnerable to such a placebo effect (see page 148 in Gosden, R. 1996. *Cheating Time: Science, Sex and Ageing*. London: Macmillan). Nevertheless, doctors were thrilled at the report, and within two years, organotherapy, as it was called, was being used worldwide. Brown-Sequard took particular umbrage at the charlatans making quick money using his (altogether incorrect and ineffectual) discovery, particularly the American hucksters soon selling "Dr. Brown-Sequard's Elixir of Life." He also expanded his theory a bit, noting that loss of semen resulted in loss of strength (twenty years earlier he had speculated on the rejuvenative effects of intravenous injections of sperm into men, an idea fortunately not tried), citing the well-known physical and mental weaknesses of men who masturbated frequently or who had frequent intercourse. (For the original citations and a thorough review of the subject, see Borell, M. 1976. "Brown-Sequard's organotherapy and its appearance in America at the end of the nineteenth century." *Bulletin of the History of Medicine* 50, 309, as well as the very entertaining section on the subject in Gosden's book.)

PAGE 27 The history of hypothalamic hormones (Harris's theory that the brain was an endocrine organ, and the work of Guillemin and Schally) has been well documented, especially in the aftermath of the award of the Nobel prize to the latter pair. This is because of the ferocity and colorfulness of the Guillemin-Schally race, and because the huge, "corporate" lab that each evolved in the process seemed the wave of the scientific future at the time. For a particularly readable account, see *The Nobel Duel: Two Scientists' 21-Year Race to Win the World's Most Coveted Research Prize* (N. Wade; Garden City, N.Y.: Anchor Press, 1981). The quotation from Schally about the competition with Guillemin is in Wade's book, page 7. For a dauntingly academic account of the sociology of Guilleman's lab (although it is not identified as Guilleman's by name), see Latour, B., and Woolgar, S. 1979. *Laboratory Life: The Social Construction of Scientific Facts* (Beverly Hills, Calif.: Sage Publications).

New releasing and inhibiting factors continue to be isolated, still often in sprints to the finish line by groups in frenzied competition with one another. An exception to this pattern came in 1981 with the isolation of what was perhaps the most sought-after of the brain hormones. This hormone, which will be discussed throughout the book, is the main way in which the brain controls a principal branch of the stress-response. Corticotropin releasing factor (CRF), as it is called, was the first brain hormone whose existence was inferred (in 1955) but one of the last ones isolated, because it turned out to be among the most chemically complex. In a wrinkle on the old Guillemin-Schally dichotomy, its isolation was carried out by a team headed by Wylie Vale, once Guillemin's right-hand man. Vale and his band of renegades, in a lab of their own, had the audacity to look for CRF in places none of the other researchers had tried in the 25 years of investigation, considering very unlikely chemical structures for CRF. One turned out to be the right one, and they beat the competition by miles. See Vale, W., Speiss, J., Rivier, C., and Rivier, J. 1983. "Characterization of a 41-residue ovine hypothalamic peptide that stimulates the secretions of corticotropin and beta-endorphin." *Science* 213, 1394.

PAGE 35 Hormonal "signatures" of different stressors: Henry, J. P. 1977. *Stress, Health, and the Social Environment* (New York: Springer-Verlag); Frankenhaeuser, M. 1983. "The sympathetic-adrenal and pituitary-adrenal response to challenge." In Dembroski, T., Schmidt, T., and Blumchen, G. (eds.). *Biobehavioral Basis of Coronary Heart Disease* (Basel: Karger), 91. For a nice discussion of signatures involving epinephrine versus norepinephrine, see Chapter 8 of Weiner, H. 1992. *Perturbing the Organism: The Biology of Stressful Experience* (Chicago: University of Chicago Press). For a particularly odd example of stress signatures (laboratory rats having different patterns of stress-responses depending on which human handled them), see Dobrakovova, M., Kvetnansky, R., Oprsalova, Z., and Jezova, D. 1993. "Specificity of the effect of repeated handling on sympathetic-adrenomedullary and pituitary-adrenocortical activity in rats." *Psychoneuroendocrinology* 18, 163. For a review of the hypothalamic stress signature for different types of psychological stress, see Romero, L., and

Sapolsky, R. 1996. "Patterns of ACTH secretagog secretion in response to psychological stimuli." *Journal of Neuroendocrinology* 8, 243.

CHAPTER 3: Stroke, Heart Attacks, and Voodoo Death

PAGE 37 Good general overviews of what the cardiovascular system does during stress can be found in most physiology textbooks, although the information is rarely explicitly organized under the topic of "stress." Instead, it can usually be found in a chapter on the heart itself, or on the physiological response to exercise. Those reviews typically focus on the role of the sympathetic nervous system in regulating the cardiovascular system. The role of glucocorticoids (which make cardiovascular tissue more sensitive to the sympathetic nervous system) is reviewed in Whitworth, J., Brown, M., Kelly, J., Williamson, P. 1995. "Mechanisms of cortisol-induced hypertension in humans." *Steroids* 60, 76. Also see Sapolsky, R., and Share, L. 1994. "Rank-related differences in cardiovascular function among wild baboons: role of sensitivity to glucocorticoids." *American Journal of Primatology* 32, 261.

PAGE 38 The 1833 study showing that emotional stress would shut down blood flow to the guts of the Canadian with the gunshot wound: Beaumont, W. 1833. *Experiments and Observations on the Gastric Juice and the Physiology of Digestion* (Plattsburgh, N.Y.).

PAGE 39 For a discussion of the role of kidneys in increasing blood pressure during stress, see Guyton, A. 1991. "Blood pressure control—special role of the kidneys and body fluids." *Science* 252, 1813.

PAGE 39 The bladder conundrum. One of my intrepid research assistants, Michelle Pearl, called up some of America's leading urologists to ask them why bladders evolved. One comparative urologist (as well as Jay Kaplan, whose research is discussed in this chapter) took the findings about territorial rodents having bladders to make scent trails and inverted the argument—maybe we have bladders so that we can avoid continual dribble of urine that would leave a scent trail so some predator could track us. A number of urologists suggested that maybe the bladder acts as a buffer between the kidney and the outside world, to reduce the chance of kidney infections. However, it seems odd to develop an organ exclusively for the purpose of protecting another organ from infection. Pearl suggested that it may have evolved for male reproduction—the acidity of urine isn't very healthy for sperm (in ancient times, women would use half a lemon as a diaphragm), so perhaps it made sense to evolve a storage site for the urine. A remarkable percentage of the urologists questioned said something like, "Well, it would be an extreme social liability to not have a bladder," before realizing that they had just suggested that vertebrates evolved bladders tens of millions of years ago so that we humans wouldn't inadvertently pee on our party clothes. Mostly, however, the urologists said things like, "To be honest, I've never thought about this before,"

"I don't know and I talked to everyone here and they don't know anything either," and "Beats me."

The strangest thing about it all is that many animals may not actually take advantage of their bladder's storage capacity. In my vast experience watching baboons go about their urinary business, it is apparent that they very rarely hold it in when they have to go.

Clearly, there's a lot of work to be done in this area.

PAGE 41 The difference in cardiovascular responses to overt physical stressors and to quiet vigilance: Fisher, L. 1991. "Stress and cardiovascular physiology in animals." In Brown, M., Koob, G., and Rivier, C. (eds.). *Stress: Neurobiology and Neuroendocrinology* (New York: Marcel Dekker), 2 hours, 10 minutes; black and white. With Claude Rains, Lily Pons, and the young Robert Mitchum as the descending aorta.

PAGE 41 Detailed discussions about how damage to the vascular lining, various hormones, and high levels of fat in the bloodstream interact to cause atherosclerosis: Ross, R. 1993. "The pathogenesis of atherosclerosis: a perspective for the 1990s." *Nature* 362, 801. The clumping of platelets during stress is discussed in Allen, M., and Patterson, S. 1995. "Hemoconcentration and stress: a review of physiological mechanisms and relevance for cardiovascular disease risk." *Biological Psychology* 41, 1. Also Rozanski, A., Krantz, D., Klein, J., and Gottdiener, J. 1991. "Mental stress and the induction of myocardial ischemia." In Brown, et al. (eds.) *Stress: Neurobiology and Neuroendocrinology* (New York: Marcel Dekker). Also see Fuster, V., Badimon, L., Badimon, J., and Chesebro, J. 1992. "The pathogenesis of coronary artery disease and the acute coronary syndromes." *New England Journal of Medicine* 326, 242. The work regarding social stress and the heart disease in rodents can be found in Henry, J. P., 1977. *Stress, Health, and the Social Environment* (New York: Springer-Verlag). The work regarding social stress and plaque formation in primates is reviewed in Manuck, S., Marsland, A., Kaplan, J., and Williams, J. 1995 "The pathogenicity of behavior and its neuroendocrine mediation: an example from coronary artery disease." *Psychosomatic Medicine* 57, 275. The work regarding interactions of the hormones of the metabolic stress-response in causing atherosclerosis can be found in Brindley, D. 1995 "Role of glucocorticoids and fatty acids in the impairment of lipid metabolism observed in the metabolic syndrome." *International Journal of Obesity and Related Metabolic Disorders* 19; suppl. I: S69.

PAGE 45 Myocardial ischemia, damaged heart muscle, and its subsequent vulnerability to stress: *Stress: Neurobiology and Neuroendocrinology* (Brown, M. et al.; New York: Marcel Dekker, 1991) contains a number of chapters with useful information. These include Chapters 20 (Verrier, R. "Stress, sleep and vulnerability to ventricular fibrillation"), 21 (Fisher, L. "Stress and cardiovascular physiology in animals"), 22 (Brodsky, M., and Allen, B. "Effects of psychological stress on cardiac rate and rhythm"), and 23 (Rozanski, A., Krantz, D., Klein, J., and Gottdiener, J. "Mental stress and the induction of myocardial ischemia"). Chapters 20 and 23 contain good reviews of ambulatory electrocardiography; the former chapter details Verrier's own studies showing that psychological stress in humans and dogs can cause acute ischemia in damaged heart tissue. (Also see Rozanski, A., and Berman, D. 1987. "Silent myocardial ischaemia. I. Pathophysiology, frequency of occurrence and approaches toward detection." *American Heart Journal* 114, 615.) For a review of the paradoxical vasoconstriction, rather than vasodilation, during stress in damaged coronary arteries, see Fuster, V., Badimon, L., Badimon, J., and Chesebro, J. 1992. "The pathogenesis of coronary artery disease and the acute coronary syndromes, part II." *New England Journal of Medicine* 326, 310. Also see Schwartz, C., Valente, A., and Hildebrandt, E. 1994. "Prevention of atherosclerosis and end-organ damage: a basis for antihypertensive interventional strategies." *Journal of Hypertension* 12, S3. Cardiologists are beginning to get some sense of what causes this paradoxical vasoconstriction. In healthy tissue, when the heart starts working hard, hormones called *EDRF* (endothelium-derived relaxant factors) and prostacyclin are secreted, causing the vasodilation. When cardiac tissue is made ischemic on a regular basis, it loses the capacity to release EDRF and prostacyclin for some reason. In addition, hormones called *endothelin* and *serotonin,* which cause vasoconstriction, seem to be released. As a result, epinephrine and norepinephrine now cause constriction instead of dilation. Interestingly, this paradoxical vasoconstriction is also observed in the socially stressed monkeys, discussed above, who developed atherosclerosis. One way to dilate coronary arteries during angina pectoris is to take a synthetic version of EDRF—nitroglycerin. For epidemiological evidence that stress is more likely to worsen preexisting heart disease than to cause it outright, see Greenwood, D., Muir, K., Packham, C., and Madeley, R. 1996. "Coronary heart disease: a review of the role of psychosocial stress and social support." *Journal of Public Health Medicine* 18, 221. For more examples of ischemia in heart patients being brought on by subtle psychological stressors (in this case, public speaking), see Taggert, P., Carruthers, M., and Somerville, W. 1973. "Electrocardiogram, plasma catecholamines, and their modification by oxyprenolol when speaking before an audience." *The Lancet* 2, 341. In another demonstration, patients were shown to have as much myocardial ischemia when describing a personal problem to a stranger as they did during exercise: Rozanski, A. 1988. "Mental stress and the induction of silent myocardial ischemia in patients with coronary artery disease." *New England Journal of Medicine* 318, 1005. For reviews of some of the special features linking stress and heart disease in women, see Brezinka V., Kittel, F. 1996. "Psychosocial factors of coronary heart disease in women; a review." *Social Science and Medicine,* 42, 1351, and Elliott, S. 1995. "Psychosocial stress, women and heart health; a critical review." *Social Science and Medicine* 40, 105.

PAGE 47 Instances of sudden cardiac death during stress in humans: Engel, G. 1971. "Sudden and rapid death during psychological stress: folklore or folk wisdom?" *Annals of Internal Medicine* 74, 771. A report shows a tripling in the incidence of myocardial infarctions of the Tel Aviv population during the first three days of the SCUD attacks, as compared with the same three days of January the year before: Meisel, S., Kutz, I., Dayan, K., Pauzner, H., Chetboun, I., Arbel, Y., and David, D. 1991. "Effect of Iraqi missile war on incidence of acute myocardial infarction and sudden death in Israeli civilians." *The Lancet* 338, 660. For data regarding the L.A. earthquake, see Leor, J., Poole, W., Kloner, R. 1996. "Sudden cardiac death triggered by an earthquake." *New England Journal of Medicine* 334, 413. The elderly couple is discussed in a letter from Dr. Paul Morrow, chief medical examiner, state of Vermont. The mechanisms underlying sudden cardiac death: Davis, A., Natelson, B. 1993. "Brain-heart interactions: the neurocardiology of arrhythmia and sudden cardiac death." *Texas Heart Institute Journal* 20, 158; also Meerson, F. 1994. "Stress-induced arrhythmic disease of the heart—part I." *Clinical Cardiology* 17, 362; this paper also describes stress making rat hearts more vulnerable to fibrillation.

PAGE 49 The Irven DeVore quotation is reproduced as a personal communication.

PAGE 50 Psychophysiological death: Davis, W., and DeSilva, R. "Psychophysiological death: a cross-cultural and medical appraisal of voodoo death." *Anthropologia*, in press. Walter Cannon contacted a variety of missionaries, anthropologists, and medical people working in the third world, collecting their descriptions of voodoo death in order to decide that it sounded like too much sympathetic nervous system to him (1942. "'Voodoo' death." *American Anthropologist* 44, 169). Curt Richter, by contrast, didn't gather any firsthand accounts of his own. Instead, he noted the similarity between the accounts in Cannon's paper and cases of parasympathetic-induced death in rats undergoing severe stressors in his own laboratory (he noted that the phenomenon occurred much more readily in wild rats captured and brought to his lab than in the lab-bred strains, and made comparisons between "uncivilized primitive humans" and undomesticated wild rats) (1957. "On the phenomenon of sudden death in animals and man." *Psychosomatic Medicine* 19, 191). Also see Morse, D., Martin, J., and Moshonov, J. 1991. "Psychosomatically induced death: relative to stress, hypnosis, mind control, and voodoo: review and possible mechanisms." *Stress Medicine* 7, 213. (Note: at no extra cost, this review also includes an excerpt of a scene describing a voodoo death, complete with descriptions of dancers "making obscene gestures with their buttocks" in what appears to be a fairly shlocky novel by the first author, something unique to any scientific paper I've seen.)

As he described in *The Serpent and the Rainbow* (New York: Warner Books, 1985), Wade Davis believed he had isolated the critical substance—a poison called *tetrodotoxin,* isolated from puffer fish—that the Haitian witch doctors use to put someone in a zombified state. This is the same poison found in the fugu fish, used in Japanese cooking. (When the fugu chef leaves a smidgen of the tetrodotoxin gland in the fish, the well-paying customer gets a mild buzz. When the chef leaves too much in, the well-paying customer gets put into a coma. Fugu chefs, by the way, are carefully licensed.) Davis made a fascinating argument that zombification in Haiti reflected the intersection of the biology of tetrodotoxin action and the anthropology of traditional Haitian religion: when a Japanese businessman gets major tetrodotoxin poisoning and recovers, he sues the chef and switches restaurants. When a Haitian villager gets the same tetrodotoxin poisoning and recovers, he realizes that his village had hired a shaman to poison him because he has done something terrible—he awakes as an ostracized zombie with no will, and then is often used for slave labor (although in some cases, the zombified person's passive state is promoted by continually drugging him). It's a charming story, although the isolation of tetrodotoxin remains controversial. Davis and tetrodotoxin zombification became so trendy in the 1980s that in Garry Trudeau's *Doonesbury*, Uncle Duke was zombified at one point, and *Miami Vice* used the zombie motif in an episode about drug runners from Haiti.

CHAPTER 4: Stress, Metabolism, and Liquidating Your Assets

PAGE 53 Energy storage and mobilization: the basics of this vastly complicated subject—involving storage tissues throughout the body, a variety of different hormonal messengers, and the liver as Grand Central Station for various nutrients coming and going—is covered in any physiology textbook. A fairly lucid presentation of the subject on an introductory college level can be found in Vander, A., Sherman, J., and Luciano, D. 1994. *Human Physiology: The Mechanisms of Body Function*, 6th ed. (New York: McGraw-Hill). For a discussion of how stress causes energy mobilization, see Mizock, B. 1995. "Alterations in carbohydrate metabolism during stress; a review of the literature." *American Journal of Medicine* 98, 75. Note that this discusses big-time stressors in humans (sepsis, burns, and trauma); the same principles hold for the more subtle ones that dominate this book.

PAGE 58 The inefficiency of the repeated activation of the metabolic stress-response: this is horrendously complicated. The introductory reference given above will teach the general principle that it is inefficient to repeatedly store away energy and then reverse the process by mobilizing it. However, in order to gain a detailed, quantitative understanding of it, one must become something of an accountant—learning what the currency of energy is in the body and how much it costs to make all those deposits and withdrawals in the body's metabolic banks. For this, one must consult biochemistry texts (typically, of the early graduate school level of difficulty); among the best is Stryer, L. 1995. *Biochemistry*, 4th ed. (New York: W. H. Freeman).

PAGE 58 Chronic glucocorticoid exposure causes muscle wastage: for a classic demonstration of this, see Kaplan, S. and Nagareda Shimizu, C. 1963. "Effects of cortisol on amino acid in skeletal muscle and plasma." *Endocrinology* 72, 267. (Cortisol is the glucocorticoid found in humans and primates.) For some recent findings, see Hong, D., and Forsberg, N. 1995. "Effects of dexamethasone on protein degradation and protease gene expression in rat L8 myotube cultures." *Molecular and Cellular Endocrinology* 108, 199.

PAGE 58 The workings of the two types of diabetes mellitus dominate chapters of every endocrinology textbook. For a review of the autoimmune features of insulin-dependent diabetes, see Andre, I., Gonzalez, A., Wang, B., Katz, J., Benoist, C., Mathis, D. 1996. "Checkpoints in the progression of autoimmune disease: lessons from diabetes models." *Proceedings of the National Academy of Sciences* USA 93, 2260. For a classic demonstration that type 2 (adult-onset) diabetes involves impaired sensitivity to insulin, rather than impaired secretion of insulin, see: Reaven, G., Bernstein, R., Davis, B., and Olefsky, J. 1976. "Nonketotic diabetes mellitus: insulin deficiency or insulin resistance?" *American Journal of Medicine* 60, 80. For demonstrations that the insulin resistance arises from a loss of insulin receptors see: Gavin, J., Roth, J., Neville, D., DeMeyts, P., and Buell, D. 1974. "Insulin-dependent regulation of insulin receptor concentrations: a direct demonstration in cell culture." *Proceedings National Academy of Sciences USA* 71, 84. For a discussion of how the insulin resistance also arises from the remaining insulin receptors' not working properly (what is called a "postreceptor" defect), see Flier, J. 1983. "Insulin receptors and insulin resistance." *Annual Review of Medicine* 34, 145. Finally, despite the primary defect of target tissue resistance to insulin's actions, a subset of patients also has a defect in the secretion of insulin. The mechanisms underlying this are reviewed by Unger, R. 1991. "Role of impaired glucose transport by cells in the pathogenesis of diabetes." *Journal of NIH Research* 3, 77.

In western societies, rates of glucose intolerance and insulin resistance rise with age: Andres, R. 1971. "Aging and diabetes." *Medical Clinics of North America* 55, 835; Davidson, M. 1979. "The effect of aging on carbohydrate metabolism: a review of the English literature and a practical approach to the diagnosis of diabetes mellitus in the elderly." *Metabolism* 28, 687. For a discussion of the increasing rate of insulin-resistant diabetes with age in western societies, see Harris, M. 1982. "The prevalence of diabetes, undiagnosed diabetes and impaired glucose tolerance in the United States." In Melish, H., Hanna, J., and Baba, S. (eds.). *Genetic Environmental Interaction in Diabetes Mellitus* (Amsterdam: Excerpta Medica), 70.

Despite this trend, insulin-resistant diabetes seems not to be an obligatory part of aging: aging rats and aging humans in our own society do not become more glucose-intolerant with age, so long as they are active and lean: Reaven, G., and Reaven, E. 1985. "Age, glucose intolerance and non-insulin-dependent diabetes mellitus." *Journal of the American Geriatrics Society* 33, 286. Also see Goldberg, A., and Coon, P. 1987. "Non-insulin-dependent diabetes mellitus in the elderly: influence of obesity and physical inactivity." *Endocrinology and Metabolism Clinics* 16, 843.

For a demonstration of extremely low rates of insulin-resistant diabetes in nonwesternized populations (for example, the Inuit and other Native Americans, New Guinea islanders, inhabitants of rural India, and North African nomads), see table 5 in Eaton, S., Konner, M., and Shostak,

M. 1988. "Stone agers in the fast lane: chronic degenerative diseases in evolutionary perspective." *American Journal of Medicine* 84, 739.

The low rates of insulin-resistant diabetes in nonwesternized populations pose a fascinating mystery. If these people begin eating westernized diets, they get astonishingly high rates of insulin-resistant diabetes. Part of this has an obvious explanation; once these various groups gain entrée into our world of packaged food and processed sugars, they tend to eat themselves into obesity (and, thus, high rates of this diabetes). However, the mystery is that given the *same* diet and degree of obesity, most people in the developing world are at greater risk for such diabetes than people in western societies. Diabetes rates soar among Mexicans and Japanese after they emigrate to the United States, among Asian Indians moving to Britain, and among Yemenite Jews moving to Israel. In the most striking cases, about half the adult residents of the Pacific island of Nauru have diabetes (fifteen times the rate in the United States), while more than 70 percent of the Pima people of Arizona over age 55 have diabetes. And in the absence of a western diet, there is virtually no diabetes.

Why should those in the developing world be at such risk for diabetes once they start consuming a western diet? One fascinating theory is that the gene for a propensity to diabetes is adaptive in nonwesternized settings. Normally, westerners are inefficient at handling dietary sugar: not all of it is absorbed from the circulation, getting lost in the urine. The notion is that people of the developing world are more efficient at utilizing sugar; the second they get any in their circulation, they have a burst of insulin secretion and every bit of the sugar gets stored, instead of urinated away. This makes sense, given tough environments with intermittent food sources, where every little bit must be exploited. And it is easy to imagine this as a genetic trait—for example, genes might alter the sensitivity with which the pancreas senses circulating glucose concentrations and releases insulin, or the sensitivity with which target tissues respond to insulin. These have been termed "thrifty genes," and at least one such candidate in fat cells has been found to have a mutation among Pima Indians (reviewed in Ezzell, C. 1995. "Fat times for obesity research." *Journal of NIH Research*, 7 (10), 39).

With traditional diets in the developing world, this trigger-happy insulin secretion keeps the body from wasting any sugar. Once people begin eating a westernized, high-sugar diet, this tendency leads to constant bursts of insulin secretion, which is more likely to cause storage tissues to become insulin resistant, leading to insulin-resistant diabetes. People in western countries, in contrast, are theorized to have more sluggish insulin responses to sugar; the net result is less efficient storing of sugar from the circulation, but lower risk of diabetes. And why are people in westernized societies theorized to be genetically less efficient in handling blood sugar? Because a few centuries back, as we first began eating typical westernized diets, those people with the greatest tendency toward insulin secretion failed to survive and pass on their genes. This predicts that populations like the Nauru islanders and Pima are undergoing the same process now; in a few centuries, most of their descendants will be the offspring of the rare individuals now with the lower diabetes risk. In support of this prediction, the rate of diabetes has already peaked among the Nauru islanders. But at present the existence of thrifty genes, and their differential presence in different human populations, is mostly speculative. For a nontechnical discussion of these ideas, see Diamond, J. 1992. "Sweet death." *Natural History* (February), 2. For technical discussions from the originator of the idea, see Neel, J. 1962. "Diabetes mellitus: a 'thrifty' genotype rendered detrimental by 'progress'?" *American Journal of Human Genetics* 14, 353; Neel, J. 1982. "The thrifty genotype revisited." In Kobberling, J., and Tattersall, R. (eds.). *The Genetics of Diabetes Mellitus* (London: Academic Press, Proceedings of the Serono Symposia, vol. 47), 283. For some technical discussions of the change in the incidence of diabetes with westernization, see Bennett, P., LeCompte, P., Miller, M., and Rushforth, N. 1976. "Epidemiological studies of diabetes in the Pima Indians." *Recent Progress in Hormone Research* 32, 333; O'Dea, K., Spargo, R., and Nestle, P. 1982. "Impact of westernization on carbohydrate and lipid metabolism in Australian Aborigines." *Diabetologia* 22, 148; Cohen, A., Chen, B., Eisenberg, S., Fidel, J., and Furst, A. 1979. "Diabetes, blood lipids, lipoproteins and change of environment. Restudy of the 'new immigrant Yemenites' in Israel." *Metabolism* 28, 716. For information on the rate of diabetes having peaked among the Nauru islanders, see Diamond, J. 1992. "Diabetes running wild." *Nature* 357, 362. For a discussion of other cases of thrifty genes, see the chapter entitled "The dangers of fallen soufflés in the developing world" in Sapolsky, R. 1997. *"The Trouble with Testosterone" and Other Essays on the Biology of the Human Predicament* (New York: Scribner). And for evidence of the "thriftiness" of metabolism among people such as Nauru islanders, see Robinson, S., Johnston, D. 1995. "Advantage of diabetes?" *Nature* 375, 640.

PAGE 58 One of the puzzles of how diabetes affects your health has been solved. It is relatively easy to understand how extra glucose in the bloodstream can clog blood vessels and cause damage. One of the mysteries, however, is why high levels of circulating glucose damage the eye (diabetes is the leading cause of blindness in this country). It turns out that glucose can stick to all sorts of proteins, causing them to form aggregates; indeed, because of its structure, glucose can stick onto proteins without the aid of enzymes to mediate the process, something called *nonenzymatic modification*. Once glucose fuses these proteins, they have to be broken apart and replaced. However, in some tissues—such as the lens of the eye—proteins are not recycled very frequently, and those cells are stuck with the fused mess. For a discussion of the nonenzymatic chemistry of sugars, focusing on its implications for aging and adult-onset diabetes, see Lee, A., and Cerami, A. 1990. "Modifications of proteins and nucleic acids by reducing sugars: possible role in aging." In Schneider, E., and Rowe, J. *Handbook of the Biology of Aging*, 3d ed. (New York: Academic Press).

Hyperglycemia can cause vascular damage even in nondiabetics: this is because of the nonenzymatic modification of glucose just discussed. See: Schmidt, A., Hori, O., Brett, J., Yan, S., Wautier, J., and Stern, D. 1994.

"Cellular receptors for advanced glycation end products. Implications for induction of oxidant stress and cellular dysfunction in the pathogenesis of vascular lesions." *Arteriosclerosis and Thrombosis* 14, 1521.

PAGE 61 Glucocorticoids promote insulin resistance: Rizza, R., Mandarino, L., and Gerich, J. 1982. "Cortisol-induced insulin resistance in man: impaired suppression of glucose production and stimulation of glucose utilization due to a postreceptor defect of insulin action." *Journal of Clinical Endocrinology and Metabolism* 54, 131. Stress promotes insulin resistance: Brandi, L., Santoro, D., Natali, A., Altomonte, F., Baldi, S., Frascerra, S., Ferrannini, E. 1993. "Insulin resistance of stress: sites and mechanisms." *Clinical Science* 85, 525.

PAGE 61 Stress disrupts metabolic control in insulin-dependent diabetics: Moberg, E., Kollind, M., Lins, P., Adamson, U. 1994. "Acute mental stress impairs insulin sensitivity in IDDM patients." (*IDDM* means "insulin-dependent diabetes mellitus.") *Diabetologia*, 37, 247. This presents a special challenge, in terms of stress management, for adolescents with insulin-dependent diabetes: Davidson, M., Boland, E., and Grey, M. 1997. "Teaching teens to cope: coping skills training for adolescents with insulin-dependent diabetes mellitus." *Journal of the Society of Pediatric Nurses* 2, 65. Controlled versus uncontrolled diabetics and stress: Dutour, A., Boiteau, V., Dadoun, F., Feissel, A., Atlan, C., and Oliver, C. 1996. "Hormonal response to stress in brittle diabetes." *Psychoneuroendocrinology* 21, 525.

PAGE 61 Glucocorticoids and stress can exacerbate the symptoms of insulin-resistant diabetes: Surwit, R., Ross, S., and Feingloss, M. 1991. "Stress, behavior, and glucose control in diabetes mellitus." In McCabe, P., Schneidermann, N., Field, T., and Skyler, J. (eds.). *Stress, Coping and Disease* (Hillsdale, N.J.: L. Erlbaum Assoc.), 97; Surwit, R., and Williams, P. 1996. "Animal models provide insight into psychosomatic factors in diabetes." *Psychosomatic Medicine* 58, 582. For a study that does not show an association between stress and worsening of symptoms, see Pipernik-Okanovic, M., Roglic, G., Prasek, M., and Metelko, Z. 1993. "War-induced prolonged stress and metabolic control in Type 2 diabetic patients." *Psychological Medicine* 23, 645.

PAGE 62 Stress causes insulin resistance and metabolic imbalances even in nondiabetics: Raikkonen, K., Keltikangas-Jarvinen, L., Adlercreutz, H., and Hautanen, A. 1996. "Psychosocial stress and the insulin resistance syndrome." *Metabolism: Clinical and Experimental* 45, 1533; Nilsson, P., Moller, L., Solstad, K. 1995. "Adverse effects of psychosocial stress on gonadal function and insulin levels in middle-aged males." *Journal of Internal Medicine* 237, 479.

PAGE 62 Stress worsens metabolic control in nondiabetics who are at genetic risk for diabetes: Esposito-Del Puente, A., Lillioja, S., Bogardus, C., McCubbin, J., Feinglos, M., Kuhn, C., and Surwit, R. 1994. "Glycemic response to stress is altered in euglycemic Pima Indians." *International Journal of Obesity and Related Metabolic Disorders* 18, 766.

PAGE 62 Statistics regarding the incidence and impact of diabetes: Rifkin, H., and Porte, D. 1990. *Diabetes Mellitus. Theory and Practice*, 4th ed. (New York: Elsevier).

CHAPTER 5: Ulcers, Colitis, and the Runs

PAGE 64 The cost of digestion: Secor, S., and Diamond, J. 1995. *Journal of Experimental Biology* 198, 1313. Those authors also report that animals that really do some energetic digesting (such as pythons and boa constrictors, who may swallow up some antelope far larger than themselves and spend the next week digesting it) use a third of their calories on the process.

PAGE 64 Stressors tend to inhibit gastrointestinal function: Desiderato, O., MacKinnon, J., and Hissom, R. 1974. "Development of gastric ulcers following stress termination." *Journal of Comparative and Physiological Psychology* 87, 208; Hess, W. 1957. *Diencephalon; Autonomic and Extra-pyramidal Functions* (New York: Grune and Stratton); Kiely, W. 1977. "From the symbolic stimulus to the pathophysiological response." In Lipowski, Z., Lipsitt, D., and Whybrow, P. (eds.), *Current Trends and Clinical Applications* (New York: Oxford University Press); Murison, R., and Bakke, H. 1990. "The role of corticotropin-releasing factor in rat gastric ulcerogenesis." In Hernandez, D., and Glavin, G. (eds.). *Neurobiology of Stress Ulcers* (*Annals of the New York Academy of Sciences*, vol. 597), 71; Tache, Y. 1991. "Effect of stress on gastric ulcer formation." In Brown, M., Koob, G., and Rivier, C. (eds.). *Stress: Neurobiology and Neuroendocrinology* (New York: Marcel Dekker), 549.

PAGE 65 Selye was the first to note that stress could cause peptic ulcers (1936. "A syndrome produced by diverse nocuous agents." *Nature* 138, 32). The first researchers to systematically explore the role of psychological stress in causing ulcers were Brady, J., Porter, D., Conrad, D., and Mason, J. 1958. "Avoidance behavior and the development of gastroduodenal ulcers." *Journal of Experimental Analysis of Behavior* I, 69; and Weiss, J. 1968. "Effects of coping responses on stress." *Journal of Comparative and Physiological Psychology* 65, 251.

The evidence that major and short-term traumas in humans can cause rapidly emerging stress ulcers can be found in Skillman, J., Bushnell, L., Goldman, H., and Silen, W. 1969. "Respiratory failure, hypotension, sepsis, and jaundice. A clinical syndrome associated with lethal hemorrhage from acute stress ulceration of the stomach." *American Journal of Surgery* 117, 523; Lucas, C., Sugawa, C., Riddle, J., Rector, F., Rosenberg, B., and Walt, A. 1971. "Natural history and surgical dilemma of 'stress' gastric bleeding." *Archives of Surgery* 102, 266; Butterfield, W. 1975. "Experimental stress ulcers: a review." *Surgical Annual* 7, 261. For evidence that more subtle psychological stress can cause gradually emerging peptic ulcers in humans see Feldman, M., Walker, P., Green, J., and Weingarden, K. 1986. "Life events, stress and psychosocial factors in men with peptic ulcer disease: a multidimensional case-controlled study." *Gastroenterology* 91, 1370.

Also see Weiner, H. 1992. *Perturbing the Organism: The Biology of Stressful Experience* (Chicago: University of Chicago Press).

The bacteria-ulcer revolution: Warren, J., Marshall, B. 1983. "Unidentified curved bacilli on gastric epithelium in active chronic gastritis." *Lancet,* i, 1273. Also, Wyatt, J., Rathbone, B., Dixon, M., and Heatley, R. 1987. "*Campylobacter pylorides* and acid induced gastric metaplasia in the pathogenesis of duodenitis." *Journal of Clinical Pathology* 40, 841. (*Campylobacter pylorides* was an earlier name for *Helicobacter*). Dooley, C., and Cohen, H. 1988. "The clinical significance of *Campylobacter pylori*." *Annals of Internal Medicine* 108, 70. For a breathless account of the discovery with Marshall (and occasionally Warren) as the heroic underdogs, see Monmaney, T. 1993. "Marshall's hunch." *The New Yorker*, September 20, 64. The bacteria's resistance to acidity: Doolittle, R. 1997. "A bug with excess gastric avidity." *Nature* 388, 515; Tom, J., White, O., and Kerlavage, A. et al. (there's a total of 42 authors—no kidding) 1997. "The complete genome sequence of the gastric pathogen *Helicobacter pylori*." *Nature* 388, 539.

PAGE 67 The efficacy of antibiotics with duodenal ulcers is reviewed in Konturek, P, 1997. "Physiological, immunohistochemical and molecular aspects of gastric adaptation to stress, aspiring and to *H. pylori*-derived gastrotoxins." *Journal of Physiology and Pharmacology* 48, 3.

PAGE 67 The decline of ulcer-stress research: Melmed, R., and Gelpin, Y. 1996. "Duodenal ulcer: the helicobacterization of a psychosomatic disease?" *Israeli Journal of Medical Science* 32, 211.

Ulcers, but no bacteria: McColl, K., El-Nujami, A., and Chittajallu, R. 1993. "A study of the pathogenesis of *Helicobacter pylori* negative chronic duodenal ulceration." *Gut* 34, 762. Bacteria but no ulcers: Tompkins, L., and Falkow, S. 1995. "The new path to preventing ulcers." *Science* 267, 1621.

The rodent studies showing no bacteria, no stress ulcers: Pare, W., Burken, M., Allen, E., and Kluczynski, J. 1993. "Reduced incidence of stress ulcer in germ-free Sprague Dawley rats." *Life Sciences* 53, 1099. The interactions of stress, bacterial load, and other risk factors in human ulcer cases: Levenstein, S., Prantera, C., Varvo, V., Scribano, M., Berto, E., Spinella, S., and Lanari, G. 1995. "Patterns of biologic and psychologic risk factors in duodenal ulcer patients." *Journal of Clinical Gastroenterology* 21, 110.

PAGE 68 Stress-related ulcers are predominantly formed during the post-stress recovery period, rather than during the stressor itself: Overmier, J., Murison, R., and Ursin, H. 1986. "The ulcerogenic effect of a rest period after exposure to water-restraint stress." *Behavioral and Neural Biology* 46, 372; Vincent, G., and Pare, W. 1982. "Post stress development and healing of supine-restraint induced stomach lesions in the rat." *Physiology and Behavior* 29, 721; Desiderato, O., MacKinnon, J., and Hissom, H. 1974. "Development of gastric ulcers in rats following stress termination." *Journal of Comparative and Physiological Psychology* 87, 208; Glavin, G. 1980. "Restraint ulcer: history, current research and future implications." *Brain Research Bulletin* 5, suppl. 1, 51. For the evidence for how this is due to rebound of the parasympathetic nervous system, see the Glavin paper just cited; also see Klein, H., Gheorghiu, T., and Hubner, G. 1975. "Morphological and functional gastric changes in stress ulcer." In Gheorghiu, T. (ed.). *Experimental Ulcer: Models, Methods and Clinical Validity* (Baden-Baden: Witzstrock).

Back to hydrochloric acid digesting the stomach in which it is secreted: if the mucous layer prevents the acid from penetrating it, how can acid, first secreted by the stomach wall, ever get through the mucous layer to digest food? This conundrum is answered by Bhaskar, K., Garik, P., Turner, B., Bradley, J., Bansil, R., Stanley, H., and Lamont, J. 1992. "Viscous fingering of hydrochloric acid through gastric mucin." *Nature* 360, 458.

Bicarbonate secretion decreases in ulcer patients: Isenberg, J., Selling, J., Hogan, D., and Koss, M. 1987. "Impaired proximal duodenal mucosal bicarbonate secretion in duodenal ulcer patients." *New England Journal of Medicine* 316, 374. Bicarbonate secretion decreases with sustained stress in an animal ulcer model: Takeuchi, K., Furukawa, O., and Okabe, S. 1986. "Induction of duodenal ulcers in rats under water-immersion stress conditions. Influence on gastric acid and duodenal alkaline secretion." *Gastroenterology* 91, 554. Mucus secretion decreases with stress and with glucocorticoid administration: Schuster, M. 1989. "Irritable bowel syndrome." In Sleisenger, M., and Fordtron, J. (eds.) *Gastrointestinal Disease: Pathophysiology, Diagnosis, Management,* 4th ed. (Philadelphia: Saunders), 1402. In approximately half the cases, during this rebound period the amount of gastric acid secreted is normal, implying that the problem is that the stomach wall is relatively more vulnerable, since the acidic attack is not stronger than usual: Dayal, Y., and DeLellis, R. 1989. "The gastrointestinal tract." In Robbins, S., Cotran, R., and Kumar, V. *Pathologic Basis of Disease*, 4th ed. (Philadelphia: Saunders), 827; also Weiner, H. 1991. "From simplicity to complexity (1950–1990): the case of peptic ulceration—I. Human studies." *Psychosomatic Medicine* 53, 467; and Weiner, H. 1991. "From simplicity to complexity (1950–1990): the case of peptic ulceration—II. Animal studies." *Psychosomatic Medicine* 53, 491; Grossman, M. 1978. "Abnormalities of acid secretion in patients with duodenal ulcer." *Gastroenterology* 75, 524; also Brodie, D., Marshall, R., and Moreno, O. 1962. "The effect of restraint on gastric acidity in the rat." *American Journal of Physiology* 202, 812.

As noted, an interesting implication of the rebound phenomenon is that in a person at risk for a stress ulcer, continuous stress may protect against the formation of an ulcer (although, as noted, this is not a good prescriptive idea for many other reasons). As a building block of that idea, sustained administration of CRF will protect against ulcer formation: Murison, R., and Bakke, H. 1990. "The role of corticotropin-releasing factor in rat gastric ulcerogenesis." In Hernandez, D., and Glavin, G. (eds.). *Neurobiology of Stress Ulcers (Annals of the New York Academy of Sciences,* vol. 597), 71.

PAGE 69 Ulcers form as a result of decreased blood flow to the stomach, causing ischemic lesions, due to both acid accumulation and to formation of oxygen radicals. These ideas are reviewed in Tsuda, A., and Tanaka, M. 1990. "Neurochemical characteristics of rats exposed to activity stress." In Hernandez, D., and Glavin, G. (eds.). *Neurobiology of Stress Ulcers* (*Annals* *of the New York Academy of Sciences,* vol. 597), 146; also Yabana, T., and Yachi, A. 1988. "Stress-induced vascular damage and ulcer." *Digestive Disease Science* 33, 751; also Menguy, R. 1980. "The prophylaxis of stress ulceration." *New England Journal of Medicine* 302, 461; also Robert, A., and Kauffman, G. 1989. "Stress ulcers, erosions and gastric motility injury." In Sleisenger, M., and Fordtron, J. (eds.). *Gastrointestinal Disease: Pathophysiology, Diagnosis, Management,* 4th ed. (Philadelphia: Saunders), 1402. Original data regarding how hemorrhage stress can cause oxidative damage: Itoh, M., and Guth, P. 1985. "Role of oxygen-derived free radicals in hemorrhagic shock-induced gastric lesions in the rat." *Gastroenterology* 88, 1162.

PAGE 70 Although glucocorticoids may cause ulcer formation by suppressing the immune system during stress via this route, it is not clear how important this is for mild stressors. During mild or infrequent stressors, the levels of glucocorticoids secreted do not predict whether ulcers form: Murison, R., and Overmeir, J. 1988. "Adrenocortical activity and disease, with reference to gastric pathology in animals." In Hellhammer, D., Florin, I., and Weiner, H. (eds.). *Neurobiological Approaches to Human Disease* (Toronto: Hans Huber), 335. Moreover, removal of glucocorticoids by adrenalectomizing a rat actually protects against ulcers: Brodie, D. 1968. "Experimental peptic ulcer." *Gastroenterology* 55, 125.

All this suggests that glucocorticoids are unlikely to be the cause of ulcers during stress. However, with more sustained or repeated stressors, the amount of glucocorticoids secreted does predict the severity of ulceration: Weiss, J. 1980. "Somatic effects of predictable and unpredictable shock." *Psychosomatic Medicine* 32, 397; Weiss, J. 1981. "Effects of coping behavior in different warning signal conditions on stress pathology in rats." *Journal of Comparative and Physiological Psychology* 77, 1; Murphy, H., Wideman, C., and Brown, T. 1979. "Plasma corticosterone levels and ulcer formation in rats with hippocampal lesions." *Neuroendocrinology* 28, 123. In addition, supraphysiological levels of glucocorticoids (levels that are higher in the bloodstream than the body can normally generate, even during stress, but are induced by taking glucocorticoid medication) can cause ulcers: Robert, A., and Nezmis, J. 1964. "Histopathology of steroid-induced ulcers: an experimental study in the rat." *Archives of Pathology* 77, 407.

PAGE 70 The role of prostaglandins in ulcerogenesis: the protective effects of prostaglandins are discussed in Kauffman, G., Zhang, L., Xing, L., Seaton, J., Colony, P., and Demers, L. 1990. "Central neurotensin protects the mucosa by a prostaglandin-mediated mechanism and inhibits gastric acid secretion in the rat." In Hernandez, D., and Glavin, G. (eds.) *Neurobiology of Stress Ulcers* (*Annals of the New York Academy of Sciences,* vol. 597), 175. See also Schepp, W., Steffen, B., Ruoff, H., Schusdziarra, V., and Classen, M. 1988. "Modulation of rat gastric mucosal prostaglandin E2 release by dietary linoleic acid: effects on gastric acid secretion and stress-induced mucosal damage." *Gastroenterology* 95, 18.

Aspirin is ulcerogenic by blocking prostaglandin synthesis: Adcock, J., Hernandez, D., Nemeroff, C., and Prang, A. 1983. "Effect of prostaglandin synthesis inhibitors on neurotensin and sodium salicylate-induced gastric cytoprotection in rats." *Life Science* 32, 2905. Glucocorticoids block prostaglandin synthesis: Flowers, R., and Blackwell, G. 1979. "Anti-inflammatory steroids induce biosynthesis of a phospholipase A2 inhibitor which prevents prostaglandin generation." *Nature* 278, 456.

PAGE 70 The role of stomach contractions in causing ulcers is discussed at length in Weiner, H. 1991. "From simplicity to complexity (1950–1990): the case of peptic ulceration—II. Animal studies." *Psychosomatic Medicine* 53, 491.

PAGE 71 Stress decreases contractions in the small intestines: Thompson, D., Richelson, E., and Malagelada, J. 1982. "Perturbation of gastric emptying and duodenal motility through the central nervous system." *Gastroenterology* 83, 1200; Thompson, D., Richelson, E., and Malagelada, J. 1983. "Perturbation of upper gastrointestinal function by cold stress." *Gut* 24, 277; O'Brien, J., Thompson, D., Holly, J., Burnham, W., and Walker, E. 1985. "Stress disturbs human gastrointestinal transit via a beta-1 adrenoreceptor mediated pathway." *Gastroenterology* 88, 1520. Stress increases contractions in the large intestines: Almy, T. 1951. "Experimental studies on irritable colon." *American Journal of Medicine* 10, 60; Almy, T., and Tulin, M. 1947. "Alterations in colonic function in man under stress: experimental production of changes simulating the 'irritable colon.'" *Gastroenterology* 8, 616; Narducci, F., Snape, W., Battle, W., London, R., and Cohen, S. 1985. "Increased colonic motility during exposure to a stressful situation." *Digestive Disease Science* 30, 40.

Chemical mediators of the sympathetic stress-response bring about the changes in contractions: Williams, C., Peterson, J., Villar, R., and Burks, T. 1987. "Corticotropin-releasing factor directly mediates colonic responses to stress." *American Journal of Physiology* 253, G582. Also Burks, T. 1990. "Central nervous system regulation of gastrointestinal motility." In Hernandez, D., and Glavin, G. (eds.). *Neurobiology of Stress Ulcers* (*Annals of the New York Academy of Sciences* 597), 36. Glucocorticoids are not mediators of the contractions: Williams, C., Villar, R., Peterson, J., and Burks, T. 1988. "Stress-induced changes in intestinal transit in the rat: a model for irritable bowel syndrome." *Gastroenterology* 94, 611.

PAGE 72 For a review of the general subject, see Lynn, R., and Friedman, L. 1993. "Irritable bowel syndrome." *New England Journal of Medicine* 329, 1940. IBS can be stress-related: Kumar, D., and Wingate, D. 1988. "Irritable bowel syndrome." In Kumar, D., and Gustavsson, S. (eds.). *An Illustrated Guide to Gastrointestinal Motility* (Chichester: John Wiley), *401.* (Note: unique Christmas gift.) Also Mendeloff, A., Monk, M., Siegel, C., and Lilienfeld, A. 1970. "Illness, experience and life stress in patients with irritable colon syndrome and with ulcerative colitis. An epidemiological study of ulceration and regional enteritis in Baltimore, 1960–1964." *New*

England Journal of Medicine 282, 14. Chaudhary, N., and Truelove, S. 1962. "The irritable colon syndrome: a study of the clinical features, predisposing causes, and prognosis in 130 cases." *Quarterly Journal of Medicine* 31, 307; Drossman, D., Sandler, R., McKee, D., and Lovitz, A. 1982. "Bowel patterns among subjects not seeking health care: use of a questionnaire to identify a population with bowel dysfunction." *Gastroenterology* 83, 529; Scribano, M., Berto, E., Andreoli, A., and Luzi, C. 1994. "Psychological stress and disease activity in ulcerative colitis: a multidimensional cross-sectional study." *American Journal of Gastroenterology* 89, 1219; Whitehead, W., Crowell, M., Robinson, J., Heller, B., and Schuster, M. 1992. "Effects of stressful life events on bowel symptoms: subjects with irritable bowel syndrome compared with subjects without bowel dysfunction. *Gut* 33, 825; Greene, B., Blanchard, E., and Wan, C. 1994. "Long-term monitoring of psychosocial stress and symptomatology in inflammatory bowel disease." *Behaviour Research and Therapy* 32, 217; Welgan, P., Meshkinpour, H., and Beller, M. 1988. Effect of anger on colon motor and myoelectric activity in irritable bowel syndrome." *Gastroenterology* 94, 1150.

A recent study showing that stress worsens experimental colitis in animals: Gue, M., Bonbonne, C., Fioramonti, J., More, J., Del Rio-Lacheze, C., Comera, C., and Bueno, L. 1997. "Stress-induced enhancement of colitis in rats; CRF and arginine vasopressin are not involved." *American Journal of Physiology* 272, G84. A demonstration of an interaction between inflammation and stress in irritable bowel syndrome: Collins, S., McHugh, K., Jacobson, K., Khan, I., Riddell, R., Murase, K., and Weingarten, H. 1996. "Previous inflammation alters the response of the rat colon to stress." *Gastroenterology* 111, 1509.

The symptoms of gut motility disorders: Burks, T. 1991. "Gastrointestinal motility disorders." In Brown, M., Koob, G., and Rivier, C. (eds.). *Stress: Neurobiology and Neuroendocrinology* (New York: Marcel Dekker), 565.

PAGE 73 People with colitis have guts that are often hyperreactive: experimental stressors induce greater motility changes in them than in control subjects: reviewed in Burks, T. 1991. "Gastrointestinal motility disorders." In Brown, M., Koob, G., and Rivier, C. (eds.). *Stress: Neurobiology and Neuroendocrinology* (New York: Marcel Dekker), 565. Original report: Wangle, A., and Deller, D. 1965. "Intestinal motility in man." *Gastroenterology* 48, 69.

PAGE 74 For the classic psychoanalytic view of these diseases, see Alexander, F. 1950 *Psychosomatic Medicine* (New York: Norton). See also Aronowitz, R., and Spiro, H. 1988. "The rise and fall of the psychosomatic hypothesis in ulcerative colitis." *Journal of Clinical Gastroenterology* 10, 298; Ramchandani, D., Schindler, B., and Katz, J. 1994. "Evolving concepts of psychopathology in inflammatory bowel disease." *Medical Clinics of North America* 78, 1321.

PAGE 74 Some studies that have not found a stress link in colitis (note that the first two studies are from the same group): Helzer, J., Stillings, W., and Chammas, S. 1982. "A controlled study of the association between ulcerative colitis and psychiatric diagnoses." *Digestive Disease Science* 27, 513. North, C., Alpers, D., and Helzer, J. 1991. "Do life events or depression exacerbate inflammatory bowel disease? A prospective study." *Annals of Internal Medicine* 114, 381. Tartar, R., Switala, J., and Carra, J. 1987. "Inflammatory bowel disease. Psychiatric status of patients before and after disease onset." *International Journal of Psych Medicine* 17, 173; Drossman, D., McKee, D., and Sandler, R. 1988. "Psychosocial factors in the irritable bowel syndrome: a multivariate study of patients and nonpatients with irritable bowel syndrome." *Gastroenterology* 95, 701; Camilleri, M., and Neri, M. 1989. "Motility disorders and stress" *Digestive Disease Science* 34, 1777.

PAGE 74 A study using time-series analysis to show a stress-symptom link: Greene, B., Blanchard, E., and Wan, C. 1994. "Long-term monitoring of psychosocial stress and symptomatology in inflammatory bowel disease." *Behaviour Research and Therapy* 32, 217. A discussion of some of the methodological problems in stress research in this area: Whitehead, W. 1994 "Assessing the effects of stress on physical symptoms." *Health Psychology* 13, 99. People are bad at accurately reporting events more than three months old: Jenkins, C., Hurst, W., and Rose, R. 1979. "Life changes: do people really remember?" *Archives of General Psychiatry* 36, 379.

PAGE 76 The effects of CRF in the brain, including the effect on appetite and feeding: Turnbull, A., and Rivier, C. 1997. "CRF and endocrine responses to stress; CRF receptors, binding protein, and related peptides." *Proceedings of the Society for Experimental Biology and Medicine 215,* 1. The urocortin story: Spina, M., Merlo-Pich, E., Chan, R., Basso, A., Rivier, J., Vale, W., and Koob, G. 1996. "Appetite-suppressing effects of urocortin, a CRF-related neuropeptide." *Science* 273, 1561. The effects of glucocorticoids on appetite are discussed in McEwen, B., de Kloet, E., and Rostene, W. 1986. "Adrenal steroid receptors and actions in the nervous system." *Physiological Reviews* 66, 1121. I am not aware of any publication in which the opposing effects of CRF and glucocorticoids on appetite are analyzed in the manner done in this chapter. However, a similar flavor (viewing some glucocorticoid actions as mediating the "recovery" from the stress-response, rather than the "mediation" of the stress-response) can be found in a very influential paper: Munck, A., Guyre, P., and Holbrook, N. 1984. "Physiological functions of glucocorticoids during stress and their relation to pharmacological actions." *Endocrine Reviews* 5, 25. Some examples of glucocorticoids increasing transcription of the *ob* gene and increase circulating leptin levels: Reul, B., Ongemba, L., Pottier, A., Henquin, J., and Brichard, S. 1997. "Insulin and insulin-like growth factor I antagonize the stimulation of *ob* gene expression by dexamethasone in cultured rat adipose tissue." *Biochemical Journal* 324, 605; Considine, R., Nyce, M., Kolaczynski, J., Zhang, P., Ohannesian, J., Moore, J., Fox, J., and Caro, J. 1997. "Dexamethasone stimulates leptin release from human adipocytes: unexpected inhibition by insulin." *Journal of Cellular Biochemistry* 65, 254; Miell, J., Englaro, P., and Blum, W. 1996. "Dexamethasone induces an acute and sustained rise in circulating leptin levels in normal human subjects."

Hormone and Metabolic Research, 28, 704. Glucocorticoids blunt the efficacy of leptin: Zakrzewska, K., Cusin, I., Sainsbury, A., Rohner-Jeanrenaud, F., and Jeanrenaud, B. 1997 "Glucocorticoids as counterregulatory hormones of leptin: toward an understanding of leptin resistance." *Diabetes* 46, 717. Chronic glucocorticoid exposure might cause leptin resistance: Ur, E., Grossman, A., and Despres, J. 1996. "Obesity results as a consequence of glucocorticoid induced leptin resistance." *Hormones and Metabolic Research* 28, 744.

CHAPTER 6: Dwarfism and the Importance of Mothers

PAGE 81 The mechanisms of growth and its regulation by various hormones can be found in any basic endocrine or physiology textbook. A relatively accessible version for nonspecialists can be found in Vander, A., Sherman, J., and Luciano, D. 1994. *Human Physiology: The Mechanisms of Body Function*, 6th ed. (New York: McGraw-Hill).

PAGE 82 Short summaries of stress dwarfism and of failure to thrive can be found in most endocrine or pediatric textbooks. A relatively recent technical summary of the subject can be found in Green, W., Campbell, M., and David, R. 1984. "Psychosocial dwarfism: a critical review of the evidence." *Journal of the American Academy of Child Psychiatry* 23, 1. A somewhat dated but very readable nontechnical account can be found in Gardner, L. 1972. "Deprivation dwarfism." *Scientific American* 227, 76. A specific discussion of the intellectual impairments found in such children can be found in Dowdney, L., Skuse, D., Heptinstall, E., Puckering, C., and Zur-Szpiro, S. 1987. "Growth retardation and developmental delay amongst inner-city children." *Journal of Child Psychology and Psychiatry* 28, 529. A demonstration that removing stress dwarfism children from their stressful environments normalizes growth and growth hormone: Albanese, A., Hamill, G., Jones, J., Skuse, D., Matthews, D., and Stanhope, R. 1994. "Reversibility of physiological growth hormone secretion in children with psychosocial dwarfism." *Clinical Endocrinology* 40, 687.

PAGE 83 Catch-up growth after stress dwarfism: Boersma, B., and Wit, J. 1997. "Catch-up growth." *Endocrine Reviews* 18, 646.

PAGE 83 Fairly consistent versions of the King Frederick story are reported by a number of his biographers, including T. Kingston, *History of Frederick the Second, Emperor of the Romans* (Cambridge, England: Macmillan, 1862), L. Allshorn, *Stupor Mundi: The Life and Times of Frederick II, Emperor of the Romans, King of Sicily and Jerusalem 1194–1250* (London: Martin Secker, 1912), and E. Kantorowicz, *Frederick the Second, 1194–1250* (London: Constable, 1931). The quotation by Salimbene comes from A. Montagu, *Touching: The Human Significance of the Skin* (New York: Harper and Row, 1978). Another story about the intersection of the monarch's scientific curiosity and barbarity concerned his interest in rates of digestion. Frederick wondered whether digestion was faster when you rested after eating or if you exercised. He had two men fed identical and sumptuous dinners and sent one off to nap afterward, while the other went for a strenuous hunt. That phase of the experiment completed, he had both men returned to his court, disemboweled in front of him, and their innards examined. The sleeper had digested his food better.

PAGE 85 The tale of the two orphanages: Widdowson, E. 1951. "Mental contentment and physical growth." *Lancet* (16 June), 1316. The information on the appalling survivorship in orphanages comes from Chapin, H. 1915. "A plea for accurate statistics in children's institutions." *Transactions of the American Pediatric Society* 27, 180. The quotation comes from Gardner, L. 1972. "Deprivation dwarfism." *Scientific American* 227, 76.

PAGE 85 The discussion of child-rearing practices at the time can be found in A. Montagu, *Touching: The Human Significance of the Skin,* cited above. The authoritative "expert" who advised against such unscientific practices as handling infants too much was Dr. Luther Holt, professor of pediatrics at Columbia University and the author of *The Care and Feeding of Children* (East Norwalk, Conn.: Appleton-Century), which went through 15 editions between 1894 and 1915.

PAGE 86 J. M. Barrie and stress dwarfism: the discussion of Barrie that so caught my attention during my student days can be found in J. Martin and S. Reichlin, *Clinical Neuroendocrinology,* 1st ed. (Philadelphia: Davis Company, 1977). I am particularly grateful to Seymour Reichlin, one of the giants of endocrinology and my teacher at the time, for remembering this source.

In preparing this book, I decided to read up a bit more on Barrie. I was surprised to discover a vast number of Barrie biographies; this now fairly obscure man was once the most popular author and playwright in Britain. The details of his life are both fascinating and grotesque. He retained a lifelong obsession with his mother, forever attempting to win her love. In one remarkable passage that encapsulated both his Oedipal wooing of her and his pathological identification with her, he predicted that in his later years, "when age must dim my mind and the past comes sweeping back like the shades of night over the bare road of the present, it will not, I believe, be my youth I shall see but hers, not a boy clinging to his mother's skirt and crying, 'Wait till I'm a man, and you'll lie on feathers,' but a little girl in a magenta frock and a white pinafore." He also had a lifelong obsession with young boys, and his private writing includes passages of sadomasochism and pedophilia.

What is perhaps most fascinating is the transition of Barrie from a rather pathetic and sympathetic loner as a young man to a far-from-sympathetic manipulator in his later years, all because his writing success brought him the power and wealth to disrupt lives around him. As he grew older, alone and childless, he inveigled his way into the lives of a succession of young couples, appearing as a generous benefactor and gradually coming to dominate them more and more, especially the fates of the sons in these families (one boy, Peter Davies, became the model for Peter Pan; he loathed the association all his adult life and, probably unrelated to that, threw himself in front of a London subway at age 63). For the most interesting of the Barrie biographies (from which the above quotation was taken), I recommend A. Birkin, *J. M. Barrie and the Lost Boys* (London: Constable, 1979). Also see the elegiac piece by Alison Lurie, "The boy who couldn't grow up," *New York Review of Books*, 6 February 1975, 11.

PAGE 87 In addition to the references given above on the clinical profiles of kids with various deprivation syndromes, the following could be checked for details of the endocrinology of the disruption of growth: Chapter 8, "Regulation of growth hormone secretion and its disorders." In Martin, J., and Reichlin, S. (eds.). 1987. *Clinical Neuroendocrinology*, 2d ed. (Philadelphia: Davis Company); chapters 8 (Underwood, L., and Van Wyk, J. "Normal and aberrant growth") and 20 (Rose, R. "Psychoendocrinology") in Wilson, J., and Foster, D. (eds.). 1985. *Williams Textbook of Endocrinology*, 7th ed. (Philadelphia: Saunders); Reichlin, S. 1988. "Prolactin and growth hormone secretion in stress." In Chrousos, G., Loriaux, D., and Gold, P. (eds.). *Mechanisms of Physical and Emotional Stress* (New York: Plenum Press). These references also discuss the differences between growth hormone regulation in the adult versus the developing child, and in primates and humans versus rodents.

PAGE 88 For a review of the regulation of ODC levels by psychological factors, see Schanberg, S., Evoniuk, G., and Kuhn, C. 1984. "Tactile and nutritional aspects of maternal care: specific regulators of neuroendocrine function and cellular development." *Proceedings of the Society for Experimental Biology and Medicine* 175, 135. Information on the requirement of active contact with the mother to normalize growth hormone levels in infant rats can be found in Kuhn, C., Paul, J., and Schanberg, S. 1990. "Endocrine responses to mother-infant separation in developing rats." *Developmental Psychobiology* 23, 395. For a discussion of the effects of maternal separation on glucocorticoid levels, see the Kuhn et al. paper just cited, plus the earlier work by Stanton, M., Guitierrez, Y., and Levine, S. 1988. "Maternal deprivation potentiates pituitary-adrenal stress responses in infant rats." *Behavioral Neuroscience* 102, 692. For the classic demonstration of the effects of neonatal handling in rats on growth rates see any of the following three reports by V. Denenberg and G. Karas: 1959,"Effects of differential handling upon weight gain and mortality in the rat and mouse." *Science* 130, 629; 1960, "Interactive effects of age and duration of infantile experience on adult learning." *Psychological Reports* 7, 313; 1961, "Interactive effects of infant and adult experience upon weight gain and mortality in the rat." *Journal of Comparative and Physiological Psychology* 54, 658.

PAGE 89 The data from the study of the child with stress dwarfism whose nurse went on vacation comes from Saenger, P., Levine, L., Wiedemann, E., Schwartz, E., Korth-Schutz, S., Pareira, J., Heinig, B., and New, M. 1977. "Somatomedin and growth hormone in psychosocial dwarfism." *Padiatr. Padol. Suppl.* 5, 1.

PAGE 90 The importance of touch in rat development: Hofer, M. 1984. "Relationships as regulators." *Psychosomatic Medicine* 46, 183.

PAGE 90 The work on touching of premature human infants is described in Field, T., Schanberg, S., Scarfidi, F., Bauer, C., Vega-Lahr, N., Garcia, R., Nystrom, J., and Kuhn, C. 1986. "Tactile/kinesthetic stimulation effects on preterm neonates." *Pediatrics* 77, 654. Also Scarfidi, F., Field, T., Schanberg, S., Bauer, C., Vega-Lahr, N., Garcia, R., Poirier, J., Nystrom, J., and Kuhn, C. 1986. "Effects of tactile-kinesthetic stimulation on the clinical course and sleep-wake behavior of pre-term infants." *Infant Behavior and Development* 9, 71. A similar experiment was carried out some years earlier in a much sketchier form with only five infants, as reported in Sokoloff, N., Yaffe, S., Weintraub, D., and Blase, G. 1969. "Effects of handling on the subsequent development of premature infants." *Developmental Psychology* 1, 765. That work, in turn, was inspired by the research of some pioneers in the field: the developmental biologist Rene Spitz and the famed pediatrician T. Berry Brazelton.

PAGE 92 The estimate of 1 billion dollars in savings is based on the following (admittedly very crude) analysis. A federal report in 1987 ("Neonatal Intensive Care for Low Birthweight Infants: Costs and Effectiveness." *Health Technology Case Study 38,* Office of Technology Assessment, Washington, D.C.) reported 150,000–200,000 infants admitted annually to neonatal intensive care units, of whom approximately 20 percent were of very low weight (less than 3 pounds). The average length of stay for that most vulnerable group was approximately 48 days at a cost of $41,000; for the other 80 percent, average stay was approximately 28 days at a cost of $24,000. Total hospitalization bills thus come to something in excess of $5 billion, based on an average individual stay of approximately 32 days (weighting the two different groups). Thus an average reduction of one week's stay in the intensive care unit constitutes an approximate 20 percent reduction and (assuming, probably incorrectly, that the rate of expense incurred is constant over time) produces savings something in excess of a billion dollars. This does not even count savings on the considerable costs of outpatient care lasting months to years for preemies after discharge (discussed in Blackman, J. 1991. "Neonatal intensive care: is it worth it?" *Pediatric Clinics of North America,* vol. 38, no. 6).

PAGE 92 Most basic physiology textbooks include descriptions of bone growth and resorption in the adult and its hormonal regulation. A particularly clear discussion can be found in Rhoades, R., and Pflanzer, R. 1989. *Human Physiology* (Philadelphia: Saunders College Publishing). A good recent review of how glucocorticoids cause osteoporosis: Canalis, E. 1996 "Mechanisms of glucocorticoid action in bone: implications to glucocorticoid-induced osteoporosis." *Journal of Clinical Endocrinology and Metabolism,* 81, 3441. The first report of bone fractures in patients with Cushing's syndrome came, of course, from Dr. Harvey Cushing himself: 1932. "The basophil adenomas of the pituitary body and their clinical manifestations as basophilism." *Bulletin of the Johns Hopkins Hospital* 1, 137. A report on how patients being treated with glucocorticoids to control a disease (in this case, asthma) will get osteoporosis: Adinoff, A., and Hollister, J. 1983. "Steroid-induced fractures and bone loss in patients with asthma." *New England Journal of Medicine* 309, 265. Sustained social stress is associated with loss of bone mass in female primates: Kaplan, J., and Manuck, S. 1989. "Behavioral and evolutionary considerations in predicting disease susceptibility in nonhuman primates." *American Journal of Physical Anthropology* 78, 250; and Shively, C., Jayo, M., Weaver, D., and Kaplan, J. 1991. "Reduced vertebral bone mineral density in socially subordinate female cynomolgus macaques." *American Journal of Primatology* 24, 135.

PAGE 94 The ability of glucocorticoids (and stress) to stimulate growth hormone release in humans in the short run, yet inhibit it in the long run, is reviewed in: Thakore, J., and Dinan, T. 1994 "Growth hormone secretion: the role of glucocorticoids." *Life Sciences* 55, 1083.

PAGE 96 The cross-cultural studies of the stressfulness of developmental rituals: Landauer, T., and Whiting, J. 1964. "Infantile stimulation and adult stature of human males." *American Anthropologist* 66, 1007. A similar theme emerges from their later studies showing that the physical stressor of immunization (and the subsequent brief illness) of children under two years of age resulted in taller adults. The population was Americans who had been children in the 1930s, a time when immunization was far from universal: Whiting, J., Landauer, T., and Jones, T. 1968. "Infantile immunization and adult stature." *Child Development* 39, 59.

PAGE 98 The quotation by Harry Harlow comes from "The nature of love." 1958. *American Psychologist* 13, 673. More technical reports of his work can be found in Harlow, H., and Zimmerman, R. 1959. "Affectional responses in the infant monkey." *Science* 130, 421; Harlow, H., Harlow, M., Dodsworth, R., and Arling, G. 1966. "Maternal behavior of rhesus monkeys deprived of mothering and peer associations in infancy." *Proceedings of the American Philosophical Society* 110, 58.

CHAPTER 7: Sex and Reproduction

PAGE 101 Basic male reproductive endocrinology and the effects of the various hormonal changes described during stress on reproduction are covered in most basic textbooks.

General reviews of male reproductive physiology during stress: Rivier, C. 1995 "Luteinizing-hormone-releasing hormone, gonadotropins, and gonadal steroids in stress." *Annals of the New York Academy of Sciences*, 771, 187; Negro-Vilar, A. 1993 "Stress and other environmental factors affecting fertility in men and women: overview." *Environmental Health Perspectives*, 101, S2, 59.

PAGE 101 Some of the original papers showing how physical stressors (such as surgery, immobilization, drought for a wild primate population, foot shock, or forced swimming) will suppress hormones of the male reproductive system: Bardin, C., and Peterson, R. 1967. "Studies of androgen production by the rat: Testosterone and androstenedione content of blood." *Endocrinology* 80, 38; Free, M., and Tillson, S. 1973. "Secretion rate of testicular steroids in conscious and halothane-anesthetized rat." *Endocrinology* 93, 874; Matsumoto, K., Takeyasu, K., Mizutani, S., Hamanaka, Y., and Uozumi, T. 1970. "Plasma testosterone levels following surgical stress in male patients." *Acta Endocrinology* 65, 11; Sapolsky, R. 1986. "Endocrine and behavioral correlates of drought in the wild baboon." *American Journal of Primatology* II, 217. Some more recent papers: Jain, S., Bruot, B., and Stevenson, J. 1996 "Cold swim stress leads to enhanced splenocyte responsiveness to concanavalin A, decreased serum testosterone, and increased serum corticosterone, glucose and protein." *Life Sciences*, 59, 209; Ellison, P., and Panter-Brick, G. 1996 "Salivary testosterone levels among Tamang and Kami males of central Nepal." *Human Biology* 68, 955.

PAGE 103 Psychological stressors will also suppress these hormones. Examples follow. A drop in social rank for a male primate: Rose, R., Bernstein, I., and Gordon, T. 1975. "Consequences of social conflict on plasma testosterone levels in rhesus monkeys." *Psychosomatic Medicine* 37, 50; Mendoza, S., Coe, C., Lowe, E., and Levine, S. 1979. "The physiological response to group formation in adult male squirrel monkeys." *Psychoneuroendocrinology* 3, 221. A difficult learning task for a primate: Mason, J., Kenion, C., and Collins, D. 1968. "Urinary testosterone response to 72-hour avoidance sessions in the monkey." *Psychosomatic Medicine* 30, 721. A first parachute jump: Davidson, J., Smith, E., and Levine, S. 1978. "Testosterone." In Ursin, H., Baade, E., and Levine, S. (eds.). *Psychobiology of Stress* (New York: Academic Press), 57. Social instability for primates: Sapolsky, R. 1983. "Endocrine aspects of social instability in the olive baboon." *American Journal of Primatology* 5, 365; Curtin, F., and Steimer. T. 1996 "Lower sex hormones in men during anticipatory stress." *Neuroreport* 7, 3, 101. The suppressive effects of Officer Candidate School on testosterone levels: Kreuz, L., Rose, R., and Jennings, J. 1972. "Suppression of plasma testosterone levels and psychological stress." *Archives of General Psychiatry* 26, 479.

PAGE 103 Opiate drugs and opioidlike hormones (for instance, beta-endorphin) block the release of LHRH: Delitala, G., Devilla, L., and Arata, L. 1981. "Opiate receptors and anterior pituitary hormone secretion in man. Effect of naloxone infusion." *Acta Endocrinology (Copenhagen)* 97, 150; Jacobs, M., and Lightman, S. 1980. "Studies in the opioid control of anterior pituitary hormones." *Journal of Physiology (London)* 300, 53; Rasmussen, D., Liu, J., Wolf, P., and Yen, S. 1983. "Endogenous opioid regulation of gonadotropin-releasing hormone release from the human fetal hypothalamus in vitro." *Journal of Clinical Endocrinology and Metabolism* 57, 881; Hulse, G., and Coleman, G. 1983. "The role of endogenous opioids in the blockade of reproductive function in the rat following exposure to acute stress." *Pharmacology, Biochemistry, and Behavior* 19, 795.

Exercise stimulates beta-endorphin release: Colt, E., Wardlaw, S., and Frantz, A. 1981. "The effect of running on plasma beta-endorphin." *Life Science* 28, 1637. For an interesting demonstration of the potential for this release to disrupt reproduction, see McArthur, J., Bellen, B., Beitins, T., Pagaon, M., Badger, T., and Klibanski, A. 1980. "Hypothalamic amenorrhea in runners of normal body composition." *Endocrine Research Communications* 7, 13. This study examined an amenorrheic runner with low LH levels; when she was given a drug (naloxone) that blocked beta-endorphin's actions, LH levels rose. Also see Samuels, M., Sanborn, C., Hofeldt, F., and Robbins, R. 1991. "The role of endogenous opiates in athletic amenorrhea." *Fertility and Sterility* 55, 507.

A moderate amount of exercise will stimulate testosterone levels: Elias, M. 1981. "Cortisol, testosterone and testosterone-binding globulin responses to competitive fighting in human males." *Aggressive Behavior* 7, 215. By contrast, sustained major exercise suppresses the system: Dessypris, A., Kuoppasalmi, K., and Adlercreutz, H. 1976. "Plasma cortisol, testosterone, androstenedione and luteinizing hormone (LH) in a noncompetitive marathon run." *Journal of Steroid Biochemistry* 7, 33; MacConnie, S., Barkan, A., Lampman, R., Schorok, M., and Beitins, I. 1986. "Decreased hypothalamic gonadotropin releasing hormone secretion in male marathon runners." *New England Journal of Medicine* 315, 411; Grandi, M., and Celani, M. 1990. "Effects of football on the pituitary-testicular axis: differences between professional and non-professional soccer players." *Experimental and Clinical Endocrinology* 96, 253; De Souza, M., Arce, J., Pescatello, L., Scherzer, H., and Luciano, A. 1994. "Gonadal hormones and semen quality in male runners. A volume threshold effect of endurance training." *International Journal of Sports Medicine* 15, 383.

PAGE 103 Similarly, major amounts of exercise suppress reproductive physiology in women. As one example, highly active ballet dancers have their onset of puberty delayed: Warren, M. 1980. "The effects of exercise on pubertal progression and reproductive function in girls." *Journal of Clinical Endocrinology and Metabolism* 51, 1150; Frisch, R., Wyshak, G., and Vincent, L. 1980. "Delayed menarche and amenorrhea in ballet dancers." *New England Journal of Medicine* 303, 17; Bale, P., Doust, J., and Dawson, D. 1996. "Gymnasts, distance runners, anorexics body composition and menstrual status." *Journal of Sports Medicine and Physical Fitness*, 36, 49. Amenorrhea occurs among women who exercise heavily: Kiningham, R., Apgar, B., and Schwenk, T. 1996. "Evaluation of amenorrhea." *American Family Physician* 53, 1185; Dale, E., Gerlach, D., and Wilhite, A. 1979. "Menstrual dysfunction in distance runners." *Obstetrics and Gynecology* 54, 47. In these cases, the degree of dysfunction is very tightly coupled to body weight or body fat content: Sanborn, C., Martin, B., and Wagner, W. 1982. "Is athletic amenorrhea specific to runners?" *American Journal of Obstetrics and Gynecology* 143, 859; Shangold, M., and Levine, H. 1982. "The effect of marathon training upon menstrual function." *American Journal of Obstetrics and Gynecology* 143, 862.

Some of the additional effects of overexercising. A moderate amount of exercise will increase bone density, particularly in the bones most heavily utilized in the exercise: Nilsson, B., and Westlin, N. 1971. "Bone density in athletes." *Clinical Orthopedics* 77, 179; Lanyon, L. 1989. "Bone loading, exercise, and the control of bone mass; the physiological basis for the prevention of osteoporosis." *Bone* 6, 19. Nevertheless, extremes of exercise can reverse this trend, leading to bone thinning, increased risk of osteoporosis, scoliosis, and stress fractures: Myburgh, K., Hutchins, J., Fataar, A., Hough, S., and Koakes, T. 1990. "Low bone density is an etiologic factor for stress fractures in athletes." *Annals of Internal Medicine* 113, 754; Drinkwater, B., Nilson, K., and Chesnut, C. 1984. "Bone mineral content of amoenorrheic and eumenorrheic athletes." *New England Journal of Medicine* 311, 277; Marcus, R., Cann, C., Madvig, P., Minkoff, J., Goddard, M., Bayer, M., Martin, M., Gaudiani, L., Haskell, W., and Genant, H. 1985. "Menstrual function and bone mass in elite women distance runners: endocrine and metabolic factors." *Annals of Internal Medicine* 102, 158; Barrow, G., and Saha, S. 1988. "Menstrual irregularity and stress fractures in collegiate female distance runners." *American Journal of Sports Medicine* 16, 209. The same occurs in male athletes: Bennell, K., Brukner, P., and Malcolm, S. 1996. "Effect of altered reproductive function and lowered testosterone levels on bone density in male endurance athletes." *British Journal of Sports Medicine* 30, 205. In prepubescent female athletes, the risks also include scoliosis: Warren, M., Brooks-Gunn, J., Hamilton, J., Warren, L., and Hamilton, G. 1986. "Scoliosis and fractures in young ballet dancers: relation to delayed menarche and secondary amenorrhea." *New England Journal of Medicine* 314, 1348.

These deleterious effects may be due, in part, to the elevated levels of glucocorticoids seen in serious athletes: Luger, A., Deuster, P., Kyle, S., Gallucci, W., Montgomery, L., Gold, P., Loriaux, L., and Chrousos, G. 1987. "Acute hypothalamic-pituitary-adrenal responses to the stress of treadmill exercise." *New England Journal of Medicine* 316, 1309; Willaneuva, A., Schlosser, C., Hopper, B., Liu, J., Hoffman, D., and Rebar, R. 1986. "Increased cortisol production in women runners." *Journal of Clinical Endocrinology and Metabolism* 63, 133; Loucks, A., Mortola, J., Girton, L., and Yen, S. 1989. "Alterations in the hypothalamic-pituitary-ovarian and the hypothalamic-pituitary-adrenal axes in athletic women." *Journal of Clinical Endocrinology and Metabolism* 68, 402. These cases documented pretty substantial increases in the levels of these hormones.

PAGE 104 Glucocorticoids work at the pituitary and the testes to block LH and testosterone release, respectively: Cummings, D., Quigley, M., and Yen, S. 1983. "Acute suppression of circulating testosterone levels by cortisol in men." *Journal of Clinical Endocrinology and Metabolism* 57, 671; Bambino, T., and Hseuh, A. 1981. "Direct inhibitory effect of glucocorticoids upon testicular luteinizing hormone receptors and steroidogenesis in vivo and in vitro." *Endocrinology* 108, 2142; Johnson, B., Welsh, T., and Juniewicz, P. 1982. "Suppression of luteinizing hormone and testosterone secretion in bulls following adrenocorticotropin hormone treatment." *Biology of Reproduction* 26, 305; Vierhapper, H., Waldhausl, W., and Nowotny, P. 1982. "Gonadotropin-secretion in adrenocortical insufficiency: impact of glucocorticoid substitution." *Acta Endocrinology (Copenhagen)* 580; Sapolsky, R. 1985. "Stress-induced suppression of testicular function in the wild baboon: role of glucocorticoids." *Endocrinology* 116, 2273.

Prolactin inhibits multiple steps in the male reproductive system: Bartke, A., Smith, M., Michael, S., Peron, F., and Dalterio, S. 1977. "Effects of experimentally-induced chronic hyperprolactinemia on testosterone and gonadotropin levels in male rats and mice." *Endocrinology* 100, 182; Bartke, A., Goldman, B., Bex, F., and Dalterio, S. 1977. "Effects of prolactin on pituitary and testicular function in mice with hereditary prolactin deficiency." *Endocrinology* 1760; McNeilly, A., Sharpe, R., and Fraser, H. 1983. "Increased sensitivity to the negative feedback effect of testosterone induced by hyperprolactinemia in the adult male rat." *Endocrinology* 112, 22.

PAGE 105 A good introductory summary of the basic workings of erections and ejaculation can be found in Previte, J. 1983. *Human Physiology* (New York: McGraw-Hill). A more detailed version is found in Guyton, A. 1986. *Textbook of Medical Physiology*, 7th ed. (Philadelphia: Saunders), 959. The parasympathetic neurotransmitter acetylcholine promotes erections: Saenz de Tejada, I., Blanco, R., Goldstein, I., Azadzoi, K., De Las Morenas, A., and Krane, R. 1988. "Cholinergic neurotransmission in human corpus cavernosum. I. Responses of isolated tissue." *American Journal of Physiology* 254, H459. The sympathetic neurotransmitter noradrenaline (norepinephrine) inhibits erections: Saenz de Tejada, I., Kim, N., Lagan, I., Krane, R., and Goldstein, I. 1989. "Regulation of adrenergic activity in penile corpus cavernosum." *Journal of Urology* 142, 1117. Just to make life and sex more complicated, researchers are coming to recognize that there are mechanisms for inducing erections that do not involve the parasympathetic nervous system. These are poorly understood, but it appears that these nerve endings make the arteries into the penis dilate (and thus engorge the penis with blood) by way of nitric oxide, a newly identified gaseous neurotransmitter that is closely related to nitrous oxide (laughing gas): Ignarro, L. 1992. "Nitric oxide as the physiological mediator of penile erection." *Journal of NIH Research* 4, 59.

PAGE 107 Incidences of psychogenic impotency: It remains controversial just how common this disorder is. Older studies reported that 90 to 95 percent of all cases of impotency were psychogenic in origin. For example, see Strauss, E. 1950. "Impotence from a psychiatric standpoint." *British Medical Journal* I, 697; or Kaplan, H. 1974. *The New Sex Therapy: Active Treatment of Dysfunctions* (New York: Brunner-Mazel). These numbers are almost certainly high, as they come from a time when many subtle organic causes of impotency were not yet understood. Some more recent studies report extremely low rates (perhaps 10 to 15 percent) of psychogenic impotency. For example, see Spark, R., White, R., and Connolly, P. 1980. "Impotence is not always psychogenic." *Journal of the American Medical Association* 243, 750. In general, recent studies indicate rates ranging from 14 percent of cases of impotency being psychogenic in nature to one study showing that 55 percent were, with another 15 percent being of unknown origin. These are summarized in Leiblum, S. and Rosen, R. 1989. *Principles and Practices of Sex Therapy* (New York: Guilford Press).

PAGE 107 For an introduction to revisionist ecology about hyenas (their role as hunters, rather than just scavengers) see Kruuk, H. 1972. *The Spotted Hyena: A Study of Predation and Social Behavior* (Chicago: University of Chicago Press). For studies of their anatomy, physiology, and behavior, see Frank, L. 1986. "Social organization of the spotted hyena: II. Dominance and reproduction." *Animal Behavior* 35, 1510; Frank, L., Glickman, S., and Licht, P. 1991. "Fatal sibling aggression, presocial development and androgens in neonatal spotted hyenas." *Science* 252, 702; Frank, L. 1997 "The evolution of female masculinization in hyenas: why does a female hyena have such a large penis?" *Trends in Ecology and Evolution*, 12, 58.

The final reference discusses the possible evolution of the unique hyena anatomy and social system. The most plausible scenario revolves around the fact that most large carnivores in Africa, such as lions, have large litters; relatively few of the offspring survive. Most starve to death, and this is because a lioness and her cubs are usually excluded from feeding on a kill until the males are sated (despite the fact that the females do the bulk of the hunting—there, one less feature to admire lions for).

By contrast, hyenas tend to have fewer progeny than these other carnivores. Suddenly the pressure is on to get those few to survive. Somewhere back when, a female hyena had a wondrous mutation—her ovaries started secreting huge amounts of the male sex hormone androstenedione, in addition to the normal estrogen. As a result, when she was pregnant, her female fetuses were exposed to the hormone, and as a result, they grew up more muscular and aggressive than typical female mammals; and the tables got turned. Within a few generations, the starving, intimidated male hyenas go and kill something, and just as they are about to gorge, the females boot them off. The kids of high-ranking moms eat before any adult males do; they survive. Thus the tendency in females toward secreting large amounts of androstenedione is highly adaptive, likely to be passed on over the generations.

There is one problem with this, however. Your average female mammal, exposed to those sorts of male sex hormone levels at birth, wouldn't be having kids. The androstenedione would have "masculinized" her hypothalamus, which is to say that as an adult, her hypothalamus would secrete LHRH at a roughly constant rate (as males do) instead of in the cyclic pattern that females need to ovulate. In any other species, this "perinatal androgenization" (masculinization around the time of birth) would make it impossible to reproduce.

Therefore, female hyenas are speculated to have a second mutation, one that protects the reproductive part of the hypothalamus from the masculinizing effects of the hormones. (By contrast, the "aggressive" part of the brain—a phrase that is obviously simplistic—is plenty sensitive to the adrostenedione: the female hyenas are terrifyingly aggressive.) At present, no one has a clue what that second mutation may be.

PAGE 110 General reviews of stress and female reproduction: Rivier, C. 1995. "Luteinizing-hormone-releasing hormone, gonadotropins, and gonadal steroids in stress." *Annals of the New York Academy of Sciences* 771,

187; Negro-Vilar, A. 1993. "Stress and other environmental factors affecting fertility in men and women: overview." *Environmental Health Perspectives* 101, S2, 59.

The subject of the effects of starvation, fat depletion, and muscle-to-fat ratios on female reproduction is reviewed in Frisch, R. 1991. "Body weight, body fat and ovulation." *Trends in Endocrinology and Metabolism* 2, 191. This review also gives a good introduction to the reproductive abnormalities seen in anorexia nervosa. Anorexia and the related eating disorder bulimia are peculiar in that more is going on than just loss of weight. Specifically, reproductive suppression occurs even before there is substantial weight loss; in other words, the reproductive systems of anorexics and bulimics are more vulnerable to such suppression than those of healthy women and girls.

PAGE 112 Opiates and opioids inhibit LHRH release in the female: Pfeiffer, A., and Herz, A. 1984. "Endocrine actions of opioids." *Hormone and Metabolic Research* 16, 386; Ching, M. 1983. "Morphine suppresses the proestrus surge of GnRH in pituitary portal plasmas of rats." *Endocrinology* 112, 2209. (GnRH, LHRH, and LHRF all refer to the same hypothalamic hormone, which causes release of LH and FSH from the pituitary.) For an interesting example of how this is relevant to female athletes, see McArthur, J., Bullen, B., Beitins, T., Pagaon, M., Badger, T., and Klibanski, A. 1980. "Hypothalamic amenorrhea in runners of normal body composition." *Endocrine Research Communications* 7, 13. This study examined an amenorrheic runner with low LH levels; when she was given a drug (naloxone) that blocked the action of beta-endorphin, LH levels rose. See immediately above in the male section for additional references regarding disrupted reproductive physiology in female athletes. An additional neurotransmitter seems to be implicated in stress-induced suppression of LHRH release: Akema, T., Chiba, A., Shinozaki, R., Oshida, M., Kimura, F., and Toyoda, J. 1995. "Acute stress suppresses the *N*-methyl-D-aspartate-induced LH release in the ovariectomized estrogen-primed rat." *Neuroendocrinology*, 62, 270. (The authors did not measure LHRH directly, but were able to infer it indirectly through a complicated trick in their LH measurements.)

Glucocorticoids suppress the responsiveness of the pituitary to LHRH: Suter, D., and Schwartz, N. 1985. "Effects of glucocorticoids on secretion of luteinizing hormone and follicle-stimulating hormone by female rat pituitary cells in vitro." *Endocrinology* 117, 849. References above show how glucocorticoid levels are elevated in heavily-exercising female athletes.

The follicular stage of the menstrual cycle is more vulnerable to disruption than the luteal phase: This is reported in many places. For an accessible version, see Hatcher, R. 1984. *Contraceptive Technology, 1984–85* (New York: Irvington Publishers). For a more detailed account, see Speroff, L., Glass, R., and Kase, N. 1989. *Clinical Gynecologic Endocrinology and Infertility* (Baltimore: Williams and Wilkins).

PAGE 113 The assertion that breast feeding prevents more pregnancies than any other type of contraception comes from Carl Djerassi, the chemist who invented the pill and has spent much of the rest of his extraordinary career studying the social, economic, and political consequences of the revolution he caused, in *The Politics of Contraception* (San Francisco: W. H. Freeman, 1979).

PAGE 114 Nursing, prolactin, and the Kalahari bushmen: Konner, M., and Worthman, C. 1980. "Nursing frequency, gonadal function, and birth spacing among !Kung hunter-gathers." *Science* 207, 788. The paper reviews what is known about how quickly prolactin rises in response to breast feeding and how long it stays up following the end of an episode of nursing. The Kalahari !Kung have been the darlings of anthropologists for decades, and they are often considered to be the prototypical hunter-gatherer society. Their "affluent" preagricultural life has been described in Lee, R. 1979. *!Kung San: Men, Women and Work in a Foraging Society* (New York: Cambridge University Press); Lee, R., and DeVore, I. 1976. *Kalahari Hunter-Gatherers* (Cambridge, Mass.: Harvard University Press); Jenkins, T., and Nurse, G. 1978. *Health and the Hunter-Gatherers* (Basel: Karger); Marshall, L. 1976. *The !Kung of Nyae Nyae* (Cambridge, Mass.: Harvard University Press); Shostak, M. 1981. *Nisa: The Life and Words of a !Kung Woman* (Cambridge, Mass.: Harvard University Press). There has been some questioning of just how typical they are of hunter-gatherers: Lewin, R. 1988. "New views emerge on hunters and gatherers." *Science* 240, 1146. The link among westernized women between a large number of menstrual cycles and a proclivity toward gynecological diseases is discussed by MacDonald, P., Dombroski, R., and Casey, M. 1991. "Recurrent secretion of progesterone in large amounts: an endocrine/metabolic disorder unique to young women?" *Endocrine Reviews* 12, 372.

PAGE 116 The effects of stress on female libido are discussed in two chapters by Sue Carter, "Neuroendocrinology of sexual behavior in the female" and "Hormonal influences on human sexual behavior," both in Becker, J., Breedlove, S., and Crews, D. (eds.), *Behavioral Endocrinology* (Cambridge, Mass.: MIT Press, 1992). Also see Rose, R. 1985. "Psychoendocrinology." In Wilson, J., and Foster, D. (eds.). *Williams Textbook of Endocrinology*, 7th ed. (Philadelphia: Saunders).

PAGE 118 Stressfulness of infertility: Domar, A., Zuttermeister, P., and Friedman, R. 1993. "The psychological impact of infertility: a comparison with patients with other medical conditions." *Journal of Psychosomatic Obstetrics and Gynaecology* 14, S45. These authors found depression rates equal to those seen in women with cancer, although less than in those with AIDS. Also see: van Balen, F., and Trimbos-Kemper, T. 1993. "Long-term infertile couples: a study of their well-being." *Journal of Psychosomatic Obstetrics and Gynaecology* 14, S53;

Stressfulness of IVF procedures: Boivin, J., and Takefman, J. 1996. "Impact of the in vitro fertilization process on emotional, physical and relational variables." *Human Reproduction* 11, 903; Harlow, C., Fahy, U., Talbot, W., Wardle, P., and Hull, M. 1996. "Stress and stress-related hormones during in vitro fertilization treatment." *Human Reproduction* 11, 274.

More stressed or depressed women are less likely to have successful IVFs: Facchinetti, F., Matteo, M., Artini, G., Volpe, A., and Genazzani, A. 1997. "An increased vulnerability to stress is associated with a poor outcome of in vitro fertilization-embryo transfer treatment." *Fertility and Sterility* 67, 309; Boivin, J., and Takefman, J. 1995. "Stress level across stages of in vitro fertilization in subsequently pregnant and nonpregnant women." *Fertility and Sterility* 64, 802; Thiering, P., Beaurepaire, J., Jones, M., Saunders, D., and Tennant, C. 1993. "Mood state as a predictor of treatment outcome after in vitro fertilization/embryo transfer technology." *Journal of Psychosomatic Research* 37, 481; Demyttenaere, K., Nijs, P., Evers-Kieybooms, G., Koninckx, P. 1994. "Personality characteristics, psychoendocrinological stress and outcome of IVF depend upon the etiology of infertility." *Gynecological Endocrinology* 8, 233. This last study was the one that showed that the stress/success link depended on the type of infertility. No relationship between stress and IVF outcome: Harlow, C., Fahy, U., Talbot, W., Wardle, P., and Hull, M. 1996. "Stress and stress-related hormones during in vitro fertilization treatment." *Human Reproduction* 11, 274.

PAGE 120 Hippocrates' advice to pregnant women is noted in Huisjes, H. 1984. *Spontaneous Abortion* (Edinburgh: Churchill Livingstone), 108. Anne Boleyn's attribution is found in Ives, E. 1986. *Anne Boleyn* (Oxford: Basil Blackwell, Ltd.). George Eliot's *Middlemarch* (London: Zodiac Press, 1982 ed.), 557. Also see Hansteen, I. 1990. "Occupational and lifestyle factors and chromosomal aberrations of spontaneous abortions." In *Mutation and the Environment*, Part B (New York: Wiley-Liss, Inc.), 467. Much of this paper reviews the links between various occupational hazards and increased risk of miscarriage; however, it also presents epidemiologic data linking stressful lifestyles to increased rates of miscarriage. For further links between stress and either pregnancy complications or miscarriage, see Vartiainen, H., Suonio, S., Halonen, P., and Rimon, R. 1994. "Psychosocial factors, female fertility and pregnancy: a prospective study—Part II: Pregnancy." *Journal of Psychosomatic Obstetrics and Gynaecology* 15, 77; O'Hare, T., and Creed, F. 1995. "Life events and miscarriage." *British Journal of Psychiatry* 167, 799; Lederman, R. 1995. "Relationship of anxiety, stress and psychosocial development to reproductive health." *Behavioral Medicine* 21, 101.

PAGE 121 Competitive infanticide in animals is reviewed in Hausfater, G., and Hrdy, S. 1984. *Infanticide: Comparative and Evolutionary Perspectives* (Hawthorne, N.Y.: Aldine). Harassment and abortion: Berger, J. 1983. "Induced abortion and social factors in wild horses." *Nature* 303, 59; Pereira, M. 1983. "Abortion following the immigration of an adult male baboon (*Papio cynephalus*)." *American Journal of Primatology* 4, 93; Alberts, S., Sapolsky, R., and Altmann, J. 1992. "Behavioral, endocrine, and immunological correlates of immigration by an aggressive male into a natural primate group." *Hormones and Behavior*, in press. Olfactory-induced abortions in rodents: Bruce, H. 1959. "An exteroceptive block to pregnancy in the mouse." *Nature* 184, 105; de Cantanzaro, D., Muir, C., O'Brien, J., and Williams, S. 1995. "Strange-male-induced pregnancy disruption in mice: reduction of vulnerability by 17 beta-estradiol antibodies." *Physiology and Behavior* 58, 401.

PAGE 122 Miscarriages typically occur many days to weeks after the death of the fetus: Chapter 24, "Abortions," in Pritchard, J., MacDonald, P., and Gant, N. 1985. *Williams Obstetrics*, 17th ed. (East Norwalk, Conn.: Appleton-Century-Crofts). For a good review of the possible mechanisms of stress-induced miscarriage, see Myers, R. 1979. "Maternal anxiety and fetal death." In Ziochella, L., and Pancheri, P. (eds.). *Psychoneuroendocrinology in Reproduction* (New York: Elsevier). The notion that decreased blood flow to the fetus can be the mechanism underlying miscarriage is found in Lapple, M. 1988. "Stress as an explanatory model for spontaneous abortions and recurrent spontaneous abortions." *Zentralblatt fur Gynakologie* 110, 325 (in German).

PAGE 123 The Kenyan birth rate: Hatcher, J., Kowal, N., Guest, S., Trussell, J., Stewart, M., Stewart, N., Bowen, T., and Cates, J. *Contraceptive Technology: International Edition* (Atlanta, Ga.: Printed Matter, Inc.), 21. Hutterite studies: Eaton, J., and Mayer, A. 1953. "The social biology of very high fertility among the Hutterites: the demography of a unique population." *Human Biology* 25, 206 (for an estimate of 9 children per family). See Frisch, R. 1978. "Population, food intake and fertility." *Science* 199, 22 (for an estimate of 10 to 12 kids per family).

PAGE 124 The Nazi studies of the women in the Theresienstadt death camp are discussed, without attribution, in Reichlin, S. 1974. "Neuroendocrinology." In Williams, R. (ed.). *Textbook of Endocrinology*, 6th ed. (Philadelphia: Saunders).

CHAPTER 8: Immunity, Stress, and Disease

PAGE 126 For an introduction to psychoimmunology, or psychoneuroimmunology (the study of the links among the nervous, endocrine, and immune systems), there is the multisyllabically titled review by A. Dunn. 1989. "Psychoneuroimmunology for the psychoneuroendocrinologist: a review of animal studies of nervous system–immune system interactions." *Psychoneuroendocrinology* 14, 251. The bible in the field is Ader, R., Felten, D., and Cohen, N. 1991. *Psychoneuroimmunology*, 2d ed. (San Diego: Academic Press)

Projections from the autonomic nervous system to immune organs, and presence of receptors for autonomic hormones in immune cells: Bulloch, K. 1985. "Neuroanatomy of lymphoid tissue: a review." *Neural Modulation of Immunity* 111.

Psychoimmunology of trained actors: Futterman, A., Kemeny, M., Shapiro, D., and Fahey, J. 1994. "Immunological and physiological changes associated with induced positive and negative mood." *Psychosomatic Medicine* 56, 499.

PAGE 128 Most college physiology textbooks will have introductions to the workings of the immune system. For those who want even more, a good introductory text for immunology is Benjamini, E., and Leskowitz, S. 1991. *Immunology: A Short Course*, 2d ed. (New York: Wiley-Liss).

PAGE 132 Recent reviews on the ability of stress to inhibit the immune system: Cohen, S., and Herbert, T. 1996. "Health psychology: psychological factors and physical disease from the perspective of human psychoneuroimmunology." *Annual Review of Psychology* 47, 113; Coe, C. 1993. "Psychosocial factors and immunity in nonhuman primates: a review." *Psychosomatic Medicine* 55, 298; Herbert, T., and Cohen, S. 1993. "Stress and immunity in humans: a meta-analytic review." *Psychosomatic Medicine* 55, 364.

Effects of glucocorticoids on the immune system: the most up-to-date and masterly of reviews can be found in McEwen, B., Biron, C., Brunson, K., Bulloch, K., Chambers, W., Dhabhar, F., Goldfarb, R., Kitson, R., Miller, A., Spencer, R., and Weiss, J. 1997. "The role of adrenocorticoids as modulators of immune function in health and disease: neural, endocrine and immune interactions." *Brain Research Reviews* 23, 79. For some of the most recent molecular findings regarding how glucocorticoids suppress the release of immune messengers, see Scheinman, R., Cogswell, P., Lofquist, A., and Baldwin, A. 1995. "Role of transcriptional activation of IkNFkappaB in mediation of immunosuppression by glucocorticoids." *Science* 270, 283; and Auphan, N., DiDonato, J., Rosette, C., Helmberg, A., and Karin, M. 1995. "Immunosuppression by glucocorticoids: inhibition of NF-KB activity through induction of IkB synthesis." *Science* 270, 286. (Note that this is another case of a pair of papers—from two groups on opposite sides of the globe—reporting the same novel finding in the same week).

Glucocorticoids kill cells of the immune system in many species and do so by causing the DNA to be chopped into small pieces. This has been shown in many studies; some of the classic ones are Wyllie, A. 1980. "Glucocorticoid-induced thymocyte apoptosis is associated with endogenous endonuclease activation." *Nature* 284, 555; Cohen, J., and Duke, R. 1984. "Glucocorticoid activation of a calcium-dependent endonuclease in thymocyte nuclei leads to cell death." *Journal of Immunology* 132, 38; Compton, M., and Cidlowski, J. 1986. "Rapid in vivo effects of glucocorticoids on the integrity of rat lymphocyte genomic DNA." *Endocrinology* 118, 38. As noted throughout the chapter, a frequent question runs along the line of "Okay, so if you inject an animal with a ton of glucocorticoids and you mess up its immune system in some way (in this case, by killing lymphocytes), is that a 'physiological' effect—will the smaller amounts of glucocorticoids secreted during stress (or stress itself) do the same thing?" The last paper also presents a small amount of data suggesting that stress will damage lymphocytes in the same way: Compton, M., Haskill, J., and Cidlowski, J. 1988. "Analysis of glucocorticoid actions on rat thymocyte DNA by fluorescence-activated flow cytometry." *Endocrinology* 122, 2158.

PAGE 133 Sympathetic role in suppressing immunity: Hori, T., Katafuchi, T., Take, S., Shimizu, N., and Nijima, A. 1995. "The autonomic nervous system as a communication channel between the brain and the immune system." *Neuroimmunomodulation* 2, 203; role of beta-endorphin: Shavit, Y., Lewis, J., and Terman, G. 1984. "Opioid peptides mediate the suppressive effect of stress on natural killer cell cytotoxicity." *Science* 3 233, 188; role of CRF: Irwin, M., Vale, W., and Rivier, C. 1990. "Central CRF mediates the suppressive effect of footshock stress on natural cytotoxicity." *Endocrinology* 126, 2837. Glucocorticoids don't play a role in some instances of immunosuppression: Gust, D., Gordon, T., and Wilson, M. 1992. "Removal from natal social group to peer housing affects cortisol levels and absolute numbers of T cell subsets in juvenile rhesus monkeys." *Brain Behavior and Evolution* 6, 189; Manuck, S., Cohen, S., Rabin, B., Muldoon, M., and Bachen, E. 1991. "Individual differences in cellular immune response to stress." *Psychological Sciences*, 2, 111; Keller, S., Weiss, J., Schleifer, S., Miller, N., and Stein, M. 1983. "Stress-induced suppression of immunity in adrenalectomized rats." *Science* 221, 1301.

PAGE 134 The idea that not all the traits of an organism are necessarily sculpted by evolution to be adaptive runs through much of Stephen Jay Gould's writing. It is most succinctly presented in "The spandrels of San Marco and the Panglossian paradigm: a critique of the adaptationist programme." Written with the geneticist Richard Lewontin, 1979. *Proceedings of the Royal Society of London* B 205.

Interleukin-1 causes the release of CRF from the hypothalamus: Sapolsky, R., Rivier, C., Yamamoto, G., Plotsky, P., and Vale, W. 1987. "Interleukin-1 stimulates the secretion of hypothalamic corticotropin-releasing factor." *Science* 238, 522; Berkenbosch, F., van Oers, J., del Rey, A., Tilders, F., and Besedovsky, H. 1987. "Corticotropin-releasing factor-producing neurons in the rat activated by interleukin-1." *Science* 238, 524. Just to make things worse, the same issue contained a report that IL-1 works at the level of the pituitary, rather than the hypothalamus in the brain, to stimulate the stress-response: Bernton, E., Beach, J., Holaday, J., Smallridge, R., and Fein, H. 1987. "Release of multiple hormones by a direct action of interleukin-1 on pituitary cells." *Science* 238, 519. I think a vague consensus is emerging in the field that the effect on the hypothalamus occurs reproducibly in an animal, whereas the pituitary effect depends on the use of pituitary cells in a petri dish (rather than in the living animal) and on the conditions under which the cells are grown in the dish.

PAGE 135 Short-term stress stimulates immunity: Berkenbosch, F., Heijnen, C., and Croiset, G. 1986. "Endocrine and immunological responses to acute stress." In: Plotnikoff, N., Faith, R., Murgo, A., and Good, R. (ed.). *Enkephalins and Endorphins: Stress and the Immune System.* (New York: Plenum Press); Croiset, G., Heijnen, C., and Veldhuis, H. 1987. "Modulation of the immune response by emotional stress." *Life Sciences* 40, 775; Dhabhar, F., and McEwen, B. 1996. "Stress-induced enhancement of antigen-specific cell-mediated immunity." *Journal of Immunology* 156, 2608; Weiss, J., Sundar, S., Becker, K., and Cierpial, M. 1989. "Behavioral and neural influences on cellular immune responses: effects of stress and interleukin-1." *Journal of Clinical Psychiatry* 50, 43; Herbert, T., Cohen, S.,

Marsland, A., Bachen, E., and Rabin, B. 1994. "Cardiovascular reactivity and the course of immune response to an acute psychological stressor." *Psychosomatic Medicine* 56, 337; Herbert, T., and Cohen, S. 1993. "Stress and immunity in humans: a meta-analytic review." *Psychosomatic Medicine* 55, 364 (see Tables 2 and 3).

PAGE 135 This short-term increase is mediated by sympathetic hormones: Bachen, E., Manuck, S., Cohen, S., Muldoon, M., and Raible, R. 1995. "Adrenergic blockage ameliorates cellular immune responses to mental stress in humans." *Psychosomatic Medicine,* in press; Landmann, R., Muller, F., and Perini, C. 1984. "Changes of immunoregulatory cells induced by psychological and physical stress: relationship to plasma catecholamines." *Clinical and Experimental Immunology* 58, 127; Ernstrom, U., and Sandberg, G. 1973. "Effects of alpha- and beta-receptor stimulation on the release of lymphocytes and granulocytes from the spleen." *Scandinavian Journal of Hematology* 11, 275. Involvement of glucocorticoids: Bateman, A., Singh, A., Kral, T., and Solomon, S. 1989. "The immune hypothalamic-pituitary-adrenal axis." *Endocrine Reviews* 10, 92. McEwen, B., Biron, C., Brunson, K., Bulloch, K., Chambers, W., Dhabhar, F., Goldfarb, R., Kitson, R., Miller, A., Spencer, R., and Weiss, J. 1997. "The role of adrenocorticoids as modulators of immune function in health and disease: neural, endocrine and immune interactions." *Brain Research Reviews* 23, 79.

PAGE 136 Munck's ideas: Munck, A., Guyre, P., and Holbrook, N. 1984. "Physiological actions of glucocorticoids in stress and their relation to pharmacological actions." *Endocrine Reviews* 5, 25. This is probably the most influential paper written about glucocorticoids in the last quarter-century.

Blockade of the glucocorticoid-mediated recovery from stress-induced activation of the immune system is associated with autoimmunity: Wick, G., Hu, Y., Schwarz, S., and Kroemer, G. 1993. *Endocrine Reviews* 14, 539; Sternberg, E., Chrousos, G., Wilder, R., and Gold, P. 1992. "The stress response and regulation of inflammatory disease." *Annals of Internal Medicine* 117, 854; Rose, N., Bacon, L., and Sundick, R. 1976. "Genetic determinants of thyroiditis in the OS chicken." *Transplantation Reviews* 31, 264–270; Takasu, N., Komiya, I., Nagasawa, Y., Aaswa, T., and Yamada, T. 1990. "Exacerbation of autoimmune thyroid dysfunction after unilateral adrenalectomy in patients with Cushing's syndrome due to adrenocortical adenoma." *New England Journal of Medicine* 322, 1708–1712; Harbuz, S., and Lightman, S. 1992. "Stress and the hypothalamo-pituitary-adrenal axis: acute, chronic and immunological activation." *Journal of Endocrinology* 134, 327–339; Green, M., and Lim, K. 1971. "Bronchial asthma with Addison's disease." *Lancet* 1, 1159–1165.

The notion of glucocorticoids sculpting the immune response can be found in Besedovsky, H., DelRay, S., Sorkin, E., and Dinarello, C. 1986. "Immunoregulatory feedback between interleukin-1 and glucocorticoid hormones." *Science* 233, 652; Besedovsky, H., and Del, Ray A. 1996. "Immuno-neuro-endocrine interactions: facts and hypotheses." *Endocrine Reviews* 17, 64.

PAGE 137 Glucocorticoids as causing a salutary redistribution of lymphocytes: Dhabhar, F., and McEwen, B. 1996. "Stress-induced enhancement of antigen-specific cell-mediated immunity." *Journal of Immunology* 156, 2608; McEwen, B., Biron, C., Brunson, K., Bulloch, K., Chambers, W., Dhabhar, F., Goldfarb, R., Kitson, R., Miller, A., Spencer, R., and Weiss, J. 1997. "The role of adrenocorticoids as modulators of immune function in health and disease: neural, endocrine and immune interactions." *Brain Research Reviews* 23, 79.

Reports that stress can exacerbate some autoimmune diseases: Leclere, J., and Weryha, G. 1989. "Stress and auto-immune endocrine diseases." *Hormone Research* 31, 90. Weiner, H. 1991. "Social and psychobiological factors in autoimmune diseases." In Ader, R., Felten, D., and Cohen, N. (eds.). *Psychoneuroimmunology*, 2d ed. (San Diego: Academic Press); Chiovato, L., and Pinchera, A. 1996. "Stressful life events and Graves' disease." *European Journal of Endocrinology* 134, 68; Rosch, P. 1993. "Stressful life events and Graves' disease." *Lancet* 342, 566; Rimon, R., Belmaker, R., and Ebstein, R. 1977. "Psychosomatic aspects of juvenile rheumatoid arthritis." *Scandinavian Journal of Rheumatology* 6, 1; Homo-Delarche, F., Fitzpatrick, F., Christeff, N., Nunez, E., Bach, J., and Dardenne, M. 1991. "Sex steroids, glucocorticoids, stress and autoimmunity." *Journal of Steroid Biochemistry and Molecular Biology* 40, 619. A report of no effect of stress: Nispeanu, P., and Korczyn, A. 1993. "Psychological stress as risk factor for exacerbations in multiple sclerosis." *Neurology* 43, 1311. In contrast: Warren, S., Greenhill, S., and Warren, K. 1982. "Emotional stress and the development of multiple sclerosis: case-control evidence of a relationship." *Journal of Chronic Disease* 35, 821.

PAGE 142 Social relationships are associated with decreased mortality rates: House, J., Landis, K., and Umberson, D. 1988. "Social relationships and health." *Science* 241, 540. Social stressors are particularly immunosuppressive: Herbert, T., and Cohen, S. 1993. "Stress and immunity in humans: a meta-analytic review." *Psychosomatic Medicine* 55, 364. Berkman, L. 1983. *Health and Ways of Living: Findings from the Alameda County Study* (New York: Oxford University Press). Lonelier individuals had less natural killer cell activity: Kiecolt-Glaser, J., Garner, W., Speicher, C., Penn, G., and Glaser, R. 1984. "Psychosocial modifiers of immunocompetence in medical students." *Psychosomatic Medicine* 46, 7.

Factors like divorce or marital discord are associated with suppressed aspects of immune function. Reviewed in Kiecolt-Glaser, J., and Glaser, R. 1991. "Stress and immune function in humans." In Ader, R., Felten, D., and Cohen, N. (eds.). *Psychoneuroimmunology*, 2d ed. (San Diego: Academic Press).

Some subtle confounds in lifestyle-disease relationships: some particularly thoughtful discussions can be found in House, J., Landis, K., and Umberson, D. 1988. "Social relationships and health." *Science* 241, 540. Less medical compliance among the socially isolated: Williams, C. 1985 "The Edgecomb County high blood pressure control program: III. Social support, social stressors, and treatment dropout." *American Journal of Public Health* 75, 483.

Social support helps the immune system of primates: Cohen, S., Kaplan, J., Cunnick, J., Manuck, S., and Rabin, B. 1992. "Chronic social stress, affiliation, and cellular immune response in nonhuman primates." *Psychological Science*, 3, 301. Social isolation suppresses primate immunity: Laudenslager, M., Capitanio, J., and Reite, M. 1985. "Possible effects of early separation experiences on subsequent immune function in adult macaque monkeys." *American Journal of Psychiatry* 142, 862; Coe, C. 1993. "Psychosocial factors and immunity in nonhuman primates: a review." *Psychosomatic Medicine* 55, 298.

PAGE 144 Bereavement: Bereavement decreases immune function and increases the risk for mortality: Kiecolt-Glaser, J., and Glaser, R. 1991. "Stress and immune function in humans." In Ader, R., Felten, D., and Cohen, N. (eds.). *Psychoneuroimmunology*, 2d ed. (San Diego: Academic Press); Levav, I., Friedlander, Y., Kark, J., and Peritz, E. 1988. "An epidemiological study of mortality among bereaved parents." *New England Journal of Medicine* 319, 457.

Bereavement among HIV patients: Goodkin, K., Feaster, D., Tuttle, R., Blaney, N., Kumar, M., Baum, M., Shapshak, P., and Fletcher, M. 1996. "Bereavement is associated with time-dependent decrements in cellular immune function in asymptomatic HIV type 1-seropositive homosexual men." *Clinical and Diagnostic Laboratory Immunology* 3, 109; Kemeny, M., Weiner, H., Duran, R., Taylor, S., Visscher, B., and Fahey, J. 1995. "Immune system changes after the death of a partner in HIV-positive gay men." *Psychosomatic medicine* 57, 547; Kemeny, M., and Dean, L. 1995. "Effects of AIDS-related bereavement on HIV progression among New York City gay men." *AIDS Education and Prevention* 7, 36.

PAGE 145 Stress and the common cold: Cohen, S., Tyrrell, D., and Smith, A. 1991. "Psychological stress and susceptibility to the common cold." *New England Journal of Medicine* 325, 606. This paper came out of the famed Common Cold Unit of the Medical Research Council in Salisbury, England, which recruited volunteers for their frequent two-week experiments about various aspects of coming down with and recovering from the common cold. Apparently quite an experience: all expenses covered plus a small salary, many recreational activities in the peaceful Salisbury countryside, daily blowing of noses into collection tubs for the staff, questionnaires to fill out, and being spritzed up the nose with either placebo or a cold-causing virus. One in three chance, on the average, of getting a cold while there. People would compete for slots as volunteers; couples have met there, married, returned for honeymoons; folks with connections would maneuver for return visits, making it an annual paid vacation. (All was not idyllic sniffling heaven at the Cold Unit, however. An occasional group would be involved in studies showing that, for example, being chilled and damp does not cause colds, and would have to stand around for hours in wet socks.) Unfortunately, because of budget limitations, the unit has been closed. For a particularly amusing account of the place, see

Roach, M. 1990. "How I blew my summer vacation." *Health* (January–February), 73.

Stress and the common cold in nonhuman primates: Cohen, S., Line, S., Manuck, S., Rabin, B., Heise, E., and Kaplan, J. 1997. "Chronic social stress, social status and susceptibility to upper respiratory infections in nonhuman primates." *Psychosomatic Medicine* 59, 213.

PAGE 146 Studies showing stress-cancer links with laboratory animals: Stress increases the rate of spontaneous tumors in mice: Henry, J., Stephens, V., and Watson, F. 1975. "Forced breeding, social disorder, and mammary tumor formation in CBA/USC mouse colonies: a pilot study." *Psychosomatic Medicine* 37, 277. Stress accelerates tumor growth in rats: Sklar, L., and Anisman, H. 1979. "Stress and coping factors influence tumor growth." *Science* 205, 513. Riley, V. 1981. "Psychoneuroendocrine influences on immunocompetence and neoplasia." *Science* 212, 1100. Visintainer, M., Volpicelli, J., and Seligman, M. 1982. "Tumor rejection in rats after inescapable or escapable shock." *Science* 216, 437. Sapolsky, R., and Donnelly, T. 1985. "Vulnerability to stress-induced tumor growth increases with age in rats: role of glucocorticoids." *Endocrinology* 117, 662. Tumor growth rates in rodents can be accelerated by housing them in stressful conditions, giving them rotational stress, and giving them glucocorticoids: Riley, V. 1981. "Psychoneuroendocrine influences on immunocompetence and neoplasia." *Science* 212, 1100. Tumor growth can also be accelerated by inescapable shock: Visintainer, M., Volpicelli, J., and Seligman, M. 1982. "Tumor rejection in rats after inescapable or escapable shock." *Science* 216, 437. Discussions of some of the limits of this literature: the links between stress and cancer mostly involve induced tumors, acceleration of tumor growth rather than initial establishment of tumors, and virally derived tumors: Fitzmaurice, M. 1988. "Physiological relationships among stress, viruses, and cancer in experimental animals." *International Journal of Neuroscience* 39, 307; Justice, A. 1985. "Review of the effects of stress on cancer in laboratory animals: importance of time of stress application and type of tumor." *Psychological Bulletin* 98, 108.

The effects of stress make sense in the context of the biology of tumorigenesis. Stress and glucocorticoid effects on natural killer cell activity: Munck, A., and Guyre, P. 1991. "Glucocorticoids and immune function." In Ader, R., Felten, D., and Cohen, N. (eds.). *Psychoneuroimmunology*, 2d ed. (San Diego: Academic Press); effects on angiogenesis: Folkman, J., Langer, R., Linhardt, R., Haudenschild, C., and Taylor, S. 1983. "Angiogenesis inhibition and tumor regression caused by heparin or a heparin fragment in the presence of cortisone." *Science* 221, 719. Effects of glucocorticoids on tumor metabolism: Romero, L., Raley-Susman, K., Redish, K., Brooke, S., Horner, H., and Sapolsky, R. 1992. "A possible mechanism by which stress accelerates growth of virally-derived tumors." *Proceedings of the National Academy of Sciences, USA* 89, 11084.

PAGE 148 No link between a history of stress and cancer in humans: the Western Electric study: Shekelle, R., Raynor, W., Ostfeld, A., Garron, D., Bieliauskas, L., Liu, S., Maliza, C., and Paul, O. 1981. "Psychological depression and 17-year risk of death from cancer." *Psychosomatic Medicine* 43, 117; Persky, V., Kempthorne-Rawson, J., and Shekelle, R. 1987. "Personality and risk of cancer: 20-year follow-up of the Western Electric Study." *Psychosomatic Medicine* 49, 435. It's revisionist debunking: Fox, B. 1989. Depressive symptoms and risk of cancer. *Journal of the American Medical Association* 262, 1231. Other studies showing no depression-cancer link: Kaplan, G., and Reynolds, P. 1988. Depression and cancer mortality and morbidity: prospective evidence from the Alameda County Study. *Journal of Behavioral Medicine* 11, 1; Hahn, R., and Petitti, D. 1988. "Minnesota Multiphasic Personality Inventory-rated depression and the incidence of breast cancer." *Cancer* 61, 845. One review (McGee, R., Williams, S., and Elwood, M. 1993. "Depression and the development of cancer; a meta-analysis." *Social Science and Medicine* 38, 187) surveyed all the studies in the field at the time and concluded that there was a tiny but significant link between depression and cancer, with depression increasing the cancer risk about 14 percent. However, one of the strongest effects in that meta-analysis was found in the since-discredited Western Electric study; once it is removed, any effect disappears.

The lack of links between other types of stressors and subsequent cancer is reviewed in Hilakivi-Clarke. L., Rowland, J., Clarke, R., and Lippman, M. 1993. "Psychosocial factors in the development and progression of breast cancer." *Breast Cancer Research and Treatment* 29, 141. This business of trying to find links between lifestyle or personality and some disease outcome years later is extremely tricky. For example, it was widely reported that cancer rates went up in the surrounding areas following the Three Mile Island nuclear accident in 1979. However, it involved all types of cancers, including those not thought to be sensitive to the radiation released. In that case, the subtle confound was probably that people were more anxious and thus more vigilant, taking more trips to the doctors, who, aware of the accident history, checked things more carefully—and thus spotted more cancer. (Pool, R. 1991. "A stress-cancer link following accident?" *Nature* 351, 429.)

PAGE 148 Cancer and personality. One of the most influential papers showing a link is: Temoshok, L., Heller, B., Sagebiel, R., Blois, M., Sweet, D., and DiClemente, R. 1985. "The relationship of psychosocial factors to prognostic indicators in cutaneous malignant melanoma." *Journal of Psychosomatic Research* 29, 139. Other papers on the subject are reviewed carefully in Spiegel, D., and Kato, P. 1996. "Psychosocial influences on cancer incidence and progression." *Harvard Review of Psychiatry* 4, 10; also Bryla, C. 1996 "The relationship between stress and the development of breast cancer: literature review." *ONF* 23, 441; Hilakivi-Clarke, L., Rowland, J., Clarke, R., and Lippman, M. 1993. "Psychosocial factors in the development and progression of breast cancer." *Breast Cancer Research and Treatment* 29, 141.

Once cancer has occurred: the helpful effects of a fighting spirit: Temoshok, L., and Fox, B. 1984. "Coping styles and other psychosocial factors related to medical status and to prognosis in patients with cutaneous malignant melanoma." In Fox, B., and Newberry, B. (eds.). *Impact of Psychoneurocrine System in Cancer and Immunity* (Toronto: Hogrefe), 86. These authors talk about the depression-prone individuals who collapse in the face of cancer as having a "type C" personality, a term that has caught on in this field to some extent. In many ways, these closely resemble the repressive individuals who appear to be somewhat cancer-prone in the first place. Just to add to the confusion, in addition to a fighting spirit, denial has been shown in some studies to be helpful as well (reviewed in Bauer, S. 1994. "Psychoneuroimmunology and cancer: an integrated review." *Journal of Advanced Nursing* 19, 1114).

PAGE 149 The Spiegel study concerning cancer survival and being in a support group: Spiegel, D., Bloom, J., and Kraemer, H. 1989. "Effect of psychosocial treatment on survival of patients with metastatic breast cancer." *Lancet* 2, 888. Other studies (both pro and con) on this issue are reviewed in Spiegel, D., and Kato, P. 1996. "Psychosocial influences on cancer incidence and progression." *Harvard Review of Psychiatry* 4, 10. The review also points out the literature showing the high rates of nonadherence to treatment regimes among cancer patients.

PAGE 150 Some very similar sentiments to those in Bernie Siegel's 1986 magnum opus, *Love, Medicine and Miracles,* can be found in other books, including one by one of Siegel's mentors: Simonton, O., Matthews-Simonton, S., and Creighton, J. 1978. *Getting Well Again* (Los Angeles: Tarcher, Inc.). The lack of an effect of Siegel's program on survivorship can be found in Morgenstern, H., Gellert, G., Walter, S., Ostfeld, A., and Siegel, B. 1984. "The impact of a psychosocial support program on survival with breast cancer: the importance of selection bias in program evaluation." *Journal of Chronic Disease* 37, 273; and Gellert, G., Maxwell, R., and Siegel, B. 1993. "Survival of breast cancer patients receiving adjunctive psychoso cial support therapy: a 10-year follow-up study." *Journal of Clinical Oncology* 11, 66. The program's lack of efficacy is pointed out in a 1992 debate between Siegel and David Spiegel (the physician whose work is discussed earlier in this chapter and who owns up to having sustained a fair amount of discomfort by having a name readily confused with Siegel's): "Psychosocial interventions and cancer." *Advances* 8, 2. The Herbert Weiner quotation comes from his 1992 book, *Perturbing the Organism: The Biology of Stressful Experience* (Chicago: University of Chicago Press).

PAGE 153 The lapsarian in the Reagan administration: in an extraordinary episode, a top appointee in the Department of Education turned out to hold lapsarian views. "There is no injustice in the universe," she wrote. "As unfair as it may seem, a person's external circumstances do fit his level of inner spiritual development. . . . [The handicapped] falsely assume that the lottery of life has penalized them at random. This is not so. 'Nothing comes to an individual that he has not [at some point in his development] summoned.'" (Second set of brackets her own.) She extended this philosophy to explaining why James Brady, Reagan's press secretary, had been seriously wounded in John Hinckley's assassination attempt. Her policy advice included terminating any special educational programs for the handicapped. Fortunately, she lasted three days in her new position before being returned to the conservative fundamentalist think tank from which she had emerged.

Testimony by and about the woman, Eileen Gardner of the conservative Heritage Foundation, can be found in the Senate Hearings Before the Committee on Appropriations, 99th Congress, First Session, 1986, HR 3424, part 3, Appropriations Hearings for the Departments of Labor, HHS, and Education, pages 74 and 177. The fiery hearings were reported in newspapers throughout the country (for example, *The New York Times,* 17–19 April 1985; *Washington Post,* 17 May 1985). In the Senate she expressed her view that sometimes congenital illnesses are visited upon newborn infants not so much on account of their own sinfulness as because of the sinfulness of the parents—all the while apparently unaware that the senator presiding over her hearings, Lowell Weicker of Connecticut, is the father of a congenitally retarded, institutionalized child and a passionate supporter of research into retardation and congenital abnormalities. Weicker, a veteran politician who probably is as knowledgeable as anyone can be about the corridors of power, described her testimony as "the most incredible thing I have read in my career in the United States Senate. I have never seen such callousness" (*The New York Times* 17 April 1985).

PAGE 157 The history of status thymicolymphaticus was originally published by me under the title "Poverty's remains." 1991. *The Sciences* (September–October), 8. The original observation of "enlarged" thymuses in SID infants was reported by Kopp, J. 1830. "Denkwurdigkeiten in der artzlichen Praxis," and was greatly expanded upon by Paltauf, A. 1889. *Plotzlicher Thymus Tod.* Wiener klin. Woechesucher, Berlin 46 and 9. The supposed disease was named a few years later in Escherich, T. 1896. *Status thymico-lymphaticus.* Berlin klin. Woechesucher 29. By the late 1920s it was in all the textbooks, complete with radiation advice (how much to administer, where to aim it, and so on). See, for example, Lucas, W. 1927. *Modern Practise of Pediatrics* (New York: Macmillan). Amid this generally grim story, I was amused to note that by the time of this textbook, the "disease" was so well established that the author now broke ground by describing the distinctive and striking behavioral features of infants who would later be found to have died of thymicolymphaticus. They were characterized as having "phlegmatic" dispositions—presumably because these were normal kids and thus were phlegmatic about their imaginary illnesses. It is a chilling experience to wander the dusty lower floor of a medical library, reading these forgotten texts and their confident discussions of this supposed disease. Page after page of errors. What similar mistakes are we making now?

Lost amid this consensus of the savants was a 1927 study by E. Boyd ("Growth of the thymus, its relation to status thymicolymphaticus and thymic symptoms." *American Journal of Diseases of Children* 33, 867), which should have put the whole thing to rest. Boyd showed for the first time that a stressor (malnutrition, in this case) caused thymic shrinking. She demonstrated, moreover, that some children who died of accidents turned out, upon autopsy, to be "suffering" from thymicolymphaticus, and suggested for the first time that the whole thing might be an artifact. It was not until the 1930s that the first of the pediatric textbooks began to voice the opinion that this conclusion might be correct; not until 1945 did the leading textbook in the field emphatically state that treating this "disease" was a disastrous thing to do [Nelson, W. 1945. *Nelson's Textbook of Pediatrics*, 4th ed. (Philadelphia: Saunders)]. In researching this subject, I had the pleasure of talking with the same Dr. Nelson, now in his nineties, still seeing inner-city children at the University of Pennsylvania Hospital every day and basking in the positive reviews of the recent edition of his classic textbook. He recalled how, by the early 1930s, the Young Turk pediatricians (one of whom he most surely was) were already contemptuous of the old guard for advocating something as crazy and outdated as radiating kids to prevent an imaginary disease. Despite that, the practice continued widely well into the 1950s.

For a discussion of how status thymicolymphaticus was a "progressive" advance in medicine in the 1800s (by substituting for simply blaming the parents), see Guntheroth, W. 1993. "The thymus, suffocation, and sudden infant death syndrome—social agenda or hubris?" *Perspectives in Biology and Medicine* 37, 2.

CHAPTER 9: Stress-Induced Analgesia

PAGE 159 The extended quotation comes from page 178 of Joseph Heller's *Catch-22* (New York: Simon and Schuster, 1955).

PAGE 160 Pain asymbolia (the inability to feel pain): Appenzeller, O., and Kornfeld, M. 1972. "Indifference to pain: a chronic peripheral neuropathy with mosaic Schwann cells." *Archives of Neurology* 27, 322; Murray, T. 1973. "Congenital sensory neuropathy." *Brain* 96, 387; Fox, J., Belvoir, F., and Huott, A. 1974. "Congenital hemihypertrophy with indifference to pain." *Archives of Neurology* 30, 490.

PAGE 161 A general review of pain pathways can be found in Hopkin, K. 1997. "Show me where it hurts: tracing the pathways of pain." *Journal of the National Institutes of Health Research* 9, (10) 37.

PAGE 162 The interactions of fast and slow pain fibers were first described in the classic paper by Melzack, R., and Wall, P. 1965. "Pain mechanisms: a new theory." *Science* 150, 971. They are elaborated in Wall, P., and Melzack, R. 1989. *Textbook of Pain*, 2d ed. (Edinburgh: Churchill Livingstone).

PAGE 165 Pain medication requests by gallbladder surgery patients: Ulrich, R. 1984. "View through a window may influence recovery from surgery." *Science* 224, 420.

PAGE 166 Most clinicians concerned with chronic pain syndromes are anecdotally familiar with stress-induced analgesia, and many basic neurology, neuroscience, or physiological psychology texts cover the subject—for example, see the chapter on pain by Dennis Kelly in Kandel, E., and Schwartz, J. (eds.). 1985. *Principles of Neural Science* (New York: Elsevier). This also contains the famous description of the phenomenon by Dr. David Livingstone upon the occasion of his being mauled by a lion. Also see Fields, H. 1987. *Pain* (New York: McGraw-Hill).

Requests for morphine by soldiers versus civilians: Beecher, H. 1956. "Relationship of significance of wound to pain experienced." *Journal of the American Medical Association* 161, 17.

PAGE 167 Stress-induced analgesia in animals: Terman, G., Shavit, Y., Lewis, J., Cannon, J., and Liebeskind, J. 1984. "Intrinsic mechanisms of pain inhibition: activation by stress." *Science* 226, 1270.

Opiates, opiate receptors, and opioids: for technical reviews on this subject, see Akil, H., Watson, S., Young, E., Lewis, M., Khachaturian, H., and Walker, J. 1984. "Endogenous opioids: biology and function." *Annual Review of Neuroscience* 7, 223; Basbaum, A., and Fields, H. 1984. "Endogenous pain control systems: brain stem spinal pathways and endorphin circuitry." *Annual Review of Neuroscience* 7, 309. For a surprisingly readable account of the history of this field, see Snyder, S. 1989. *Brainstorming: The Science and Politics of Opiate Research* (Cambridge: Harvard University Press). Snyder, one of the discoverers of the opiate receptor and a leading figure in the field, is an excellent nontechnical writer.

PAGE 168 The effects of acupuncture are mediated by opiate receptors: Mayer, D., Price, D., Barber, J., and Rafii, A. 1976. "Acupuncture analgesia: evidence for activation of a pain inhibitory system as a mechanism of action." In Bonica, J., and Albe-Fessard, D. (eds.) *Advances in Pain Research and Therapy*, vol. 1. (New York: Raven Press), 751; Mayer, D., and Hayes, R. 1975. "Stimulation-produced analgesia: development of tolerance and cross-tolerance to morphine." *Science* 188, 941.

PAGE 169 First demonstration of endorphin release during stress: Guillemin, R., Vargo, T., and Rossier, J. 1977. "Beta-endorphin and adrenocorticotropin are secreted concomitantly by pituitary gland." *Science* 197, 1367. Its stimulation by a variety of stressors: Colt, E., Wardlaw, S., and Frantz, A. 1981. "The effect of running on plasma beta-endorphin." *Life Sciences* 28, 1637; Cohen, M., Pickar, D., and Dubois, M. 1982. "Stress-induced plasma beta-endorphin immunoreactivity may predict postoperative morphine usage." *Psychiatry Research* 6, 7; Katz, E., Sharp, B., and Kellermann, J. 1982. "Beta-endorphin immunoreactivity and acute behavioral distress in children with leukemia." *Journal of Nervous and Mental Disease* 170, 72; Jungkunz, G., Engel, R., and King, U. 1983. "Endogenous opiates increase pain tolerance after stress in humans." *Psychiatry Research* 8, 13.

PAGE 169 Nonopioid mediated analgesia during stress: Mogil, J., Sternberg, W., Marek, P., Sadowski, B., Belknap, J., and Liebeskind, J. 1996. "The genetics of pain and pain inhibition." *Proceedings of the National Academy of*

Sciences, USA 93, 3048; Mogil, J., Marek, P., Yirmiya, R., Balian, H., Sadowski, B., Taylor, A., and Liebeskind, J. 1993. "Antagonism of the non-opioid component of ethanol-induced analgesia by the NMDA receptor antagonist MK-801." *Brain Research* 602, 126; Nakao, K., Takahashi, M., Kaneto, H. 1996. "Implications of ATP-sensitive K+ channels in various stress-induced analgesia in mice." *Japanese Journal of Pharmacology* 71, 269.

CHAPTER 10 Stress and Memory

PAGE 175 For some primers into the biology and neuropsychology of how memory works: Squire, L. 1987. *Memory and Brain* (New York: Oxford University Press); Gazzaniga, M. 1995. *The Cognitive Neurosciences* (Cambridge, Mass.: MIT Press; warning: this book is almost 1500 pages long); Hebb, D. O. 1947. *The Organization of Behavior* (New York: Wiley). This last book is something of a cult classic. Hebb was one of the great neuroscientists of all time and, in this one book, predicted how long-term potentiation and neural networks were going to work—long before any of the underlying biology had been sorted out. Basically, anything new that has come along in this field for decades was outlined somewhere in this 1947 book. The Squire book gives a good overview of H.M. and his extraordinary history.

PAGE 177 One of the Nobel prize–winning classics of Hubel and Wiesel: Hubel, D., and Wiesel, T. 1962. "Receptive fields, binocular interaction and functional architecture in the cat's visual cortex." *Journal of Physiology* (London) 160, 106.

PAGE 178 For an introduction to neural networks (and a lesson in how distorted and simplified this chapter's version of a network is) see Arbib, M. 1995. *The Handbook of Brain Theory and Neural Networks* (Cambridge, Mass.: MIT Press); also Taylor, J. 1996. *Neural Networks and Their Applications* (Chichester, England: Wiley). Also see: Fitzsimonds, R., Song, H., and Poo, M. 1997. "Propagation of activity-dependent synaptic depression in simple neural networks." *Nature* 388, 439.

PAGE 180 For introductions to long-term potentiation, see Gluck, M., and Meyers, C. 1997. "Psychobiological models of hippocampal function in learning and memory." *Annual Review of Psychology*, 48, 481. For a review of long-term depression, see Stevens, C. 1996. "Strengths and weaknesses in memory." *Nature* 381, 471; Nicoll, R., and Malenka, R. 1997. "Long-distance long-term depression." *Nature* 388, 427.

PAGE 181 For a broad review of the effects of stress on cognition, see McEwen, B., and Sapolsky, R. 1995. "Stress and cognitive function." *Current Opinion in Neurobiology* 5, 205. Stress enhancing memory in humans, and this mediated by sympathetic nervous system: Cahil, L., Prins, B., Weber, M., and McGaugh, J. 1994. "Beta-adrenergic activation and memory for emotional events." *Nature* 371, 702. Glucose, energy, and memory: reviewed in Gold, P. 1995. "Role of glucose in regulating the brain and cognition." *American Journal of Clinical Nutrition* 61, 987S. Sympathetic nervous system enhances neuronal energetics: McGaugh, J. 1989. "Involvement of hormonal and neuromodulatory systems in the regulation of memory storage." *Annual Review of Neuroscience* 12, 255. Glucocorticoids initially stimulate glucose uptake into the brain: Bryan, R., and King, J. 1988. "Glucocorticoids modulate the effect of plasma epinephrine on regional glucose utilization." *Society for Neuroscience Abstracts* 20, 427.11. A moderate increase in glucocorticoid levels enhances long-term potentiation: Diamond, D., Bennet, M., Fleshner, M., and Rose, G. 1992. "Inverted-U relationship between the level of peripheral corticosterone and the magnitude of hippocampal primed burst potentiation." *Hippocampus* 2,421.

PAGE 184 Stress levels of glucocorticoids inhibit long-term potentiation: Diamond, D., Bennet, M., Fleshner, M., and Rose, G. 1992. "Inverted-U relationship between the level of peripheral corticosterone and the magnitude of hippocampal primed burst potentiation." *Hippocampus* 2, 421; Joels, M. 1997. "Steroid hormones and excitability in the mammalian brain." *Frontiers in Neuroendocrinology* 18, 2. Stress enhances long-term depression: Xu, L., Anwyl, R., and Rowan, M. 1997. "Behavioural stress facilitates the induction of long-term depression in the hippocampus." *Nature* 387, 497.

PAGE 184 Glucocorticoids inhibit glucose utilization and transport in the hippocampus and in hippocampal neurons and glia: Kadekaro, M., Masonori, I., and Gross, P. 1988. "Local cerebral glucose utilization is increased in acutely adrenalectomized rats." *Neuroendocrinology* 47, 329; Horner, H., Packan, D., and Sapolsky, R. 1990. "Glucocorticoids inhibit glucose transport in cultured hippocampal neurons and glia." *Neuroendocrinology* 52, 57; Virgin, C., Ha, T., Packan, D., Tombaugh, G., Yang, S., Horner, H., and Sapolsky, R. 1991. "Glucocorticoids inhibit glucose transport and glutamate uptake in hippocampal astrocytes: implications for glucocorticoid neurotoxicity." *Journal of Neurochemistry* 57, 1422.

PAGE 186 Atrophy of hippocampal neuronal connections with stress: Woolley, C., Gould, E., and McEwen, B. 1990. "Exposure to excess glucocorticoids alters dendritic morphology of adult hippocampal pyramidal neurons." *Brain Research*, 531, 225; Magarinos, A., and McEwen, B. 1995. "Stress-induced atrophy of apical dendrites of hippocampal CA3c neurons: comparison of stressors." *Neuroscience* 69, 83; Magarinos, A., and McEwen, B. 1995.: "Stress-induced atrophy of apical dendrites of hippocampal CA3c neurons: involvement of glucocorticoid secretion and excitatory amino acid receptors." *Neuroscience* 69, 88; Magarinos, A., McEwen, B., Flugge, G., and Fuchs, E. 1996. "Chronic psychosocial stress causes apical dendritic atrophy of hippocampal CA3 pyramidal neurons in subordinate tree shrews." *Journal of Neuroscience* 16, 3534.

PAGE 187 Hippocampal atrophy in Cushing's disease: Starkman, M., Gebarski, S., Berent, S., and Schteingart, D. 1992. "Hippocampal formation volume, memory dysfunction, and cortisol levels in patients with Cushing's syndrome." *Biological Psychiatry* 32, 756.

PAGE 188 The concept of neuroendangerment by glucocorticoids is discussed in Sapolsky, R. 1996. "Stress, glucocorticoids, and damage to the nervous system: the current state of confusion." *Stress* 1, 1. Also, see Sapolsky, R. 1992. *Stress, the Aging Brain, and the Mechanisms of Neuron Death* (Cambridge Mass.: MIT Press).

Glucocorticoids worsen the hippocampal damage caused by seizure in the rat: Sapolsky, R. 1995. "A mechanism for glucocorticoid toxicity in the hippocampus: increased neuronal vulnerability to metabolic insults." *Journal of Neuroscience* 5, 1227; and by lack of oxygen caused by cardiac arrest: Sapolsky, R., and Pulsinelli, W. 1985. "Glucocorticoids potentiate ischemic injury to neurons: therapeutic implications." *Science* 229, 1397; vulnerability to damage caused by the amyloid fragment of Alzheimer's disease: Behl, C., Lezoualc'h, F., Trapp, T., Widmann, M., Skutella, T., and Holsboer, F. 1997. "Glucocorticoids enhance oxidative stress-induced cell death in hippocampal neurons in vitro." *Endocrinology 138, 101*; Goodman, Y., Bruce, A., Cheng, B., and Mattson, M. 1996. "Estrogens attenuate and corticosterone exacerbates excitotoxicity, oxidative injury, and amyloid beta-peptide toxicity in hippocampal neurons." *Journal of Neurochemistry* 66, 1836; gp120-induced damage in neurons: Brooke, S., Chan, R., Howard, S., and Sapolsky, R. 1997. "Endocrine modulation of the neurotoxicity of gp120 implications for AIDS-related dementia complex." *Proceedings of the National Academy of Sciences USA,* in press.

PAGE 189 Higher glucocorticoid levels associated with worse outcome to a stroke in humans: Astrom, M., Olsson, T., and Asplund, K. 1993. "Different linkage of depression to hypercortisolism early versus later after stroke." *Stroke* 24, 52.

The problem with glucocorticoid treatment in post-stroke edema reviewed in Sapolsky, R. 1992. *Stress, the Aging Brain, and the Mechanisms of Neuron Death,* (Cambridge, Mass.: MIT Press). Glucocorticoids decrease spinal cord injury: Bracken, M., Shepard, M., Collins, W., Holford, T., et al. 1990. "A randomized controlled trial of methylprednisolone or naloxone in the treatment of acute spinal cord injury." *New England Journal of Medicine* 322, 1405.

PAGE 190 Glucocorticoids and their clinical use with AIDS patients: Bozzette, S., Sattler, F., Chiu, J., Wu, A., Gluckstein, D., et al. 1990. "A controlled trial of early adjunctive treatment with corticosteroids for *Pneumocystis carinii* pneumonia in the acquired immunodeficiency syndrome." *New England Journal of Medicine* 323, 1451; Gagnon, S., Boota, A., Fischl, M., Baier, H., Kirksey, O., La Voie, L. 1990. "Corticosteroids as adjunctive therapy for severe *Pneumocystis carinii* pneumonia in the acquired immunodeficiency syndrome. a double-blind, placebo-controlled trial." *New England Journal of Medicine* 323, 1444.

Huge stress response after neurological insults in humans: Feibel, J., Hardi, P., Campbell, M., Goldstein, N., and Joynt, R. 1977. "Prognostic value of the stress response following stroke." *Journal of the American Medical Association* 238, 1374. Blocking glucocorticoid secretion after a stroke or seizure in a rat is neuroprotective: Stein, B., and Sapolsky, R. 1988.

"Chemical adrenalectomy reduces hippocampal damage induced by kainic acid." *Brain Research* 473, 175; Morse, J., and Davis, J. 1989. "Chemical adrenalectomy protects hippocampal cells following ischemia." *Society for Neuroscience Abstracts* 15, 149.4.

PAGE 191 The first report of glucocorticoid neurotoxicity: Aus der Muhlen, K., and Ockenfels, H. 1969. "Morphologische veranderungen im diencephalon und telenceaphlin nach storngen des regelkreises adenohypophyse-nebennierenrinde III. Ergebnisse beim meerschweinchen nach verabreichung von cortison und hydrocortison." *Z Zellforsch* 93, 126. The first report of the hippocampus as a glucocorticoid target site: McEwen, B., Weiss, J., and Schwartz, l. 1968. "Selective retention of corticosterone by limbic structures in rat brain." *Nature* 220, 911. Glucocorticoids and stress and accelerating hippocampal neuron loss: Sapolsky, R., Krey, L., and McEwen, B. 1985. "Prolonged glucocorticoid exposure reduces hippocampal neuron number: implications for aging." *Journal of Neuroscience* 5, 1221; Kerr, D., Campbell, L., Applegate, M., Brodish, A., and Landfield, P. 1991. "Chronic stress-induced acceleration of electrophysiologic and morphometric biomarkers of hippocampal aging." *Journal of Neuroscience* 11, 1316. Removing glucocorticoids or decreasing their secretion delays hippocampal neuron loss: Landfield, P., Baskin, R., and Pitler, T. 1981. "Brain-aging correlates: retardation by hormonal–pharmacological treatments." *Science* 214, 581; Meaney, M., Aitken, D., Bhatnager, S., van Berkel, C., and Sapolsky, R. 1988. "Effect of neonatal handling on age-related impairments associated with the hippocampus." *Science* 239, 766.

PAGE 192 Stress and glucocorticoids damage the primate hippocampus: Uno, H., Tarara, R., Else, J., Suleman, M., and Sapolsky, R. 1989. "Hippocampal damage associated with prolonged and fatal stress in primates." *Journal of Neuroscience* 9, 1705; Sapolsky, R., Uno, H., Rebert, C., and Finch, C. 1990. "Hippocampal damage associated with prolonged glucocorticoid exposure in primates." *Journal of Neuroscience* 10, 2897; Uno, H., Eisele, S., Sakai, A., Shelton, S., Baker, E., DeJesus, O., and Holden, J. 1994. "Neurotoxicity of glucocorticoids in the primate brain." *Hormones and Behavior* 28, 336.

PAGE 193 Hippocampal atrophy in PTSD: Bremner, J., Randall, P., Scott, T., Bronen, R., et al. 1995. "MRI-based measurement of hippocampal volume in patients with combat-related PTSD." *American Journal of Psychiatry* 152, 973; Gurvits, T., Shenton, M., Hokama, H., Ohta, H., Lasko, N., Gilbertson, M., et al. 1996 "Magnetic resonance imaging study of hippocampal volume in chronic, combat-related posttraumatic stress disorder." *Biological Psychiatry* 40, 1091; Bremner, J., Randall, P., Vermetten, E., Staib, L., Bronen, A., et al. 1997. "Magnetic resonance imaging-based measurement of hippocampal volume in PTSD related to childhood physical and sexual abuse—a preliminary report." *Biological Psychiatry* 41, 23. Hippocampal atrophy in human depression: Sheline, Y., Wang, P., Gado, M., Csernansky, J., and Vannier, M. 1996. "Hippocampal atrophy in recurrent major depression." *Proceedings of the National Academy of Sciences* USA 93, 3908. Also see Sapolsky, R. 1996. "Why stress is bad for your brain." *Science* 273, 749.

PAGE 193 Newcomer, J., Craft, S., and Askins, K. 1994. "Glucocorticoid-induced impairment in declarative memory performance in adult humans." *Journal of Neuroscience* 14, 2047; Wolkowitz, O., Reuss, V., and Weingartner, H. 1990. "Cognitive effects of corticosteroids." *American Journal of Psychiatry* 147, 1297. Studies of memory problems in patients receiving glucocorticoids: Keenan, P., Jacobson, M., Soleymani, R., Mayes, M., Stress, M., and Yaldoo, D. 1996. "The effect on memory of chronic prednisone treatment in patients with systemic disease." *Neurology* 47, 1396.

PAGE 194 The Woody Allen quote is from *Sleeper*.

CHAPTER 11: Aging and Death

PAGE 199 Functioning in old organisms is disrupted by stress more than in young ones: the classic studies on temperature dysregulation during aging can be found in Shock, N. 1977. "Systems integration." In Finch, C., and Hayflick, L. (eds.). *Handbook of the Biology of Aging*, 1st ed. (New York: Van Nostrand). A detailed discussion of all the ways in which the cardiovascular system functions similarly in young and old healthy subjects in the absence of stress can be found in Lakatta, E. 1990. "Heart and circulation." In Schneider, E., and Rowe, J. (eds.). *Handbook of the Biology of Aging*, 3d ed. (New York: Academic Press). Decrease in maximal heart rate and work capacity with age: Gerstenblith, G., Lakatta, E., and Weisfeldt, M. 1976. "Age changes in myocardial function and exercise response." *Progress in Cardiovascular Disease* 19, 1. Decreased ejection volume during exercise with age: Rodeheffer, R., Gerstenblith, G., Becker, L., Fleg, J., Weisfeldt, M., and Lakatta, E. 1984. "Exercise cardiac output is maintained with advancing age in healthy human subjects: cardiac dilatation and increased stroke volume compensate for diminished heart rate." *Circulation* 69, 203. Increased cardiac muscle stiffness with force as a function of age: Spurgeon, H., Thorne, P., Yin, F., Shock, N., and Weisfeldt, M. 1977. "Increased dynamic stiffness of tabeculae carneae from senescent rats." *American Journal of Physiology* 232, H373. Easier immunosuppression in aged primates: Ershler, W., Coe, C., and Gravenstein, S. 1988. "Aging and immunity in nonhuman primates. I. Effects of age and gender on cellular immune function in rhesus monkeys." *American Journal of Primatology* 15, 181.

PAGE 200 The aged brain is slower in compensatory sprouting after an injury than is the young brain: Cotman, C. 1985. *Synaptic Plasticity* (New York: Guilford). The effect of age on the vulnerability of cerebral metabolism to a metabolic stressor: Benzi, G., Pastoris, O., Vercesi, L., Gorini, A., Viganotti, C., and Villa, R. 1987. "Energetic state of aged brain during hypoxia." *Gerontology* 33, 207; Hoffman, W., Pelligrino, D., Miletich, D., and Albrecht, R. 1985. "Brain metabolic changes in young versus aged rats during hypoxia." *Stroke* 16, 860.

The effect of age on performance on intelligence tests: this vast subject is reviewed in a number of chapters in Birren, J., and Schaie, K. 1990. *Handbook of the Psychology of Aging*, 3d ed. (New York: Van Nostrand): Cerella, J. "Aging and information-processing rate"; Kausler, D. "Motivation, human aging and cognitive performance"; Hultsch, D., and Dixson, R. "Learning and memory in aging." See also Katzman, R., and Terry, R. 1983. *The Neurology of Aging* (Philadelphia: Davis).

Elevated epinephrine and norepinephrine concentrations during exercise as a function of age: Fleg, J., Tzankoff, S., and Lakatta, E. 1985. "Age-related augmentation of plasma catecholamines during dynamic exercise in healthy males." *Journal of Applied Physiology* 59, 1033. Decreased cardiovascular sensitivity to adrenaline and noradrenaline with age: Lakatta, E. 1987. "Catecholamines and cardiovascular function in aging." *Endocrinology and Metabolism Clinics of North America* 16, 877.

PAGE 201 Slower epinephrine-norepinephrine recovery after the end of stress: McCarty, R. 1986. "Age-related alterations in sympathetic-adrenal medullary responses to stress." *Gerontology* 32, 172. Slower glucocorticoid recovery after the end of stress: Sapolsky, R., Krey, L., and McEwen, B. 1983. "The adrenocortical stress-response in the aged male rat: impairment of recovery from stress." *Experimental Gerontology* 18, 55; Ida, Y., Tanaka, M., and Tsuda, A. 1984. "Recovery of stress-induced increases in noradrenaline turnover is delayed in specific brain regions of old rats." *Life Sciences* 34, 2357. The delayed glucocorticoid recovery may accelerate tumor growth: Sapolsky, R., and Donnelly, T. 1985. "Vulnerability to stress-induced tumor growth increases with age in the rat: role of glucocorticoid hypersecretion." *Endocrinology* 117, 662.

Resting epinephrine and norepinephrine levels increase with age: Fleg, J., Tzankoff, S., and Lakatta, E. 1985. "Age-related augmentation of plasma catecholamines during dynamic exercise in healthy males." *Journal of Applied Physiology* 59, 1033; also Rowe, J. and Troen, B. 1980. "Sympathetic nervous system and aging in man." *Endocrine Reviews* I, 167. Resting glucocorticoid levels rise with age in the rat: reviewed in Sapolsky, R. 1991. "Do glucocorticoid concentrations rise with age in the rat?" *Neurobiology of Aging* 13, 171. In the aged human: reviewed in Sapolsky, R. 1990. "The adrenocortical axis." In Schneider, E., and Rowe, J. (eds.). *Handbook of the Biology of Aging*, 3d ed. (New York: Academic Press). In the wild baboon: Sapolsky, R., and Altmann, J. 1991. "Incidences of hypercortisolism and dexamethasone resistance increase with age among wild baboons." *Biological Psychiatry*, 30, 1008.

PAGE 202 The elevated epinephrine-norepinephrine levels contribute to the elevated blood pressure: Lakatta, E. 1990. "Heart and circulation." In Schneider, E., and Rowe, J. (eds.). *Handbook of the Biology of Aging*, 3d ed. (New York: Academic Press). Hypertension as the most common disease of the elderly: Kannel, W., and Vokonas, P. 1986. "Primary risk factors for coronary heart disease in the elderly. the Framingham study." In Wenger, N., and Furberg, C. (eds.). *Current Heart Disease in the Elderly* (London: Elsevier).

The elevated glucocorticoid levels disrupt the ability of the brain for sprouting after injury: Scheff, S., and Cotman, C. 1982. "Chronic glucocorticoid therapy alters axon sprouting in the hippocampal dentate gyrus." *Experimental Neurology* 76, 644; DeKosky, S., Scheff, S., and Cotman, C. 1984. "Elevated corticosterone levels: a possible cause of reduced axon sprouting in aged animals." *Neuroendocrinology* 38, 33.

PAGE 203 Can stress accelerate the aging process? For those who want to go straight to the horse's mouth (in German), there is Rubner, M. 1908. *Das problem der lebensdauer und seine beziehungen zun wachstum und ernahrun* (Munchen: Oldenbourg). Also see Pearl, R. 1928. *The Rate of Living* (New York: Knopf) for the most detailed exploration of rate-of-living hypotheses. For some of Selye's ideas about stress and aging, see Selye, H., and Tuchweber, B. 1976. "Stress in relation to aging and disease." In Everitt, A., and Burgess, J. (eds.). *Hypothalamus, Pituitary and Aging* (Springfield, Ill.: Charles C. Thomas). For a scholarly discussion of the whole topic by one of the wisest thinkers in gerontology, see Chapter 5 ("Rates of living and dying: correlations of life span with size, metabolic rates, and cellular and biochemical characteristics") in Finch, C. 1990. *Longevity, Senescence, and the Genome* (Chicago: University of Chicago Press).

PAGE 205 For a discussion of why programmed aging (and aging in general) may have evolved, see Sapolsky, R., and Finch, C. 1991. "On growing old: not every creature ages, but most do. The question is why." *The Sciences* (March–April), 30. For the original demonstration of what goes wrong in the salmon, see Robertson, O., and Wexler, B. 1957. "Pituitary degeneration and adrenal tissue hyperplasia in spawning Pacific salmon." *Science* 125, 1295. For a comparison of the effects of salmon aging with the effects of glucocorticoid excess, see Wexler, B. 1976. "Comparative aspects of hyperadrenocorticism and aging." In Everitt, A., and Burgess, J. (eds.). *Hypothalamus, Pituitary and Aging* (Springfield Ill.: Thomas). For an introduction to the marsupial mouse aging literature, see McDonald, I., Lee, A., and Bradley, A. 1981. "Endocrine changes in dasyurid marsupials with differing mortality patterns." *General and Comparative Endocrinology* 44, 292, and McDonald, I., Lee, A., and Than, K. 1986. "Failure of glucocorticoid feedback in males of a population of small marsupials (*Antechinus swainsonii*) during the period of mating." *Journal of Endocrinology* 108, 63.

A brief digression: what about those children who senesce incredibly rapidly and die of old age when they are twelve? Progeria, as the disease is called, is extremely rare. Those afflicted go bald, have bony chins, beaked noses, and dry, scratchy voices; they lose their hearing, get hardening of the arteries and heart disease (which is what usually kills them). When you try to grow some of their cells in a petri dish, you have as much trouble as you would with the cells of a seventy-year-old. Despite that, not everything about progeric kids is prematurely aged: they don't become demented, nor do they get cancer—two diseases typically linked with aging (although not necessary features of it, obviously). General consensus in the field is that progeria is a disease of some *facets* of aging being accelerated, rather than the entire aging process (which indirectly demonstrates that aging involves the ticking of multiple and independent clocks in the body). For a discussion of progeria and its relationship to aging, see Finch, C. 1991. *Longevity, Senescence and the Genome* (Chicago: University of Chicago Press); Mills, R., and Weiss, A. 1990. "Does progeria provide the best model of accelerated aging in humans?" *Gerontology* 36, 84.

PAGE 207 Old humans, primates, and rats tend to become dexamethasone-resistant with age: reviewed in Sapolsky, R. 1990. "The adrenocortical axis." In Schneider, E., and Rowe, J. (eds.). *Handbook of the Biology of Aging,* 3d ed. (New York: Academic Press). In the wild baboon: Sapolsky, R., and Altmann, J. 1991. "Incidences of hypercortisolism and dexamethasone resistance increase with age among wild baboons." *Biological Psychiatry,* 30, 1008.

PAGE 208 Patterns of neuron loss during aging: for a review, see Coleman, P., and Flood, D. 1987. "Neuron numbers and dendritic extent in normal aging and Alzheimer's disease." *Neurobiology of Aging* 8, 521. Some changes in the techniques used to count neurons in the brain have led to controversy as to just how many neurons are typically lost in the aging hippocampus: Wickelgren, I. 1996. "Horizons in Aging." *Science* 274, 48; Gallagher, M., Landfield, P., McEwen, B., Meaney, M., Rapp, P., Sapolsky, R., and West, M. 1996. "Hippocampal neurodegeneration in aging." *Science* 274, 1423.

PAGE 208 The hippocampus plays a role in inhibiting glucocorticoid secretion: reviewed in Jacobson, L., and Sapolsky, R. 1991. "The role of the hipocampus in feedback regulation of the hypothalamic-pituitary-adrenocortical axis." *Endocrine Reviews* 12, 118.

PAGE 208 The interaction between the effects of glucocorticoids on the hippocampus and the effects of the hippocampus on glucocorticoid secretion: Sapolsky, R., Krey, L., and McEwen, B. 1986. "The neuroendocrinology of stress and aging: the glucocorticoid cascade hypothesis." *Endocrine Reviews* 7, 284.

PAGE 210 For a review of this entire topic in unreadably technical detail, the truly masochistic may want to buy a dozen copies of Sapolsky, R. 1992. *Stress, the Aging Brain and the Mechanisms of Neuron Death* (Cambridge: MIT Press). The final chapter is a detailed discussion of whether glucocorticoids and stress can damage the human brain, and the possible implications for aging, Alzheimer's disease, depression, and a number of neurological disorders.

CHAPTER 12: Why Is Psychological Stress Stressful?

PAGE 211 Teddy Roosevelt's childhood lament can be found in Morris, E. 1979. *The Rise of Theodore Roosevelt* (New York: Ballantine Books). For a history of the field of stress research, as well as the celebrated debate between Selye and Mason, see Selye, H. 1975. "Confusion and controversy in the stress field." *Journal of Human Stress* 1, 37; Mason, J. 1975. "A historical view of the stress field." *Journal of Human Stress* 1, 6. For a nontechnical review of Weiss's work, see Weiss, J. 1972. "Psychological factors in stress and disease." *Scientific American*, 226 (June), 104.

PAGE 214 Outlets for frustration: the difference in the stress response in patients depending on whether they expressed fear to their doctor: Greene, W., Conron, D., Schalch, S., and Schreiner, B. 1970. "Psychological correlates of growth hormone and adrenal secretory responses of patients undergoing cardiac catheterization." *Psychosomatic Medicine* 32, 599. The demonstration that social support networks are associated with lower glucocorticoid concentrations can be found in Ray, J., and Sapolsky, R. 1992. "Styles of male social behavior and their endocrine correlates among high-ranking wild baboons." *American Journal of Primatology* 28, 231; Virgin, C., and Sapolsky, R. 1997. "Styles of male social behavior and their endocrine correlates among low-ranking baboons." *American Journal of Primatology* 42, 25.

PAGE 215 Support and primates in a novel setting: Gust, D., Gordon, T., Brodie, A., and McClure, H. 1996. "Effect of companions in modulating stress associated with new group formation in juvenile rhesus macaques." *Physiology and Behavior* 59, 941.

PAGE 216 Supportive friend during human stressors: Lepore, S., Allen, K., and Evans, G. 1993. Social support lowers cardiovascular reactivity to an acute stressor." *Psychosomatic Medicine* 55, 518; Edens, J., Larkin, K., and Abel, J. 1992. "The effect of social support and physical touch on cardiovascular reactions to mental stress." *Journal of Psychosomatic Research* 36, 371; Gerin, W., Pieper, C., Levy, R., and Pickering, T. 1992. "Social support in social interaction: a moderator of cardiovascular reactivity." *Psychosomatic Medicine* 54, 324; Kamarck, T., Manuck, S., and Jennings, J. 1990. "Social support reduces cardiovascular reactivity to psychological challenge: a laboratory model." *Psychosomatic Medicine* 52, 42. Glucocorticoids in stepchildren: Flinn, M., and England, B. 1997. "Social economics of childhood glucocorticoid stress response and health." *American Journal of Physical Anthropology* 102, 33.

Social support and sympathetic nervous system: Fleming, R. 1982. "Mediating influence of social support on stress at Three Mile Island." *Journal of Human Stress* 8, 14. Social support and cardiovascular disease: Williams, R., and Littman, A. 1996. "Psychosocial factors: role in cardiac risk and treatment strategies." *Cardiology Clinics* 14, 97.

PAGE 218 Importance of predictability: Abbott, B., Schoen, L., and Badia, P. 1984. "Predictable and unpredictable shock: behavioral measures of aversion and physiological measures of stress." *Psychological Bulletin* 96, 45; Davis, H., and Levine, S. 1982. "Predictability, control, and the pituitary-adrenal response in rats." *Journal of Comparative and Physiological Psychology* 96, 393; Seligman, M., and Meyer, B. 1970. "Chronic fear and ulcers with rats as a function of the unpredictability of safety." *Journal of*

Comparative and Physiological Psychology 73, 202. Predictability: an analysis similar to mine (a warning signal tells you when to worry and, even more importantly, when you can relax) has been termed the *safety-signal hypothesis* by psychologist Martin Seligman. 1975. *Helplessness: On Depression, Development and Death* (San Francisco: W. H. Freeman and Co.).

Parachute jumpers habituating: Ursin, H., Baade, E., and Levine, S. 1978. *Psychobiology of Stress: A Study of Coping Men* (New York: Academic Press).

PAGE 219 Ulcers and bombings in World War II: Stewart, D., and Winser, D. 1942. "Incidence of perforated peptic ulcer: effect of heavy air-raids." *Lancet* (28 February), 259. Gary Gilmore and his execution were the subject of Norman Mailer's book *The Executioner's Song* (Boston: Little, Brown and Company, 1979).

PAGE 219 Control: you don't actually need to exercise control in order to get its benefits: Glass, D., and Singer, J. 1972. *Urban Stress: Experiments on Noise and Social Stressors* (New York: Academic Press). Effects of uncontrollability on tumor growth: Visintainer, M., Volpicelli, J., and Seligman, M. 1982. "Tumor rejection in rats after inescapable or escapable shock." *Science* 216, 437. General references for control being helpful: Houston, B. 1972. "Control over stress, locus of control, and response to stress." *Journal of Personality and Social Psychology* 21, 249; Lundberg, U., and Frankenhaeuser, M. 1978. "Psychophysiological reactions to noise as modified by personal control over stimulus intensity." *Biological Psychology* 6, 51; Brier, A., Albus, M., Pickar, D., Zahn, TP., Wolkowitz, O., and Paul, S. 1987. "Controllable and uncontrollable stress in humans: alterations in mood and neuroendocrine and psychophysiological function." *American Journal of Psychiatry* 144, 1419; Manuck, S., Harvey, A., Lechleiter, S., and Neal, K. 1978. "Effects of coping on blood pressure responses to threat of aversive stimulation." *Psychophysiology* 15, 544.

Primates in novel settings: Coe, C., Rosenberg, L., Fischer, M., and Levine, S. 1987. "Psychological factors capable of preventing the inhibition of antibody responses in separated infants." *Child Development* 58, 1420.

PAGE 220 The perception of things worsening or improving: baboons rising or declining in the hierarchy: Sapolsky, R. 1992. "Cortisol concentrations and the social significance of rank instability among wild baboons." *Psychoneuroendocrinology* 17, 701 (cortisol is glucocorticoid found in the bloodstream of primates and humans). Parents of children with cancer: Wolff, C., Friedman, S., Hofer, M., and Mason, J. 1964. "Relationship between psychological defenses and mean urinary 17-hydroxycorticosteroid excretion rates." *Psychosomatic Medicine* 26, 576 (17-hydroxycorticosteroids are the versions of glucocorticoids that humans excrete).

PAGE 221 Predictive information doesn't work with long lag time: Pitman, D., Natelson, B., Ottenweller, J., McCarty, R., Pritzel, T., and Tapp, W. 1995. "Effects of exposure to stressors of varying predictability on adrenal function in rats." *Behavioral Neuroscience* 109, 767; Arthur, A. 1986.

"Stress of predictable and unpredictable shock." *Psychological Bulletin* 100, 379.

PAGE 225 Subtleties of control: DeGood, D. 1975. "Cognitive control factors in vascular stress responses." *Psychophysiology* 12, 399; Houston, B. 1972. "Control over stress, locus of control, and response to stress." *Journal of Personality and Social Psychology* 21, 249; Lundberg, U., and Frankenhaeuser, M. 1978. "Psychophysiological reactions to noise as modified by personal control over stimulus intensity." *Biological Psychology* 6, 51.

PAGE 226 Executive stress syndrome and ulcerating monkeys: technical and nontechnical versions of the famous experiment with executive monkeys can be found, respectively, in Brady, J., Porter, R., Conrad, D., and Mason, J. 1958. "Avoidance behavior and the development of gastroduodenal ulcers." *Journal of the Experimental Analysis of Behavior* I, 69, and Brady, J. 1958. "Ulcers in 'executive' monkeys." *Scientific American* 199, 95. Technical and nontechnical critiques by Weiss of this experiment can be found, respectively, in Weiss, J. 1968. "Effects of coping response on stress." *Journal of Comparative and Physiological Psychology* 65, 251, and Weiss, J. 1972. "Psychological factors in stress and disease." *Scientific American* 226, 104. A technical critique is also offered by Natelson, B., Dubois, A., and Sodetz, F. 1977. "Effect of multiple stress procedures on monkey gastro-duodenal mucosa, serum gastrin and hydrogen ion kinetics." *American Journal of Digestive Diseases* 22, 888.

Many of the ideas in this chapter will be returned to in the final chapter on stress management, along with additional references.

CHAPTER 13: Stress and Depression

PAGE 229 A masterful overview of the entire field can be found in Goodwin, F., and Jamison, K. 1990. *Manic-Depressive Illness* (New York: Oxford University Press).

Five to twenty percent of the population will suffer from a major depression: Robins, L., Helzer, J., Weissman, M., Orvaschel, H., Gruenberg, E., Burke, J., and Regier, D. 1984. "Lifetime prevalence of specific psychiatric disorders in three sites." *Archives of General Psychiatry* 41, 949; Weissman, M., and Myers, J. 1978. "Rates and risks of depressive symptoms in a United States urban community." *Acta Psychiatr. Scand.* 57, 219; Helgason, T. 1979. "Epidemiological investigation concerning affective disorders." In Schor, M., and Stromgren, M. (eds.). *Origin, Presentation and Treatment of Affective Disorders* (London: Academic Press), 241.

PAGE 230 Good descriptions of the symptoms found in different depressive subtypes can be found in the bible on the subject, the *Diagnostic and Statistical Manual of Mental Disorders* (DSM-III-R), 3d ed., rev. (Washington, D.C.: American Psychiatric Association, 1987). Also see Gold, P., Goodwin, F., and Chrousos, G. 1988. "Clinical and biochemical manifestations of depression. Relation to the neurobiology of stress." *New England Journal of Medicine* 319, 348.

PAGE 232 For the classic discussion of depression as a cognitive disorder, see Beck, A. 1976. *Cognitive Therapy and the Emotional Disorders* (New York: International Universities Press).

PAGE 233 Vegetative symptoms: for the first report of sleep changes in many depressives: Diaz-Guerrero, R., Gottlieb, J., and Knott, J. 1946. "The sleep of patients with manic-depressive psychosis, depressive type: an electroencephalographic study." *Psychosomatic Medicine* 8, 399. Also see Coble, P., Foster, F., and Kupfer, D. 1976. "Electroencephalographic sleep diagnosis of primary depression." *Archives of General Psychiatry* 33, 1124; Gillin, J., Duncan, W., Pettigrew, K., Frankel, B., and Snyder, F. 1979. "Successful separation of depressed, normal and insomniac subjects by EEG sleep data." *Archives of General Psychiatry* 36, 85. Cortisol (glucocorticoid) levels are elevated in many depressives: for an early demonstration of this, see Sachar, E. 1975. "Neuroendocrine abnormalities in depressive illness." In Sachar, E. (ed.). *Topics of Psychoendocrinology* (New York: Grune and Stratton), 135. For a more recent review, see Sapolsky, R., and Plotsky, P. 1990. "Hypercortisolism and its possible neural bases." *Biological Psychiatry* 27, 937.

PAGE 234 Depressive symptomatology can follow cyclic patterns over time: a classic demonstration of this can be found in Richter, C. 1938. "Two-day cycles of alternating good and bad behavior in psychotic patients." *Archives of Neurology and Psychiatry* 39, 587. For a good review of seasonal affective disorders, see Rosenthal, N., Sack, D., Gillin, C., Lewy, A., Goodwin, F., Davenport, Y., Mueller, P., Newsome, D., and Wehr, T. 1984. "Seasonal affective disorder." *Archives of General Psychiatry* 41, 72. For demonstrations of the use of light therapy for SADs, see Rosenthal, N., Sack, D., Carpenter, C., Parry, B., Mendelson, W., and Wehr, T. 1985. "Antidepressant effects of light in seasonal affective disorder." *American Journal of Psychiatry* 142, 163; Wehr, T., Jacobsen, F., Sack, D., Arendt, J., Tamarkin, L., and Rosenthal, N. 1986. "Phototherapy of seasonal affective disorder." *Archives of General Psychiatry* 43, 870.

PAGE 235 An excellent and accessible introduction to the topic of neurotransmitters can be found in Barondes, S. 1993. *Molecules and Mental Illness* (New York: Scientific American Library, W. H. Freeman).

PAGE 235 The neurochemistry of depression is a vast subject with a dizzying number of papers, many of them contradicting one another. For an authoritative and relatively accessible discussion of the current confusion about whether it is a neuroepinephrine or serotonin problem, involving too much or too little of the neurotransmitters, or too much or too little of the receptors, see Kandel, E. 1991. "Disorders of mood." In Kandel, E., Schwartz, J., and Jessell, T. *Principles of Neural Sciences,* 3d ed. (New York: Elsevier). Also see Barondes, S. 1993. *Molecules and Mental Illness* (New York: Scientific American Library, W. H. Freeman).

PAGE 240 A brief tirade about ECT: few medical procedures of our time have a worse popular image. In the past, ECT involved sufficient amounts of electricity to cause brain damage and memory loss, and to induce convulsions, causing body injury. Far worse, ECT's use for all sorts of things besides intractable depression—behavior disorders, juvenile delinquency, and so on—smacked of medicopolitical control and punishment. However, ECT is now conducted very differently—far less electricity is used, and there is no evidence that the modern form of ECT causes brain damage or permanent memory loss. Moreover, people are now typically sedated during ECT sessions, which virtually eliminates the danger of physical injury from convulsing. Most important, when it is administered correctly, ECT can save lives. For people who have been through every type of psychotherapy, every known antidepressant, and every combination of the two, yet are still suicidally depressed, ECT may be the only known technique that will ever get them functioning again. It can be an extraordinarily helpful procedure, and many former depressives swear by it. For a discussion of the history of ECT and its rather safe record as currently used, see Fink, M. 1985. "Convulsive therapy: fifty years of progress." *Convulsive Therapy* I, 204. Mechanisms of ECT action: some papers showing effects of ECT on numbers of receptors for norepinephrine and related neurotransmitters: Kellar, K., and Stockmeier, C. 1986. "Effects of electroconvulsive shock and serotonin axon lesions on beta-adrenergic and serotonin-2 receptors in rat brain." *Annals of the New York Academy of Sciences* 462, 76; Chiodo, L., and Antelman, S. 1980. "Electroconvulsive shock: progressive dopamine autoreceptor subsensitivity independent of repeated treatment." *Science* 210, 799; Reches, A., Wagner, H., Barkai, A., Jackson, V., Yablonskaya-Alter, E., and Fahn, S. 1984. "Electroconvulsive treatment and haloperidol: effects on pre-and postsynaptic dopamine receptors in rat brain." *Psychopharmacology* 83, 155.

PAGE 241 Pleasure pathways in the brain: For a history of the start of this field by one of its two discoverers, see Milner, P. 1989. "The discovery of self-stimulation and other stories." *Neuroscience and Biobehavioral Reviews* 13, 61. For another general overview of the field, see Routtenberg, A. 1978. "The reward system of the brain." *Scientific American* (November). For a demonstration that stimulation of these pathways can be more reinforcing than food, see Routtenberg, A., and Lindy, J. 1965. "Effects of the availability of rewarding septal and hypothalamic stimulation on bar pressing for food under conditions of deprivation." *Journal of Comparative and Physiological Psychology* 60, 158. For an early study implicating norepinephrine in the pleasure pathway, see Stein, L. 1962. "Effects and interactions of imipramine, chlorpromzaine, reserpine, and amphetamine on self-stimulation: possible neurophysiological basis of depression." In Wortis, J. (ed.). *Recent Advances in Biological Psychiatry*, vol. 4 (New York: Plenum), 288. This study showed that norepinephrine depletion in the rat decreases self-stimulation of the pleasure pathways. In recent years, there has been a shift in this field away from considering norepinephrine to be the principal neurotransmitter of the pleasure pathways, much as there has been a shift away from considering it the sole culprit in depression. A neurotransmitter called *dopamine* is moving toward the forefront as the first neurotransmitter among equals involved in pleasure signaling. This makes some sense, as cocaine works mostly on dopamine synapses. However, although norepinephrine is probably not the most important neurotransmitter of pleasure perception, a defect in norepinephrine regulation in that part of the brain still has a great deal of potential for wreaking havoc. In considering how multiple neurotransmitters are used in many synaptic steps in these pleasure pathways, an analogy might help: a long cable may be made of stronger and weaker materials at different points; nevertheless, severing the cable in any place causes problems, and the norepinephrine link might be where the severing occurs. This is reviewed in Milner, P. 1991. "Brain-stimulation reward: a review." *Canadian Journal of Psychology* 45, 1.

PAGE 242 A review of the human literature regarding pleasure pathways and self-stimulation can be found in Heath, R. 1963. "Electrical self-stimulation of the brain in man." *American Journal of Psychiatry* 120, 571.

PAGE 244 For a discussion of the cons and surprising number of pros concerning cingulotomy (and for a thoughtful discussion of psychosurgical controversies in general), see Konner, M. 1988. "Too desperate a cure?" Originally published in *The Sciences* (May), 6; reprinted in Konner, M. 1990. *Why the Reckless Survive* (New York: Viking Penguin). For a technical discussion of the outcome of cingulotomies, see Ballantine, H., Bouckoms, A., Thomas, E., and Giriunas, I. 1987. "Treatment of psychiatric illness by stereotactic cingulotomy." *Biological Psychiatry* 22, 807. For a history of psychosurgery and its accompanying controversies, see Valenstein, E. 1986. *Great and Desperate Cures: The Rise and Decline of Psychosurgery and Other Radical Treatments for Mental Illness* (New York: Basic Books). Interestingly, a recent paper supports the rough picture of "the cortex whispering too many depressing thoughts to the limbic system"; this paper demonstrates that depressed patients have enhanced metabolism (relative to nondepressed patients) in the prefrontal cortex and the amygdala: Drevets, W., Videen, T., Price, J., Preskorn, S., Carmichael, S., and Raichle, M. 1992. "A functional anatomical study of unipolar depression." *The Journal of Neuroscience* 12, 3628.

PAGE 245 Thyroid hormone insufficiency can lead to depression: Denko, J., and Kaelbling, R. 1962. "Psychiatric aspects of hypoparathyroidism." *Acta Psychiatr. Scand.* (suppl. 164) 38, 7; Whybrow, P., Prange, A., and Treadway, C. 1969. "Mental changes accompanying thyroid gland dysfunction." *Archives of General Psychiatry* 20, 47. One way in which this may occur comes with the finding that thyroid hormones influence norepinephrine processing in the brain: Prange, A., Meek, J., and Lipton, M. 1970. "Catecholamines: diminished rate of synthesis in rat brain and heart after thyroxine pretreatment." *Life Sciences* 9, 901. Many patients with depression turn out to have an underlying thyroid hormone deficiency: Lipton, M., Breese, G., Prange, A., Wilson, I., and Cooper, B. 1976. "Behavioral effects of hypothalamic polypeptide hormones in animals and man." In Sacher, E. (ed.). *Hormones, Behavior and Psychopathology* (New York: Raven Press), 15.

PAGE 245 Higher rates of depression in women than in men: Murphy, M., Sobol, A., Neff, R., Olivier, D., and Leighton, A. 1984. "Stability of prevalence." *Archives of General Psychiatry* 41, 990. Sex differences in the rates of depression: the best overview of some of the nonhormonal theories can be found in Nolen-Hoeksma, S. 1987. "Sex differences in depression: theory and evidence." *Psychological Bulletin* 101, 259. Hormonal aspects of sex differences in depression: women have particularly high incidences of depression around the time of menstruation: Abramowitz, E., Baker, A., and Fleischer, S. 1982. "Onset of depressive psychiatric crises and the menstrual cycle." *American Journal of Psychiatry* 139, 475. The immediate postparturition period is one of great risk for depression: Campbell, S., and Cohn, J. 1991. "Prevalence and correlates of postpartum depression in first-time mothers." *Journal of Abnormal Psychology* 100, 594; O'Hara, M., Schlechte, J., Lewis, D., and Wright, E. 1991. "Prospective study of postpartum blues. Biologic and psychosocial factors." *Archives of General Psychiatry* 48, 801. What is generally viewed to be a heretical idea was voiced in a recent study, namely, that fathers have the same rate of postpartum depression as mothers do: Richman, J., Raskin, V., and Gaines, C. 1991. "Gender roles, social support, and postpartum depressive symptomatology." *Journal of Nervous and Mental Disease* 179, 139.

PAGE 246 Estrogen and progesterone have effects on the brain: As just some examples of these, estrogen will change the electrical excitability of the brain (Teyler, T., Vardaris, R., Lewis, D., and Rawitch, A. 1980. "Gonadal steroids: effects on excitability of hippocampal pyramidal cells." *Science* 209, 1017), and the number of receptors for some of the major neurotransmitters (Schumacher, M. 1990. "Rapid membrane effects of steroid hormones: an emerging concept in neuroendocrinology." *Trends in Neurosciences,* 13, 359; see also Weiland, N. 1990. "Sex steroids alter *N*-methyl-D-aspartate receptor binding in the hippocampus." *Society for Neuroscience Abstracts* 16, 959), as well as the number of receiving sites on dendrites ("dendritic spines") that form synapses with axon terminals. This last observation is particularly interesting, as it has been shown that the number of dendritic spines fluctuates in parts of the brain of the rat as a function of the reproductive cycle of the female (Woolley, C., Gould, E., Frankfurt, M., and McEwen, B. 1990. "Naturally occurring fluctuation in dendritic spine density on adult hippocampal pyramidal neurons." *Journal of Neuroscience* 10, 4035).

Progesterone also has effects, in that one of its breakdown products (metabolites) can bind to one of the main neurotransmitter receptor types in the brain and alter its functioning: Majewska, M., Harrison, N., Schwartz, R., Barker, J., and Paul, S. 1986. "Steroid hormone metabolites are barbiturate-like modulators of the GABA receptor." *Science* 232, 1004. This is particularly interesting for two reasons. First, the fact that the critical agent there is not progesterone but its metabolite (called *3-alpha-hydroxy-5-alpha-dihydroprogesterone* by its close friends) means that one must keep track of not only how much progesterone there is on the scene but how much of it gets converted to the metabolite. Of particular interest in terms of the menstrual cycle, progesterone, mood, and depression is the fact that these progesterone metabolites bind to the same receptor complex that binds the benzodiazepine tranquilizers (like those marketed as Valium and Librium) as well as barbiturate anesthetics ("downers"). Moreover, at proper doses, this progesterone metabolite can work as an anesthetic itself (such "steroid anesthetics" have even been used on humans during surgery). No one has quite sorted out the functional significance of this yet, but everyone assumes that something extremely interesting is going on.

Finally, for a way in which estrogen and progesterone can alter the action of antidepressant drugs in the brain, see Wilson, M., Dwuyer, K., and Roy, E. 1989. "Direct effects of ovarian hormones on antidepressant binding sites." *Brain Research Bulletin* 22, 181. For a demonstration that females break down antidepressant drugs in the bloodstream more slowly than males, so that more gets into the brain, see Biegon, A., and Samuel, D. 1979. "The in vivo distribution of an antidepressant drug (DMI) in male and female rats." *Psychopharmacology* 65, 259. For a fascinating discussion of the ways in which people of different ethnic backgrounds vary in their sensitivity to various psychoactive drugs, see Holden, C. 1991. "New center to study therapies and ethnicity." *Science* 251, 748.

PAGE 247 For broad discussions of the connections between stress and depression, see Gold, P., Goodwin, F., and Chrousos, G. 1988. "Clinical and biochemical manifestations of depression. Relation to the neurobiology of stress." *New England Journal of Medicine* 319, 348 (outlines a model, very similar to that proposed in this chapter, of the genetic defect in depression as a failure for stress to induce tyrosine hydroxylase); Zis, A., and Goodwin, F. 1979. "Major affective disorders as a recurrent illness: a critical review." *Archives of General Psychiatry* 36, 385; Anisman, H., and Zacharko, R. 1982. "Depression: the predisposing influence of stress." *Behavioral and Brain Science* 5, 89; Turner, R., and Beiser, M. 1990. "Major depression and depressive symptomatology among the physically disabled: assessing the role of chronic stress." *Journal of Nervous and Mental Disease* 178, 343.

PAGE 250 Freud's classic essay, "Mourning and melancholia," can be found in *The Collected Papers, Vol. IV* (New York: Basic Books, 1959).

PAGE 252 Psychological features of learned helplessness: the definitive book on the subject is by Martin Seligman (from which the various quotations are taken): *Helplessness: On Depression, Development and Death* (San Francisco: W. H. Freeman, 1975). This monumental (and quite readable) work is one of the most influential books ever published in psychology. The specific human experiments cited in this section are Hiroto, D. 1974. "Locus of control and learned helplessness." *Journal of Experimental Psychology* 102, 187 (uncontrollable noise induces helplessness with a noise-avoidance task); Hiroto, D., and Seligman, M. 1974. "Generality of learned helplessness in man." *Journal of Personality and Social Psychology* 31, 311 (uncontrollable noise disrupts learning of simple word puzzles, and unsolvable tasks induce helplessness); Seligman, 1975 (cited above), p. 35 (unsolvable tasks induce social helplessness).

For a discussion of learned helplessness as a cognitive or affective phenomenon, see Seligman, M. 1975. *Helplessness: On Depression, Development and Death* (San Francisco: W. H. Freeman). For a discussion of learned helplessness as a phenomenon of psychomotor retardation, see Weiss, J., Bailey, W., Goodman, P., Hoffman, L., Ambrose, M., Salman, S., and Charry, J. 1982. "A model for neurochemical study of depression." In Spiegelstein, M., and Levy, A. (eds.). *Behavioral Models and the Analysis of Drug Action* (Amsterdam: Elsevier). "Learned laziness" in animals given noncontingent reward: the use of the term "spoiled brat" is cited by Seligman, 1975 (cited above), p. 35. The published version of those findings can be found in Engberg, L., Hansen, G., Welker, R., and Thomas, D. 1973. "Acquisition of key-pecking via autoshaping as a function of prior experience: 'learned laziness?'" *Science* 178, 1002.

PAGE 254 Biological features of learned helplessness, where rats show altered grooming, social behavior, sexual behavior, feeding, plus many of the vegetative symptoms: Stone, E. 1978. "Possible grooming deficit in stressed rats." *Research Communication in Psychology, Psychiatry and Behavior* 3, 109; Weiss, J., Simson, P., Ambrose, M., Webster, A., and Hoffman, L. 1985. "Neurochemical basis of behavioral depression." In Katkin, E., and Manuck, S. (eds.). *Advances in Behavioral Medicine,* vol. 1 (Greenwich, Conn.: JAI Press); Weiss, J., Goodman, P., Losito, P., Corrigan, S., Charry, J., and Bailey, W. 1981. "Behavioral depression produced by an uncontrolled stressor: relation to norepinephrine, dopamine and serotonin levels in various regions of the rat brain." *Brain Research Reviews* 3, 167. For an explicit comparison between the symptoms of depression (DSM-III criteria) and learned helplessness, see Weiss, J., Bailey, W., Goodman, P., Hoffman, L., Ambrose, M., Salman, S., and Charry, J. 1982. "A model for neurochemical study of depression." In Spiegelstein, M., and Levy, A. (eds.). *Behavioral Models and the Analysis of Drug Action* (Amsterdam: Elsevier).

PAGE 255 Learned helplessness can be lessened by antidepressants or ECT: Dorworth, T., and Overmier, J. 1977. "On learned helplessness: the therapeutic effects of electroconvulsive shocks." *Physiological Psychology* 5, 355; Leshner, A., Remler, H., Biegon, A., and Samuel, D. 1979. "Desmethylimipramine counteracts learned helplessness in rats." *Psychopharmacology* 66, 207; Petty, F., and Sherman, A. 1980. "Reversal of learned helplessness by imipramine." *Communications in Psychopharmacology* 3, 371; Sherman, A., Allers, G., Petty, F., and Henn, F. 1979. "A neuropharmacologically-relevant animal model of depression." *Neuropharmacology* 18, 891.

PAGE 256 Rozin, P., Poritsky, S., and Sotsky, R. 1971. "American children with reading problems can easily learn to read English represented by Chinese characters." *Science* 171, 1264.

PAGE 256 Early parental loss increases the risk of adulthood depression: this subject is reviewed in Breier, A., Kelso, J., Kirwin, P., Beller, S.,

Wolkowitz, O., and Pickar, D. 1988. "Early parental loss and development of adult psychopathology." *Archives of General Psychiatry* 45, 987.

PAGE 259 Stress depletes parts of the brain of norepinephrine, while also increasing the activity of tyrosine hydroxylase: Stone, E., and McCarty, R. 1983. "Adaptation to stress: tyrosine hydroxylase activity and catecholamine release." *Neuroscience and Biobehavioral Reviews* 7, 29. Glucocorticoids have something to do with this: Dunn, A., Gildersleeve, N., and Gray, H. 1978. "Mouse brain tyrosine hydroxylase and glutamic acid decarboxylase following treatment with adrenocorticotropic hormone, vasopressin or corticosterone." *Journal of Neurochemistry* 31, 977. In addition, CRF may have something to do with this: Ahlers, S., Salander, M., Shurtleff, D., and Thomas, J. 1992. "Tyrosine pretreatment alleviates suppression of schedule-controlled responding produced by CRF in rats." *Brain Research Bulletin* 29, 567.

CHAPTER 14: Personality, Temperament and Their Stress-Related Consequences

PAGE 264 For a review of primate personality, see Clarke, A., and Boinski, S. 1995. "Temperament in nonhuman primates." *American Journal of Primatology* 37, 103. Personality in sunfish: Wilson, D., Coleman, K., Clark, A., and Biderman, L. 1993. "Shy-bold continuum in pumpkinseed sunfish (*Lepomis gibbosus*): an ecological study of a psychological trait." *Journal of Comparative Psychology* 107, 250.

PAGE 264 Baboons, personality and physiology: Sapolsky, R., and Ray, J. 1989. "Styles of dominance and their physiological correlates among wild baboons." *American Journal of Primatology* 18, 1. Ray, J., and Sapolsky, R. 1992. "Styles of male social behavior and their endocrine correlates among high-ranking baboons." *American Journal of Primatology* 28, 231. Sapolsky, R. 1996. "Why should an aged male baboon transfer troops?" *American Journal of Primatology* 39, 149. Virgin, C., and Sapolsky, R. 1997. "Styles of male social behavior and their endocrine correlates among low-ranking baboons." *American Journal of Primatology* 42, 25. Also see: Suomi, S. 1997. "Early determinants of behaviour: evidence from primate studies." *British Medical Bulletin* 53, 270.

PAGE 271 For the most detailed review of the literature on psychogenic abortions, see Huisjes, H. 1984. *Spontaneous Abortion* (New York: Churchill Livingstone).

PAGE 273 The reader is referred back to Chapter 13 for references on the cognitive and endocrine profile of depression.

PAGE 273 Anxiety and catecholamines: Friedman, B., Thayer, J., Borkovec, T., Tyrrell, R., Johnson, B., and Columbo, R. 1993. "Autonomic characteristics of nonclinical panic and blood phobia." *Biological Psychiatry* 34, 298. For a discussion of the dichotomy between still striving to cope (accompanied by catecholamine secretion) and having given up (characterized by glucocorticoid hypersecretion): Frankenhaeuser, M. 1983. "The sympathetic-adrenal and pituitary-adrenal response to challenge." In Dembroski, T., Schmidt, T., and Blumchen, G. (eds.). *Biobehavioral Basis of Coronary Heart Disease* (Basel: Karger), 91

PAGE 275 The Sisyphus pattern of personality was first identified in Bruhn, J., Paredes, A., Adsett, C., and Wolf, S. 1974. "Psychological predictors of sudden death in myocardial infarction." *Journal of Psychosomatic Research* 18, 187.

Type A: the definitive prospective study showing a link between Type A personality and coronary heart disease is Rosenman, R., Brand, R., Jenkins, C., Friedman, M., Straus, R., and Wurm, M. 1975. "Coronary heart disease in the Western Collaborative Group Study: final follow-up experience of 812 years." *Journal of the American Medical Association* 233, 872. See also Friedman, M., and Rosenman, R. 1974. *Type A Behavior and Your Heart* (New York: Knopf). The blue-ribbon panel that endorsed the Type A concept published its report as: Cooper, T., Detre, T., and Weiss, S. 1981. "Coronary prone behavior and coronary heart disease; a critical review." *Circulation* 63, 1199.

Problems with replicating the original Type A finding: the most influential study was Shekelle, R., Billings, J., and Borhani, N. 1985. "The MRFIT behavior pattern study. II. Type A behavior and incidence of coronary heart disease." *American Journal of Epidemiology* 122, 599. Others are discussed in Barefoot, J., Peterson, B., Harrell, F., et al. 1989. "Type A behavior and survival: a follow-up study of 1,467 patients with coronary artery disease." *American Journal of Cardiology* 64, 427.

The demonstration that Type A behavior is associated with better survivorship: the Barefoot study just cited, plus Ragland, D., and Brand, R. 1988. "Type A behavior and mortality from coronary heart disease." *New England Journal of Medicine* 313, 65. That finding is a good lesson in how incredibly subtle some confounds can be in epidemiology research. Why should being Type A be associated with better survivorship, once you are diagnosed with coronary heart disease? Some possibilities: Type A people, because of their driven and disciplined nature, are more likely to comply with the medicine, dieting, and exercise schedules given them by their doctors. Or certain people may immediately be recognized as being Type A by their doctors, who then think "Aha, here's this Type A patient with coronary heart disease. I know all about Friedman and Rosenman's studies; I'd better take extra good care of this person." Or Type A people may be more disciplined about annual checkups with doctors, and thus be diagnosed with coronary heart disease earlier than average, when it is still pretty mild—leading to seemingly better survivorship. This last factor has probably been ruled out, but no one is sure about the other possible confounds yet, as the studies showing better Type A survivorship are quite recent. Some of the possible sources of confounds in these finding are discussed in Matthews, K., and Haynes, S. 1986. "Type A behavior pattern and coronary disease risk." *American Journal of Epidemiology* 123, 923.

PAGE 276 The importance of hostility as a predictor of heart disease: the demonstration of this by reanalysis of the original Friedman and Rosenman data: Hecker, M., Chesney, M. N., Black, G., and Frautsch, N. 1988. "Coronary-prone behaviors in the Western Collaborative Group Study." *Psychosomatic Medicine* 50, 153. Hostility in medical students: Barefoot, J., Dahlstrom, W., and Williams, R. 1983. "Hostility, CHD incidence, and total mortality: a 25-year follow-up study of 255 physicians." *Psychosomatic Medicine* 45, 59. In lawyers: Barefoot, J., Dodge, K., Peterson, B., Dahlstrom, W., and Williams, R. 1989. "The Cook-Medley Hostility scale: item content and ability to predict survival." *Psychosomatic Medicine* 51, 46. In Finnish twins: Koskenvuo, M., Kaprio, J., Rose, R., Kesaniemi, A., Sarna, S., Heikkila, K., and Langinvainio, H. 1988. "Hostility as a risk factor for mortality and ischemic heart disease in men." *Psychosomatic Medicine* 50, 330. In Western Electric employees: Shekelle, R., Gale, M., Ostfeld, A., and Paul, O. 1983. "Hostility, risk of coronary disease, and mortality." *Psychosomatic Medicine* 45, 219. For general reviews, see Miller, T., Smith, T., Turner, C., Guijarro, M., and Hallet, A. 1996. "A meta-analytic review of research on hostility and physical health." *Psychological Bulletin* 119, 322; Williams, R., and Littman, A. 1996. "Psychosocial factors: role in cardiac risk and treatment strategies." *Cardiology Clinics* 14, 97. Hostility as a predictor of overall mortality: Houston, B., Babyak, M., Chesney, M., Black, G., and Ragland, D. 1997. "Social dominance and 22-year all-cause mortality in men." *Psychosomatic Medicine* 50, 5.

PAGE 277 Insecurity as the key to Type A-ness: Price, V., Friedman, M., Ghandour, G., and Fleischmann, N. 1995. "Relation between insecurity and type A behavior." *American Heart Journal* 129, 488.

PAGE 278 The studies of James Gross concerning the intentional inhibition of emotional expression can be found in Gross, J., and Levenson, R. 1993. "Emotional suppression: physiology, self-report, and expressive behavior." *Journal of Personality and Social Psychology* 64, 870. See also Gross, J., and Levenson, R. 1997. "Hiding feelings: the acute effects of inhibiting negative and positive emotion." *Journal of Abnormal Psychology*, 106, 95. For a vote in favor of expression of hostility as being most injurious: Siegman, A. 1993. "Cardiovascular consequences of expressing, experiencing, and repressing anger." *Journal of Behavioral Medicine* 16, 539.

PAGE 279 Hormonal and cardiovascular function in hostile versus nonhostile people. Demonstration that hostile and nonhostile people do not differ in hormone and blood pressure measures during rest or during nonsocial stressors: Sallis, J., Johnson, C., Treverow, T., Kaplan, R., and Hovell, M. 1987. "The relationship between cynical hostility and blood pressure reactivity." *Journal of Psychosomatic Research* 31, 111. See also Smith, M., and Houston, B. 1987. "Hostility, anger expression, cardiovascular responsivity, and social support." *Biological Psychology* 24, 39. Also Krantz, D., and Manuck, S. 1984. "Acute psychophysiologic reactivity and risk of cardiovascular disease: a review and methodological critique." *Psychological Bulletin* 96, 435.

Demonstration of larger responses in hostile people to social provocations: being interrupted during a task: Suarez, E., and Williams, R. 1989. "Situational determinants of cardiovascular and emotional reactivity in high and low hostile men." *Psychosomatic Medicine* 51, 404. During a rigged game against a disparaging opponent: Glass, D., Krakoff, L., and Contrada, R. 1980. "Effect of harassment and competition upon cardiovascular and catecholamine responses in Type A and Type B individuals." *Psychophysiology* 17, 453. During a role-play of social conflict: Hardy, J., and Smith, T. 1988. "Cynical hostility and vulnerability to disease: social support, life stress, and physiological response to conflict." *Health Psychology* 7, 477. Unsolvable tasks with bad instructions: Weidner, G., Friend, R., Ficarrotto, T., and Mendell, N. 1989. "Hostility and cardiovascular reactivity to stress in women and men." *Psychosomatic Medicine* 51, 36; also see Suls, J., and Wan, C. 1993. "The relationship between trait hostility and cardiovascular reactivity: a quantitative review and analysis." *Psychophysiology* 30, 615. Note that many of these studies were done when everyone was still dichotomizing between Type A and Type B people, rather than hostile and nonhostile.

PAGE 279 If you can change Type A tendencies, you decrease risk of coronary heart disease: Friedman, M., Thoresen, C., and Gill, J. 1986. "Alteration of Type A behavior and its effect on cardiac recurrences in post-myocardial infarction patients: summary results of the Recurrent Coronary Prevention Project." *American Heart Journal* 112, 653; Friedman, M., Breall, W., Goodwin, M., Sparagon, B., Ghandour, G., and Fleischmann, N. 1996. "Effect of type A behavioral counseling on frequency of episodes of silent myocardial ischemia in coronary patients." *American Heart Journal* 132, 933.

Finally, for an analysis of the medical histories and the significantly shortened life spans of the American presidents, in which the author concludes that they have died disproportionately from stress-related cardiovascular disease: Gilbert, R. 1993. "Travails of the chief." *The Sciences* (January–February), 8.

PAGE 283 Repressive personality: they really are happy: Brandtstadter, J., Balte, S., Gotz, B., Kirschbaum, C., and Hellhammer, D. 1991. "Developmental and personality correlates of adrenocortical activity as indexed by salivary cortisol: observations in the age range of 35 to 65 years." *Journal of Psychosomatic Research* 35, 173. Weinberger, D., Schwartz, G., and Davidson, R. 1979. Low-anxious, high-anxious, and repressive coping styles: psychometric patterns and behavioral and physiological responses to stress. *Journal of Abnormal Psychology* 88, 369; Shaw, R., Cohen, F., Fishman-Rosen, R., Murphy, M., Stertzer, S., Clark, D., and Myler, K. 1986. "Psychologic predictors of psychosocial and medical outcomes in patients undergoing coronary angioplasty." *Psychosomatic Medicine* 48, 582; Shaw, R., Cohen, F., Doyle, B., Palesky, J. 1985. "The impact of denial and repressive style on information gain and rehabilitation outcomes in myocardial infarction patients." *Psychosomatic Medicine* 47, 262.

Patterns of repressives: Brown, L., Tomarken, A., Orth, D., Loosen, P., Kalin, N., and Davidson, R. 1996. "Individual differences in repressive-defensiveness predict basal salivary cortisol levels." *Journal of Personality and Social Psychology* 70, 362. Such individuals have impaired immune profiles: Jamner, L., Schwartz, G., and Leigh, H. 1988. "The relationship between repressive and defensive coping styles and monocyte, eosinophile, and serum glucose levels: support for the opioid peptide hypothesis of repression." *Psychosomatic Medicine* 50, 567. Tomarken, A., and Davidson, R. 1994. "Frontal brain activation in repressors and nonrepressors." *Journal of Abnormal Psychology* 103, 339. Sociopathy and the frontal cortex: Damasio, A., Tranel, D., and Damasio, H. 1990. "Individuals with sociopathic behavior caused by frontal damage fail to respond autonomically to social stimuli." *Behavioural Brain Research* 41, 81.

CHAPTER 15: The View from the Bottom

PAGE 289 Rudolph Virchow: Rosen, G. 1972. "The evolution of social medicine." In Freeman, H., Levine, S., and Reeder, L. (eds.). *Handbook of Medical Sociology*, 2d ed. (Englewood Cliffs, N.J.: Prentice-Hall). This is also the source of the Virchow quotes.

PAGE 290 For introductions to baboon social behavior, see: Strum, S. 1987. *Almost Human* (New York: Random House); Smuts, B. 1985. *Sex and Friend In Baboons* (New York: Aldine); Ransom, T. 1981. *Beach Troop of the Gombe* (Lewisburg, Pa.: Bucknell University Press).

PAGE 293 Elevated glucocorticoids and other problems in low-ranking male baboons: Sapolsky, R. 1990. "Adrenocortical function, social rank and personality among wild baboons." *Biological Psychiatry* 28, 862; Sapolsky, R. 1993. "Endocrinology alfresco: psychoendocrine studies of wild baboons." *Recent Progress in Hormone Research*, 48, 437; Sapolsky, R., and Spencer, E. 1997. "Social subordinance is associated with suppression of insulin-like growth factor I in a population of wild primates." *American Journal of Physiology*, 273, R1346.

PAGE 294 A review of rank-related differences in the stress-response of other species: Sapolsky, R. "The physiological and pathophysiological implications of social stress in mammals." In McEwen, B. (ed.). *Coping with the Environment. Handbook of Physiology* (Washington, D.C.: American Physiological Association Press), in press.

PAGE 294 Marmoset review: Abbott, D., Saltzman, W., Schultz-Darken, N., and Smith, T. 1997. "Specific neuroendocrine mechanisms not involving generalized stress mediate social regulation of female reproduction in cooperatively breeding marmoset monkeys." In Carter, C., Kirpatrick, B., Liederhendler, I. (eds.): *The Integrative Neurobiology of Affiliation. Annals of the New York Academy of Sciences*, in press.

PAGE 294 Life for dominant wild dogs and mongooses: Creel, S., Creel, N., and Monfort, S. 1996. "Social stress and dominance." *Nature* 379, 212.

PAGE 295 For an introduction to primate culture: Wrangham, R. 1994. *Chimpanzee Cultures* (Cambridge, Mass.: Harvard University Press).

PAGE 296 Reconciliative rhesus: Gust, D., Gordon, T., Hambright, K., and Wilson, M. 1993. "Relationship between social factors and pituitary-adrenocortical activity in female rhesus monkeys." *Hormones and Behavior* 27, 318. Baboons during a drought: Sapolsky, R. 1986. "Endocrine and behavioral correlates of drought in the wild baboon." *American Journal of Primatology* 11, 217.

PAGE 296 Social instability as a primate stressor: Sapolsky, R. 1993. "The physiology of dominance in stable versus unstable social hierarchies." In Mason, W., and Mendoza, S. (eds.). *Primate Social Conflict* (New York: SUNY Press); Cohen, S., Kaplan, J., and Cunnick, J. 1992. "Chronic social stress, affiliation and cellular immune response in nonhuman primates." *Psychological Sciences* 3, 301.

PAGE 297 The immunosuppressive effects of the aggressive transfer male: Alberts, S., Altmann, J., and Sapolsky, R. 1992. "Behavioral, endocrine and immunological correlates of immigration by an aggressive male into a natural primate group." *Hormones and Behavior* 26, 167.

PAGE 297 A review of primate personality: Clarke, A., and Boinski, S. 1995. "Temperament in nonhuman primates." *American Journal of Primatology* 37, 103.

PAGE 298 Some studies of human rank: Elias, M. 1981. "Cortisol, testosterone and testosterone-binding globulin responses to competitive fighting in human males." *Aggressive Behavior* 7, 215. Meyerhoff, J., Leshansky, M., and Mougey, E. 1988. "Effects of psychological stress on pituitary hormones in man." In Chrousos, G., Loriaux, D., Gold, P. (eds.). *Mechanisms of Physical and Emotional Stress* (New York: Plenum Press); Houston, B., Babyak, M., Chesney, M., Black, G., and Ragland, D. 1997. "Social dominance and 22-year all-cause mortality in men." *Psychosomatic Medicine* 59, 5; Mazur, A., and Booth, A. 1997. "Testosterone and dominance in men." *Brain and Behavioral Sciences,* in press.

PAGE 298 Stress in the workplace: Pickering, T., Devereux, R., James, G., Gerin, W., Landsbergis, P., Schnall, P., and Schwartz, J. 1996. "Environmental influences on blood pressure and the role of job strain." *Journal of Hypertension,* 14, S179.

PAGE 300 Endocrine consequences of winning by effort versus by luck: Mazur, A., and Lamb, T. 1980. "Testosterone, status and mood in human males." *Hormones and Behavior* 14, 236; McCaul, K., Gladue, B., and Joppa, M. 1992. "Winning, losing, mood, and testosterone." *Hormones and Behavior* 26, 486.

PAGE 300 The threat of unemployment disrupts health: Beale, N., and Nethercott, S. 1985. "Job-loss and family morbidity: a study of a factory closure." *Journal of the Royal College of General Practitioners* 35, 510; Cobb, S., and Kasl, S. 1977. *Termination: The Consequences of Job Loss,* DHEW-NIOSH Publication No. 77-224 (Cincinnati: U.S. NIOSH). The study about workers not taking their diuretics because they couldn't go to the bathroom at work: cited in Adler, N., Boyce, T., Chesney, M., Folkman, S., and Syme, S. 1993. "Socioeconomic inequalities in health: no easy solution." *Journal of the American Medical Association* 269, 3140.

PAGE 301 Reviews about the SES gradient and health (these are all leaders in the field): Pincus, T., and Callahan, L. 1995. "What explains the association between socioeconomic status and health: primarily access to medical care or mind-body variables?" *Advances* 11, 4; Syme, S., and Berkman, L. 1976. "Social class, susceptibility and sickness." *American Journal of Epidemiology* 104, 1; Adler, N., Boyce, T., Chesney, M., Folkman, S., and Syme, S. 1993. "Socioeconomic inequalities in health: no easy solution." *Journal of the American Medical Association* 269, 3140; Anderson, N., and Armstead, C. 1995. "Toward understanding the association of SES and health; a new challenge for the biopsychosocial approach." *Psychosomatic Medicine* 57, 213; Evans, R., Barer, M., and Marmor, T. 1994. *Why Are Some People Healthy and Others Not? The Determinants of Health of Populations* (New York: Aldine de Gruyter); Antonovsky, A. 1968. "Social class and the major cardiovascular diseases." *Journal of Chronic Diseases* 21, 65; Marmot, M. 1983. "Stress, social and cultural variations in heart disease." *Journal of Psychosomatic Research* 27, 377; Levenstein, S., Prantera, C., Varvo, V., Arca, M., Scribano, M., Spinella, S., and Berto, E. 1996. "Long-term symptom patterns in duodenal ulcer: psychosocial factors." *Journal of Psychosomatic Research* 41, 465; Hahn, R., Eaker, E., Barker, N., Teutsch, S., Sosniak, W., and Krieger, N. 1996. "Poverty and death in the United States." *International Journal of Health Services*. 26, 673.

PAGE 302, FOOTNOTE Diseases more prevalent among the wealthy: malignant melanoma and breast cancer: Kitagawa, E., and Hauser, P. 1973. *Differential Mortality in the United States* (Cambridge, Mass.: Harvard University Press). Multiple sclerosis: Pincus, T., and Callahan, L. 1995. "What explains the association between socioeconomic status and health: primarily access to medical care or mind-body variables?" *Advances* 11, 4. Polio: Pincus, T. In Davis, B. (ed.). *Microbiology, Including Immunology and Molecular Genetics,* 3d ed. (New York: Harper and Row).

PAGE 302 SES accounts for all or most of the racial difference in health in America: Rogers, R. 1992. "Living and dying in the U.S.A.: sociodemographic determinants of death among blacks and whites." *Demography* 29, 287; Guralnik, J., Land, K., Blazer, D., Fillenbaum, G., and Branch, L. 1993. "Educational status and active life expectancy among older blacks and whites." *New England Journal of Medicine* 329, 110; Dayal, H., Power, R., and Chiu, C. 1982. "Race and socioeconomic status in survival from breast cancer." *Journal of Chronic Disability* 35, 675; Otton, M., Teutsch, S., Williamson, D., and Marks, J. 1990. "The effect of known risk factors in the excess mortality of black adults in the United States." *Journal of the American Medical Association,* 263, 845.

PAGE 302 The persistent consequences of poverty: Sennett, R., and Cobb, J. 1972. *The Hidden Injuries of Class* (New York: Vintage).

PAGE 302 The nun study: Snowdon, D., Ostwald, S., and Kane, R. 1989. "Education, survival and independence in elderly Catholic sisters 1936–1988." *American Journal of Epidemiology* 120, 999; Snowdon, D., Ostwald, S., Kane, R., and Keenan, N. 1989. "Years of life with good and poor mental and physical function in the elderly." *Journal of Clinical Epidemiology* 42, 1,055.

PAGE 303 SES and being resuscitated in an ambulance: Sudnow, D. 1967. *Passing On: The Social Organization of Dying* (Englewood Cliffs, N.J.: Prentice-Hall).

PAGE 303 People never having heard of PAP tests: Harlan, L., Bernstein, A., and Kessler, L. 1991. "Cervical cancer screening: who is not screened and why?" *American Journal of Public Health* 81, 885.

PAGE 304 The SES gradient worsening in England: Susser, M., Watson, W., and Hopper, K. 1985. *Sociology in Medicine*, 3d ed. (Oxford, England: Oxford University Press).

PAGE 305 Study where poor had made more use of a prepaid health plan and still had more disease: Oakes, T., and Syme, S. 1973. "Social factors in newly discovered elevated blood pressure." *Journal of Health, Society and Behavior*. 14, 198.

PAGE 305 Pincus's work is reviewed in Pincus, T., and Callahan, L. 1995. "What explains the association between socioeconomic status and health: primarily access to medical care or mind-body variables?" *Advances* 11, 4.

PAGE 305 Marmot's work is reviewed in Marmot, M. 1983. "Stress, social and cultural variations in heart disease." *Journal of Psychosomatic Research* 27, 377.

PAGE 306 The poor do have the most stressors: McLeod, J., and Kessler, R. 1990. "Socioeconomic status differences in vulnerability to undesirable life events." *Journal of Health Society and Behavior* 31, 162; Cohen, S., and Wills, T. 1985. "Stress, social support and the buffering hypothesis." *Psychological Bulletin* 98, 310; Brown, G., and Harris, T. 1978. *Social Origins of Depression* (London: Tavistock).

PAGE 306 A sense of coherence: Antonovsky, A. 1994. "A sociological critique of the 'Well-Being' movement." *Advances* 10, no. 3, 6.

PAGE 307 The conclusion by Adler and colleagues: Adler, N., Boyce, T., Chesney, M., Folkman, S., and Syme, S. 1993. "Socioeconomic inequalities in health: no easy solution." *Journal of the American Medical Association* 269, 3140.

CHAPTER 16: Managing Stress

PAGE 309 A technical description of alopecia areata can be found in Rook, A., and Dawber, R. 1991. *Diseases of the Hair and Scalp*, 2d ed. (Oxford:

Blackwell Scientific Publications). In actuality, though, there is not really a change in hair color under those circumstances. Alopecia areata occurs in people who already have some degree of whitening or graying of their hair. With the onset of the trauma, hair that is not white or gray falls out, probably because the immune system attacks dark-hair bulbs. Thus, all that is left is the white or gray hair. Various experts I've consulted suggest that the phenomenon represents a bit of media hype—it is extremely rare and usually takes weeks or months, rather than occurring in a single night.

A particularly amusing account of the history of the disorder and speculations about it can be found in Jelinek, J. 1972. "Sudden whitening of the hair." *Bulletin of the New York Academy of Medicine* 48, 1003. Jelinek, a professor of dermatology, recounts many tales over the centuries of people who, condemned to be executed by their king, turn white with terror the night before the scheduled execution. The now white-haired prisoner is brought before the king and assembled court for execution the next morning. Everyone is moved with wonder and pity at the transformation, and the poor wretch is pardoned. Numerous sources claim that the hair and beard of Sir Thomas More, who had fallen out of favor with King Henry VIII and was condemned to death, turned white the day before his execution. In contrast to the general pattern of these tales, Henry, unimpressed, still had him killed and his head parboiled and displayed on London Bridge. Over the course of her imprisonment prior to her execution, Marie Antoinette's hair was also reported to have turned gray. This may not have represented a true case of alopecia areata, however. "It has cynically been conjectured that the keepers of her dungeon neglected to furnish their guest's dressing table with hair dyes. The iconoclast respects nothing, not even the grey hairs of royalty," opined the mordant Dr. Jelinek.

PAGE 309 The tendency for variability to increase in aging populations is discussed in Rowe, J., Wang, S., and Elahi, D. 1990. "Design, conduct, and analysis of human aging research." In Schneider, E., and Rowe, J. (eds.) *Handbook of the Biology of Aging*, 3d ed. (San Diego: Academic Press), 63.

PAGE 311 The encouraging topic of successful aging is reviewed in Rowe, J., and Kahn, R. 1987. "Human aging: usual and successful." *Science* 237, 143; Baltes, P., and Baltes, M. 1990. *Successful Aging* (Cambridge, Eng.: Cambridge University Press).

PAGE 313 Cognitively unimpaired aged rats showed none of the usual degenerations: Issa, A., Rowe, W., Gauthier, S., and Meaney, M. 1991. "Hypothalamic-pituitary-adrenal activity in aged, cognitively impaired and cognitively unimpaired rats." *Journal of Neuroscience* 10, 3247. Neonatal handling produces similar protection in old age: Meaney, M., Aitken, D., Bhatnager, S., van Berkel, C., and Sapolsky, R. 1988. "Effect of neonatal handling on age-related impairments associated with the hippocampus." *Science* 239, 766; Meaney, M., Aitken, D., and Sapolsky, R. 1990. "Postnatal handling attenuates neuroendocrine, anatomical and cognitive dysfunctions associated with aging in female rats." *Neurobiology of Aging* 12, 31.

PAGE 314 Coping styles among parents of children with cancer: Wolff, C., Friedman, S., Hofer, M., and Mason, J. 1964. "Relationship between psychological defenses and mean urinary 17-hydroxycorticosteroid excretion rates. I. A predictive study of parents of fatally ill children." *Psychosomatic Medicine* 26, 576; Hofer, M., Wolff, E., Friedman, S., and Mason, J. 1972. "A psychoendocrine study of bereavement, parts I and II." *Psychosomatic Medicine* 34, 481.

PAGE 316 Benson, H. 1996. *Timeless Healing. The Power and Biology of Belief* (New York: Scribner). A discussion of the controversy that this has generated: Roush, W. 1997. "Herbert Benson: mind-body maverick pushes the envelope." *Science* 276, 357. Other people who have done research in this area: Kendler, K., Gardner, C., and Prescott, C. 1997. "Religion, psychopathology, and substance use and abuse: a multimeasure, genetic-epidemiologic study." *American Journal of Psychiatry* 154, 322; Williams, D., Larson, D., Buckerl, R., Heckmann, R., and Pyle, C. 1991. "Religion and psychological distress in a community sample." *Social Science and Medicine* 32, 1257; Maton, K. 1989. "The stress-buffering role of spiritual support: cross-sectional and prospective investigations." *Journal of Scientific Study of Religion* 28, 310; Krause, N., and Van Tran, T. 1987. "Stress and religious involvement among older blacks." *Journal of Gerontology* 44, S4; Pressman, P., Lyons, J., Larson, D., and Strain, J. 1990. "Religious belief, depression, and ambulation status in elderly women with broken hips." *American Journal of Psychiatry* 147, 758.

PAGE 317 Resistance to learned helplessness is discussed in Seligman, M. 1992. *Helplessness*, 2d ed. (New York: W. H. Freeman).

PAGE 318 Sapolsky, R. 1996. "Why should an aged male baboon transfer troops?" *American Journal of Primatology* 39, 149.

PAGE 321 Change in cholesterol profiles in Type A individuals receiving counseling: Gill, J., Price, V., and Friedman, M. 1985. "Reduction in Type A behavior in healthy middle-aged American military officers." *American Heart Journal* 110, 503. Also see Thoresen, C., and Powell, L. 1993. "Type A behavior pattern: new perspectives on theory, assessment and intervention." *Journal of Consulting and Clinical Psychology*, in press.

Some papers showing the salutary effects of transcendental meditation on various physiological endpoints (resting glucocorticoid levels, oxygen consumption, heart rate, and so on): Wallace, R. 1970. "Physiological effects of transcendental meditation." *Science* 167, 1751; Wallace, R., and Benson, H. 1972. "The physiology of meditation." *Scientific American* (February), 84 (these two papers cover much the same material, but the latter is more accessible and gives more of an overview of the subject); Jevning, R., Wilson, A., and Davidson, J. 1978. "Adrenocortical activity during meditation." *Hormones and Behavior* 10, 54.

PAGE 322 Changing of the stress-response over time among parachuting trainees: Ursin, H., Baade, E., and Levine, S. 1978. *Psychobiology of Stress* (San Diego: Academic Press).

PAGE 322 Self-medication among acute pain patients can be conducted safely: Norman, J., White, W., and Pearce, D. 1978. "New possibilities in analgesia: the demand analgesia computer. Round table on morphinomimetics." *5th European Congress of Anaesthesiology*, Paris; Jully, C., and Sibbald, A. 1981. "Control of postoperative pain by interactive demand analgesia." *British Journal of Anaesthesiology* 53, 385; Baumann, T., Bastenhorst, R., Graves, D., Foster, T., and Bennett, R. 1986. "Patient-controlled analgesia in the terminally ill cancer patient." *Drug Intell Clinical Pharmacology* 20, 297; Citron, M., Johnston-Early, A., Boyer, M., Krasnow, S., Hood, H., and Cohen, M. 1986. "Patient-controlled analgesia for severe cancer pain." *Archives of Internal Medicine* 146, 734. Such self-medication can be associated with an overall decrease in the amount of medication taken: Chapman, C., and Hill, H. 1989. "Prolonged morphine self-administration and addiction liability: evaluation of two theories in a bone marrow transplant unit." *Cancer* 63, 1636; Chapman, C. 1989. "Giving the patient control of opioid analgesic administration." In Hill, C., and Fields, W. (eds.). *Advances in Pain Research and Therapy*, vol. 11 (New York: Raven Press), 339; Chapman, C., and Hill, H. 1990. "Patient-controlled analgesia in a bone marrow transplant setting." In Foley, K. (ed.). *Advances in Pain Research and Therapy*, vol. 16, 231.

PAGE 323 Coping styles differ between young and aged humans: Folkman, S., Lazarus, R., Pimley, S., and Novacek, J. 1987. "Age differences in stress and coping processes." *Psychology and Aging* 2, 171.

Manipulating psychological variables in nursing home populations: This large literature is reviewed in Rodin, J. 1986. "Aging and health: effects of the sense of control." *Science* 233, 1271; Rowe, J., and Kahn, R. 1987. "Human aging: usual and successful." *Science* 237, 143. Individual studies: Schulz, J. 1976. "Effects of control and predictability on the physical and psychological well-being of the institutionalized aged." *Journal of Personality and Social Psychology* 33, 563; Schulz, J., and Hanusa, B. 1978. "Long-term effects of control and predictability-enhancing interventions: findings and ethical issues." *Journal of Personality and Social Psychology* 36, 1194.

PAGE 327 Follow-up to cancer study: Hofer, M., Wolff, E., Friedman, S., and Mason, J. 1972. "A psychoendocrine study of bereavement, parts I and II." *Psychosomatic Medicine* 34, 481.

PAGE 327 Follow-up to nursing home study: Schulz, J. 1976. "Effects of control and predictability on the physical and psychological well-being of the institutionalized aged." *Journal of Personality and Social Psychology* 33, 563

PAGE 328 Immune function tends to decline in rodents socially housed: Bohus, B., and Koolhaas, J. 1991. "Psychoimmunology of social factors in rodents and other subprimate vertebrates." In Ader, R., Felten, D., and Cohen, N. (eds.). *Psychoneuroimmunology*, 2d ed. (San Diego: Academic Press), 807. Social housing tends to elevate glucocorticoid levels in rodents and primates: Levine, S., Wiener, S., and Coe, C. 1989. "The psychoneuroendocrinology of stress: a psychobiological perspective." In Levine, S., and Brush, F. (eds.). *Psychoendocrinology* (San Diego: Academic Press). This review also discusses the study showing that infant monkeys separated from their mothers are not necessarily comforted (that is, show lowered glucocorticoid secretion) merely by being put in a social group. Also see Clarke, A., Czekala, N., and Lindburg, D. "Behavioral and adrenocortical responses of male cynomolgus and lion-tailed macaques to social stimulation and group formation." *American Journal of Primatology,* submitted. Marital discord is associated with immune suppression: Kiecolt-Glaser, J., Fisher, L., Ogrocki, P., Stout, J., Speicher, C., and Glaser, R. 1987. "Marital quality, marital disruption, and immune function." *Psychosomatic Medicine* 49, 13; Kiecolt-Glaser, J., Kennedy, S., Malkoff, S., Fisher, L., Speicher, C., and Glaser, R. 1988. "Marital discord and immunity in males." *Psychosomatic Medicine* 50, 213; Kiecolt-Glaser, J., Malar, J., Chee, M., Newton, T., Cacioppo, J., Mao, H., and Glaser, R. 1993. "Negative behavior during martial conflict is associated with immunological down-regulation." *Psychosomatic Medicine* 55, 395.

PAGE 330 Cyril Radcliffe and his terrible footnote in history are discussed in Collins, L., and Lapierre, D. 1975. *Freedom at Midnight* (New York: Simon and Schuster). Radcliffe's bizarrely detached speech, "Thoughts on India as 'The page is turned'" can be found in Radcliffe, C. 1968. *Not in Feather Beds: Some Collected Papers* (London: Hamish Hamilton).

PAGE 332 John Henryism: James, S. 1994. "John Henryism and the health of African-Americans." *Culture, Medicine and Psychiatry* 18, 163. An internal locus of control in a population of Harvard grads: Peterson, C., Seligman, M., and Waillant, G. 1988. "Pessimistic explanatory style is a risk factor for physical illness: a thirty-five-year longitudinal study." *Journal of Personality and Social Psychology* 55, 23.

PAGE 334 For a good overview of health psychology and stress management issues, see Chapter 15 in Nolen-Hoeksema, S. 1996. *Abnormal Psychology* (Brown and Benchmark).

PAGE 334 Warning against overloading aged individuals with too much control and responsibility: Rodin, J. 1986. "Aging and health: effects of the sense of control." *Science* 233, 1271; Langer, E. 1983. *The Psychology of Control* (Beverly Hills, Calif.: Sage); Langer, E. 1989. *Mindfulness* (Reading, Mass.: Addison Wesley Publishing Co.).

PAGE 334 The dangers of unrealistic anger: Williams, R. 1989. *The Trusting Heart: Great News about Type A Behavior* (New York: Random House). Also Williams, R., and Williams, V. 1993. *Anger Kills: Seventeen Strategies for Controlling the Hostility That Can Harm Your Health* (New York: Times/Random House).

PAGE 334 Spiegel, D. 1992. "Psychosocial interventions and cancer." *Advances* 8, no. 1, 3. The study he cited concerning types of beliefs in

control in cancer patients: Watson, M., Greer, S., Pruyn, J., and van den Borne, B. 1990. "Locus of control and adjustment to cancer." *Psychological Reports* 66, 39.

PAGE 335 The advantages of denial: Lazarus, R. 1983. "The costs and benefits of denial." In Breznitz, S. (ed.). *The Denial of Stress* (New York: International Universities Press), 1.

PAGE 335 Seligman, M. 1991. *Learned Optimism* (New York: Knopf).

SOURCES

PAGE 3
The National Archives

PAGE 5
Courtesy of Robert Longo and Metro Pictures

PAGE 23
Nick Downes © 1998 from The Cartoon Bank. All Rights Reserved.

PAGE 26
Merrilley Borell, 1976, *Bulletin of the History of Medicine* 50, p. 309, and Johns Hopkins Press, Baltimore, Maryland

PAGE 40
The Far Side © 1988 Farworks, Inc. Used by permission of Universal Press Syndicate. All Rights Reserved.

PAGE 43
(left) Photo Researchers
(right) Biophoto / Photo Researchers

PAGE 45
Superstock

PAGE 46
Custom Medical Stock Photo

PAGE 54
Roz Chast © 1980 from The New Yorker Collection. All Rights Reserved.

PAGE 60
P. Motta, Dept. of Anatomy, University "La Sapeinza," Rome, Science Source / Photo Researchers

PAGE 65
J. James, Science Photo Library, Photo Researchers

PAGE 76
Courtesy of Mark Daughhetee

PAGE 84
P. Saenger et al., 1977, "Somatedin growth hormone in psychosocial dwarfism," *Padiatrie und Padologie,* Supplement 5, p. 2; courtesy of Maria I. New, New York Hospital / Cornell Medical Center

PAGE 90
Data from P. Saenger et al., 1977 "Somatedin growth hormone in psychosocial dwarfism," *Padiatrie und Padologie,* Supplement 5, p. 2

PAGE 91
M. Newman / Superstock

PAGE 99
Courtesy of Harlow Primate Laboratory, University of Wisconsin

PAGE 105
(left) The Menil Collection, Houston, and © 1998 Artists Rights society (ARS), New York / ADAGP, Paris
(right) Private Collection, from W. Spies, *Max Ernst: A Retrospective,* Prestel, 1991 and © 1998 Artists Rights society (ARS), New York / ADAGP, Paris

PAGE 109
Courtesy of Laurance Frank, University of California, Berkeley

PAGE 115
Konner / Anthro-Photo

PAGE 132
Courtesy of Gilla Kaplan, The Rockefeller University

PAGE 161
Philadelphia Museum of Art, SmithKline Beckman Corporation Fund

PAGE 170
Courtesy of Vic Boff

PAGE 174
Alfred Eisenstaedt , Life Magazine

PAGE 183
The Far Side © 1984 Farworks, Inc. Used by permission of Universal Press Syndicate. All Rights Reserved.

PAGE 185
By permission of Jerry Van Amerongen and Creators Syndicate

PAGE 187
Courtesy Dr. Ana María Magariños and Dr. Bruce S. McEwen, Rockefeller University, New York

PAGE 196
Courtesy of Morris Zlapo

PAGE 197
DeVore / Anthro-Photo

PAGE 198
Anthony Taber © 1983 from The New Yorker Collection. All Rights Reserved.

PAGE 205
Rick Backlaws / Image West Photography

PAGE 210
Courtesy of Sidney Janis Gallery, New York; © 1993 George Segal / VAGA, New York

PAGE 216
Private collection; courtesy of George Tooker and Chameleon Books, Inc.

PAGE 223
Drawing by Roz Chast

PAGE 231
New Britain Museum of American Art, Gift of Olga H. Knoepke; courtesy of George Tooker and Chameleon Books, Inc.

PAGE 265
Robert Sapolsky, Stanford University

PAGE 274
Mick Stevens © 1982 from The New Yorker Collection. All Rights Reserved.

PAGE 278
Courtesy Gerald W. Friedland, Stanford University Department of Radiology

PAGE 282
Dr. Meyer Friedman, Meyer Friedman Institute

PAGE 284
Clifford Goodenough, Davidson Galleries, Seattle

PAGE 288
Vic DeLucia / The New York Times

PAGE 291
Robert Sapolsky, Stanford University

PAGE 292
Robert Sapolsky, Stanford University

PAGE 310
Hermitage Museum, St. Petersburg, Russia; Scala Art Resource, New York; © 1998 Succession H. Matisse / Artists Rights Society (ARS), New York

PAGE 315
Courtesy of Ed Speilman and E. Allen Becker and Son

PAGE 337
Edward Korne © 1993 from The New Yorker Collection. All Rights Reserved.

PAGE 338
Philadelphia Museum of Art: The Louise and Walter Arensberg Collection; © 1998 Artists Rights Society (ARS), New York / ADAGP, Paris

PAGE 414
The Far Side © 1982 Farworks, Inc. Used by permission of Universal Press Syndicate. All Rights Reserved.

INDEX